Power Quality in Modern Power Systems

Power Quality in Modern Power Systems

Edited by

P. Sanjeevikumar
Esbjerg Energy Section, Department of Energy Technology, Aalborg University, Esbjerg, Denmark

C. Sharmeela
Department of EEE, Anna University, Chennai, India

Jens Bo Holm-Nielsen
Esbjerg Energy Section, Department of Energy Technology, Aalborg University, Esbjerg, Denmark

P. Sivaraman
Leading Engineering Organisation, Chennai, India

ELSEVIER

ACADEMIC PRESS
An imprint of Elsevier

Academic Press is an imprint of Elsevier
125 London Wall, London EC2Y 5AS, United Kingdom
525 B Street, Suite 1650, San Diego, CA 92101, United States
50 Hampshire Street, 5th Floor, Cambridge, MA 02139, United States
The Boulevard, Langford Lane, Kidlington, Oxford OX5 1GB, United Kingdom

Library of Congress Cataloging-in-Publication Data
A catalog record for this book is available from the Library of Congress

British Library Cataloguing-in-Publication Data
A catalogue record for this book is available from the British Library

ISBN: 978-0-12-823346-7

For information on all Academic Press publications visit our website at https://www.elsevier.com/books-and-journals

Publisher: Joe Hayton
Acquisitions Editor: Lisa Reading
Editorial Project Manager: Naomi Robertson
Production Project Manager: Prasanna Kalyanaraman
Cover Designer: Victoria Pearson

Typeset by TNQ Technologies

Contents

Contributors xiii
Preface xv
Acknowledgment xvii

1. Power quality and its characteristics

P. Sivaraman and C. Sharmeela

1.1 How do changes in voltage, current, or frequency impact the quality of power? 2
1.2 Types of power quality problems 2
1.3 Transients 3
1.3.1 Cause 3
1.3.2 Impact 4
1.3.3 Types 4
1.3.4 Example 5
1.4 Short duration RMS variation 6
1.4.1 Voltage sag 7
1.4.2 Voltage swell 12
1.4.3 Interruption 19
1.5 Long duration voltage variation 24
1.5.1 Undervoltage 24
1.5.2 Overvoltage 27
1.5.3 Sustained interruption 29
1.6 Unbalance or imbalance 32
1.6.1 Cause 33
1.6.2 Impact 33
1.6.3 Example 33
1.7 Voltage fluctuations 34
1.7.1 Cause 37
1.7.2 Impact 37
1.7.3 Example 38
1.8 Waveform distortion 38
1.8.1 Harmonics 39
1.8.2 Interharmonics 42
1.8.3 DC offset 43
1.8.4 Notching 43
1.8.5 Noise 45

1.9	**Annexure 1.1**	47
1.10	**Annexure 1.2**	50
1.11	**Annexure 1.3: harmonics**	50
	1.11.1 Odd harmonics	50
	1.11.2 Even harmonics	53
	1.11.3 Tripplen harmonics	57
	References	59
	Further reading	60

2. Power system harmonics

P. Sivaraman and C. Sharmeela

2.1	**Introduction**	61
2.2	**Types of harmonic distortion**	62
	2.2.1 Voltage harmonics	62
	2.2.2 Current harmonics	62
2.3	**Harmonic-producing loads**	63
	2.3.1 Single-phase rectifiers	63
	2.3.2 Three-phase rectifiers	65
2.4	**Point of common coupling**	73
	2.4.1 Source-side harmonics	76
	2.4.2 Load-side harmonics	76
2.5	**Impact of harmonics on connected loads sharing the same feeder**	78
2.6	**Impact of harmonics in multitenanted premises**	79
2.7	**Evaluation of harmonics in the system**	80
	2.7.1 Total harmonic distortion	80
	2.7.2 Total demand distortion	85
	2.7.3 Total rated current distortion	87
2.8	**Resonance**	91
	2.8.1 Series resonance	91
	2.8.2 Parallel resonance	92
	2.8.3 How to prevent the system resonance condition	92
2.9	**Annexure 2.1**	93
	References	103
	Further reading	103

3. Power quality problems with renewable energy integration

Vasundhara Mahajan, Pramod Agarwal and Hari Om Gupta

3.1	**Motivation**	105
3.2	**Renewable energy sources and their integration**	107
3.3	**Power quality analysis with RES**	107
	3.3.1 Voltage fluctuations	108
	3.3.2 Voltage transients	108
	3.3.3 Voltage unbalance/sag/swell	108
	3.3.4 Origins of harmonics	109
	3.3.5 Consequences/impacts of harmonics	109

3.4 **Mathematical modeling of power systems with RES** 110
3.5 **Artificial intelligence-based controllers** 112
3.5.1 Artificial neural network-based controller 112
3.5.2 ANN controller for separation of events related to PQ of RES 115
3.5.3 ANN controller for harmonic separation 116
3.5.4 Fuzzy logic controller 118
3.5.5 Fuzzy logic controller for DC link capacitor voltage matching 119
3.5.6 Fuzzy logic controller for current adaptation and modulation 121
3.6 **Simulation results and analysis** 123
3.7 **Conclusion** 129
References 129

4. Fault ride-through (FRT) capability and current FRT methods in photovoltaic-based distributed generators

Gökay Bayrak, Davood Ghaderi and P. Sanjeevikumar

4.1 **Introduction** 133
4.2 **Fault ride-through strategies for PV power plants** 136
4.3 **Fault ride-through methods in PV power plants** 137
4.3.1 External FRT methods 139
4.3.2 External FRT methods 143
4.4 **Assessment of current FRT methods for PV power plants** 146
4.5 **Conclusion** 147
References 147

5. Power quality problems associated with electric vehicle charging infrastructure

P. Sivaraman and C. Sharmeela

5.1 **Introduction** 151
5.2 **Types of charging station** 152
5.3 **Power quality problems associated with EVSE and its impacts** 154
5.3.1 Power quality problems from EV charging station into the grid 154
5.3.2 Power quality problems from grid into EV charging station 157
5.3.3 Case study: harmonics analysis for a DC fast charger of 150 kW rating 157
5.4 **Mitigation of impact of higher penetration of EVs into distribution system** 158
5.4.1 Energy storage system 159

	5.4.2	Distributed FACTS devices	160
	5.4.3	Demand response	160
5.5	**Conclusion**		160
	References		161
	Further reading		161

6. Impact of power quality issues in residential systems

R. Senthil Kumar, I. Gerald Christopher Raj, S. Saravanan, P. Leninpugalhanthi and P. Pandiyan

6.1	**Introduction**		164
6.2	**Power quality disturbances in residential systems**		165
	6.2.1	Unpredictable disturbances	165
	6.2.2	Residential customers	166
	6.2.3	Effects of power quality in residential systems	169
6.3	**Power quality measurement**		170
6.4	**Study of power quality disturbances caused by a home appliance using MATLAB/Simulink**		172
	6.4.1	Modeling methodology	172
	6.4.2	Power quality impacts due to single-phase nonlinear loads	172
	6.4.3	Power quality impacts due to single-phase AC motors starting	174
6.5	**Mitigation techniques**		181
	6.5.1	Power factor correction (compensation)	184
	6.5.2	Harmonic filter	184
	6.5.3	Power quality-based equipment	186
	6.5.4	Surge protection devices	187
	6.5.5	Programmable protection devices for household appliances	188
6.6	**Conclusion**		190
	References		190

7. Voltage control in active distribution networks

Fco. Javier Zarco-Soto, José L. Martínez-Ramos and Pedro J. Zarco-Periñán

7.1	**Introduction**		193
7.2	**Voltage quality requirements**		194
7.3	**Traditional voltage control strategies**		196
7.4	**Voltage control strategies in active distribution networks**		199
	7.4.1	Transmission and distribution systems: two different scenarios	201
	7.4.2	Novel voltage control strategies	203
	7.4.3	Case study	206
7.5	**Conclusions**		214
	References		215

8. **Voltage dips caused by faults in a transmission system: a monitoring case study of a sensitive industrial consumer**

Roger Alves de Oliveira, Bibiana Petry Ferraz, Ruth Agustini, Renato Ferraz and Roberto Chouhy Leborgne

8.1 Introduction 219
8.2 Methodology 223
8.2.1 Proposed approach for voltage dip quantification 223
8.2.2 Proposed approach for grouping the faults and voltage dips 228
8.3 Case study and results 229
8.3.1 Quantification of voltage dips and comparison of IEEE 1564 and brazilian grid code 230
8.3.2 Grouping the main influence variables 237
8.3.3 Assessment of the voltage dip vulnerability area 241
8.4 Conclusions 242
References 243
Further reading 244

9. **Power quality improvement by a double-source multilevel inverter with reduced device and standing voltage on switches**

Mahdi Karimi, Paria Kargar, Kazem Varesi and P. Sanjeevikumar

9.1 Introduction 245
9.2 Proposed structure 248
9.2.1 Converter configuration 248
9.2.2 Operational scheme 249
9.2.3 Standing voltage on switches 252
9.2.4 Cascaded version of the proposed topology 253
9.3 Modulation technique 254
9.4 Power loss analysis 260
9.4.1 Switching losses 260
9.4.2 Conduction losses 261
9.4.3 Capacitor ripple losses 265
9.5 Capacitor design 265
9.6 Comparative analysis 265
9.7 Simulation results 270
9.8 Power quality analysis 272
9.8.1 Form factor 272
9.8.2 Ripple factor 272
9.8.3 Harmonic factor 274
9.8.4 Total harmonic distortion 276
9.9 Conclusions 279
References 279

10. E-STATCOM (energy storage + STATCOM): a solution to integrate large-scale wind farms into the grid at medium and high power levels

Anil Bharadwaj, Suman Maiti, Nirmalya Dhal, S. Chakraborty and Sreekesh K. Pillai

10.1	**Introduction**		284
10.2	**Challenges of renewable power-dominated grids**		285
10.3	**Existing methods to provide grid codes**		285
10.4	**Concept of an E-STATCOM**		286
	10.4.1	Energy storage system in an E-STATCOM	287
	10.4.2	Two-/three-level converter-based E-STATCOM	287
	10.4.3	Modular multilevel converter-based E-STATCOM	288
10.5	**System for performance study**		289
10.6	**Control methodology of an E-STATCOM**		291
	10.6.1	Operation of an MMC-based E-STATCOM	292
	10.6.2	Operation of hybrid storage system	297
10.7	**Discussion of results**		302
10.8	**Conclusion**		304
	References		309

11. PQ disturbance detection and classification combining advanced signal processing and machine learning tools

Md Shafiullah, Meer A.M. Khan and Shakir D. Ahmed

11.1	**Introduction**		311
11.2	**Methods**		313
	11.2.1	Advanced signal processing techniques	313
	11.2.2	Artificial neural network	315
11.3	**Proposed PQD detection and classification scheme**		317
	11.3.1	Mathematical models of PQD	317
	11.3.2	Feature extraction explanation	318
11.4	**Results and discussion**		322
	11.4.1	PQD detection results	322
	11.4.2	PQD classification results	326
	11.4.3	Validation of the proposed scheme	327
11.5	**Conclusions**		331
11.6	**List of abbreviations**		332
	Acknowledgments		333
	References		333

12. Recent trends and advances in power quality

Subrat Sahoo

12.1	**Introduction**	338
12.2	**Business motivation**	339

12.3	**Measurement methodologies and techniques**		340
	12.3.1	Flicker measurement	341
	12.3.2	Classification techniques using multiresolution methods	341
	12.3.3	Alternative architecture design for building power quality monitoring systems	341
12.4	**PQ data analytics: examples**		343
	12.4.1	Case 1: detection of a transient condition	344
	12.4.2	Case 2: detection of a voltage sag	345
	12.4.3	Case 3: detection of voltage dip in two phases	346
	12.4.4	Case 4: detection of a voltage sag—downstream event restored after 1.2 s	347
	12.4.5	Case 5: momentary interruption—short circuit cleared in the reclosure	347
	12.4.6	Case 6: several reclosure event	348
	12.4.7	Case 7: voltage dips in the network leading to overvoltage in the system in an EV environment	350
	12.4.8	Case 8: impact on EV charging infrastructure during various PQ events	350
12.5	**Recent trends on power quality**		350
	12.5.1	Impact of modern grid integration framework	351
	12.5.2	Impact of supraharmonics	351
	12.5.3	Impact of fast voltage variations	352
	12.5.4	Developments in monitoring and integration opportunities	353
	12.5.5	Data collection format	353
	12.5.6	IoT-based framework for power quality solutions	354
12.6	**Conclusion**		355
	References		355

Index 359

Contributors

Pramod Agarwal, Department of Electrical Engineering, Indian Institute of Technology, Roorkee, Uttarakhand, India

Ruth Agustini, Universidade Federal do Rio Grande do Sul (UFRGS), Porto Alegre, Rio Grande do Sul, Brazil

Shakir D. Ahmed, Center of Research Excellence in Renewable Energy, King Fahd University of Petroleum and Minerals, Dhahran, Eastern Province, Saudi Arabia

Roger Alves de Oliveira, Universidade Federal do Rio Grande do Sul (UFRGS), Porto Alegre, Rio Grande do Sul, Brazil

Gökay Bayrak, Bursa Technical University, Engineering and Natural Sciences, Department of Electrical and Electronics Engineering, Bursa, Turkey

Anil Bharadwaj, Department of Electrical Engineering, Indian Institute of Technology Kharagpur, Kharagpur, West Bengal, India

S. Chakraborty, Department of Electrical Engineering, Indian Institute of Technology Kharagpur, Kharagpur, West Bengal, India

I. Gerald Christopher Raj, Department of Electrical and Electronics Engineering, PSNA College of Engineering and Technology, Dindigul, Tamil Nadu, India

Nirmalya Dhal, Department of Electrical Engineering, Indian Institute of Technology Kharagpur, Kharagpur, West Bengal, India

Bibiana Petry Ferraz, Universidade Federal do Rio Grande do Sul (UFRGS), Porto Alegre, Rio Grande do Sul, Brazil

Renato Ferraz, Universidade Federal do Rio Grande do Sul (UFRGS), Porto Alegre, Rio Grande do Sul, Brazil

Davood Ghaderi, Bursa Technical University, Engineering and Natural Sciences, Department of Electrical and Electronics Engineering, Bursa, Turkey

Paria Kargar, Power Electronics Research Lab. (PERL), Electrical Engineering, Sahand University of Technology, Tabriz, East Azerbaijan, Iran

Mahdi Karimi, Power Electronics Research Lab. (PERL), Electrical Engineering, Sahand University of Technology, Tabriz, East Azerbaijan, Iran

Meer A.M. Khan, Center of Research Excellence in Renewable Energy, King Fahd University of Petroleum and Minerals, Dhahran, Eastern Province, Saudi Arabia

R. Senthil Kumar, Department of Electrical and Electronics Engineering, Sri Krishna College of Technology, Coimbatore, Tamil Nadu, India

Roberto Chouhy Leborgne, Universidade Federal do Rio Grande do Sul (UFRGS), Porto Alegre, Rio Grande do Sul, Brazil

P. Leninpugalhanthi, Department of Electrical and Electronics Engineering, Sri Krishna College of Technology, Coimbatore, Tamil Nadu, India

Vasundhara Mahajan, Department of Electrical Engineering, Sardar Vallabhbhai National Institute of Technology (SVNIT), Surat, Gujarat, India

Suman Maiti, Department of Electrical Engineering, Indian Institute of Technology Kharagpur, Kharagpur, West Bengal, India

José L. Martínez-Ramos, Department of Electrical Engineering, University of Seville, Seville, Spain

Hari Om Gupta, Jaypee Institute of Technology, Greater Noida, Uttar Pradesh, India

P. Pandiyan, Department of Electrical and Electronics Engineering, Sri Krishna College of Technology, Coimbatore, Tamil Nadu, India

Sreekesh K. Pillai, Department of Electrical Engineering, Indian Institute of Technology Kharagpur, Kharagpur, West Bengal, India

Subrat Sahoo, ABB Power Grids Research, Vasteras, Sweden

P. Sanjeevikumar, Esbjerg Energy Section, Department of Energy Technology, Aalborg University, Esbjerg, Denmark

S. Saravanan, Department of Electrical and Electronics Engineering, Sri Krishna College of Technology, Coimbatore, Tamil Nadu, India

Md Shafiullah, Center of Research Excellence in Renewable Energy, King Fahd University of Petroleum and Minerals, Dhahran, Eastern Province, Saudi Arabia

C. Sharmeela, Department of Electrical and Electronics Engineering, College of Engineering Guindy, Anna University, Chennai, Tamil Nadu, India

P. Sivaraman, Leading Engineering Organisation, Chennai, Tamil Nadu, India

Kazem Varesi, Power Electronics Research Lab. (PERL), Electrical Engineering, Sahand University of Technology, Tabriz, East Azerbaijan, Iran

Pedro J. Zarco-Periñán, Department of Electrical Engineering, University of Seville, Seville, Spain

Fco. Javier Zarco-Soto, Department of Electrical Engineering, University of Seville, Seville, Spain

Preface

The term power quality represents *new findings* and *challenges* for researchers to fulfill the power and energy demands of societal communities. Any researcher with vast experience in power engineering and research will encounter power quality disturbances such as transients, root mean square variation of both long and short duration, imbalance, voltage and current distortions, voltage fluctuations, frequency variations, and power factor compensation. This book will find an enhanced solution for power quality with real-time case studies from the editors and diverse authors.

The book covers different areas dealing with: power quality analysis in real-time analysis accompanied by practical issues, fault issues and fault monitoring in the transmission network, power quality in electric vehicles and residential buildings, fault detection using artificial intelligence, application of power electronics circuits in the power system E-STATCOM, and present/future trends in power quality.

It describes topics with theoretical-based analysis with numerical solutions and hardware experimental results from real-time cases, which will help readers to select power quality as their future research or profession. Exciting results from ABB Power Grids Research, Sweden, will help readers to understand the basics and advancements of power quality industrial innovation.

This book mainly focuses on the guidelines followed in the IEEE standards for the presented results and cross-verification from real-time case studies. The chapters cover significant and bottleneck challenges prevailing in modern power systems and enable readers to better understand IEEE standards 1159 -2019, 519-2014, 1547, 1346, etc. The book will be readily available as a reference for IEEE standards and enable the student community to create more interest and take up the challenges of the power quality profession.

Editors

Acknowledgment

Foremost, thanks to the Almighty for his everlasting love throughout this endeavor.

Acknowledgment is always appreciated and timely support and the bond of encouragement are vital tools for teachers and researchers from their institutions. In this regard, the editors express their sincere thanks to Mr. S. Muthukumaran, Director, TECH Engineering Services, Chennai, India; Mr. K. Balaji, Sr., Electrical Engineer, Sree Nandees Technologies, Chennai, India; Mr. S. Rajkumar, Executive, JLL, Bengaluru, India; Mr. K. Sasikumar, Electrical Engineer, Mott MacDonald, Noida, India; Ms. A. Thaiyal Nayagi, Lecturer, Rane Polytechnic College, Trichy, India; and the Center for Bioenergy and Green Engineering, Department of Energy Technology, Aalborg University, Esbjerg, Denmark, and the Department of Electrical and Electronics Engineering, College of Engineering Guindy, Anna University, Chennai, India. The editors appreciate the full support of their institution, which hopefully will make this book a great success.

I wish One and All for their devoted time and effort.

Editors

Chapter 1

Power quality and its characteristics

P. Sivaraman[1], C. Sharmeela[2]
[1]*Leading Engineering Organisation, Chennai, Tamil Nadu, India;* [2]*Department of Electrical and Electronics Engineering, College of Engineering Guindy, Anna University, Chennai, Tamil Nadu, India*

Chapter outline

1.1 How do changes in voltage, current, or frequency impact the quality of power? 2
1.2 Types of power quality problems 2
1.3 Transients 3
1.3.1 Cause 3
1.3.2 Impact 4
1.3.3 Types 4
1.3.3.1 Impulsive transients 4
1.3.3.2 Oscillatory transients 5
1.3.4 Example 5
1.4 Short duration RMS variation 6
1.4.1 Voltage sag 7
1.4.1.1 Causes 10
1.4.1.2 Impact 10
1.4.1.3 Example 10
1.4.2 Voltage swell 12
1.4.2.1 Cause 14
1.4.2.2 Impact 15
1.4.2.3 Example 16
1.4.3 Interruption 19
1.4.3.1 Cause 20
1.4.3.2 Impact 21
1.4.3.3 Example 21
1.5 Long duration voltage variation 24
1.5.1 Undervoltage 24
1.5.1.1 Cause 25
1.5.1.2 Impact 25
1.5.1.3 Example 26
1.5.2 Overvoltage 27
1.5.2.1 Cause 27
1.5.2.2 Impact 29
1.5.3 Sustained interruption 29
1.5.3.1 Cause 30
1.5.3.2 Impact 30
1.5.3.3 Example 30
1.6 Unbalance or imbalance 32
1.6.1 Cause 33
1.6.2 Impact 33
1.6.3 Example 33
1.7 Voltage fluctuations 34
1.7.1 Cause 37
1.7.2 Impact 37
1.7.3 Example 38
1.8 Waveform distortion 38
1.8.1 Harmonics 39
1.8.2 Interharmonics 42
1.8.2.1 Cause 43
1.8.2.2 Impacts 43
1.8.3 DC offset 43
1.8.3.1 Cause 43
1.8.3.2 Impacts 43
1.8.4 Notching 43
1.8.4.1 Cause 44
1.8.4.2 Impacts 45
1.8.5 Noise 45

Power Quality in Modern Power Systems. https://doi.org/10.1016/B978-0-12-823346-7.00001-3

1.8.5.1 Cause 45
1.8.5.2 Impacts 46
1.9 Annexure 1.1 47
1.10 Annexure 1.2 50
1.11 Annexure 1.3: harmonics 50
1.11.1 Odd harmonics 50
1.11.2 Even harmonics 53
1.11.3 Tripplen harmonics 57
References 59
Further reading 60

1.1 How do changes in voltage, current, or frequency impact the quality of power?

An ideal source should deliver proper voltage irrespective of the connected load. The voltage can be either AC or DC. In a normal scenario, there will be some impedance associated with the source called source impedance. Current flows from the voltage source and the magnitude are decided by the circuit impedance. Circuit impedance is a combination of source impedance, distribution impedance, and connected load in which source impedance would be a negligible component. The source voltage can be disturbed due to various reasons like lightning, faults, and noncompatible loads. This disturbance can be classified into various categories based on the magnitude and duration of the disturbance. Similar to voltage, current can also have disturbances that again are classified into various categories. The sinusoidal component should have a stable frequency. If there are frequency variations, these will also affect the connected load, which indirectly has an impact on the quality of power.

1.2 Types of power quality problems

Generally, power quality problems are divided into two types:

- ➢ Voltage-related problems
- ➢ Current-related problems

Voltage-related problems include sags, swells, blackouts, harmonics, etc., while current-related problems include electromagnetic field, leakage current, electromagnetic interference, radiofrequency interference, etc.

As per IEEE standard 1159–2019, power quality problems are classified into the following:

1. Transients
2. Short duration root mean square (RMS) variation
3. Long duration RMS variation
4. Imbalance
5. Waveform distortion
6. Voltage fluctuations
7. Power frequency variations

TABLE 1.1 Classifications of transients.

Parameter	Category	Type	Duration/Frequency
Transients	Impulsive	Nanosecond	<50 ns
		Microsecond	50 ns–1 ms
		Millisecond	>1 ms
	Oscillatory	Low frequency	<5 kHz
		Medium frequency	5–500 kHz
		High frequency	0.5–5 MHz

1.3 Transients

Transients are momentary changes in voltage or current or both resulting from switching operation, lightning, faults, etc. The time duration for a transient varies from nanoseconds to milliseconds. The classifications of transients as listed in IEEE 1159–2019 [2] are shown in Table 1.1.

1.3.1 Cause

Transient overvoltage is the result of the rapid change of current in an inductive circuit. Sources of transients are listed in Table 1.2.

TABLE 1.2 Sources of transients.

Category	Type	Sources of problems
Transients	External sources	Lightning strike
		Opening and closing of energized lines
		Breaker opening and closing
		Transients from other users in shared sources
	Internal sources	Switching operation
		High resistance fault
		Photocopy machines
		Welding machines
		Capacitor bank switching
		High-frequency switching in inverter or switch-mode power supply

1.3.2 Impact

1. Failures of insulation in inductive loads if the transients exceed the impulsive voltage withstand capability.
2. Failure of equipment due to lightning-induced overvoltage.
3. Failure of semiconductor devices and their associated circuits due to peak inverse voltage.
4. Failure of control wiring in electronic circuits.

1.3.3 Types

Based on the polarity change, transients are classified into two categories:

➢ Impulsive
➢ Oscillatory

1.3.3.1 Impulsive transients

An impulsive transient is a sudden change in voltage or current or both in unidirectional polarity either positive or negative [5]. Fig. 1.1 shows a positive polarity impulsive transient appearing in the negative cycle, whereas Fig. 1.2 shows a positive polarity impulsive transient appearing in the positive cycle. Similarly, a negative polarity impulsive transient can appear either in the positive cycle or in the negative cycle.

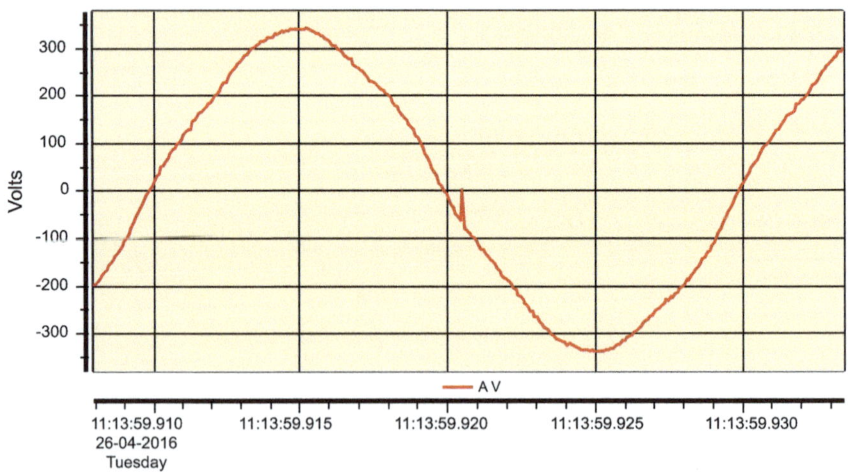

FIGURE 1.1 Positive polarity impulsive transient appearing in the negative cycle.

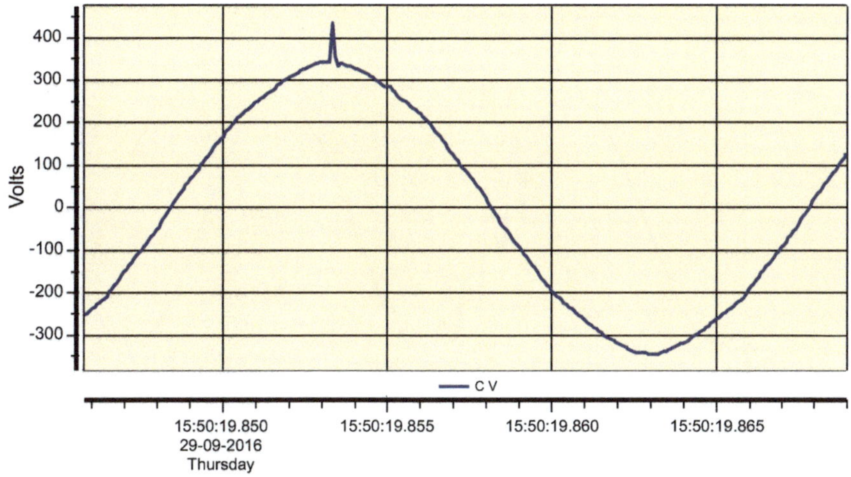

FIGURE 1.2 Positive polarity impulsive transient appearing in the positive cycle.

1.3.3.2 Oscillatory transients

An oscillatory transient is a sudden change in voltage or current or both in bidirectional polarity (positive and negative). In other words, frequency variations within the voltage cycle up to 5 MHz are called oscillatory transients.

Fig. 1.3 shows the oscillatory transients in the voltage wave shape of a three-phase circuit.

1.3.4 Example

Oscillatory transients at an 11 kV utility grid through practical measurements

The grid supply voltage of 11 kV is stepped down to 433 V by a 2.5 MVA stepdown transformer. A single-line diagram (SLD) of electrical distribution is shown in Fig. 1.4.

The power quality analyzer configured in 3φ 4 W Y mode is connected at the metering potential transformers (PTs) and current transformers (CTs) of an 11 kV supply voltage as shown in Fig. 1.4. The captured oscillatory transients in all three-phase voltage wave shapes are shown in Figs. 1.5 and 1.6.

Event 1: @09:10:29:25 h, the power quality analyzer captured the oscillatory transient in all three-phase voltage wave shapes followed by the voltage sag event as shown in Fig. 1.5.

Event 2: @09:31:10:96 h, the power quality analyzer captured another oscillatory transient in the same circuit as shown in Fig. 1.6.

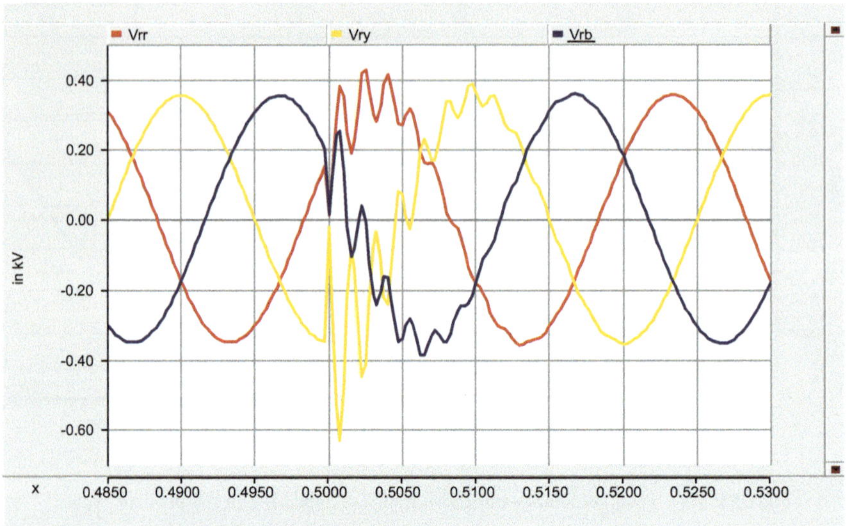

FIGURE 1.3 Oscillatory transients in a three-phase circuit.

1.4 Short duration RMS variation

Variations in RMS voltage lasting from 0.5 cycles to less than or equal to 1 min are classified as short duration RMS variations as per IEEE 1159–2019 [2]. Accordingly, reduction in voltage is termed sag, while an increase in voltage is termed swell. The absence of voltage is termed an interruption. Based on the duration of the event, short duration voltage variation can be classified into three categories, namely instantaneous, momentary, and temporary. The different classifications of short duration voltage variations are listed in Table 1.3 [2].

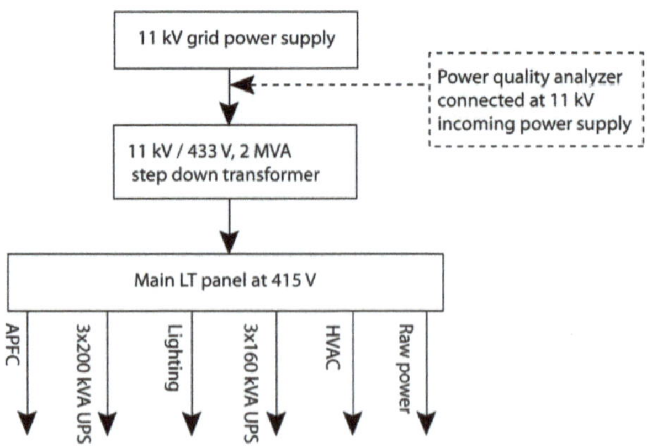

FIGURE 1.4 Schematic diagram of power distribution. *APFC*, Automatic power factor controller; *HVAC*, heating, ventilation, and air conditioning; *LT*, low tension; *UPS*, uninterruptible power supply.

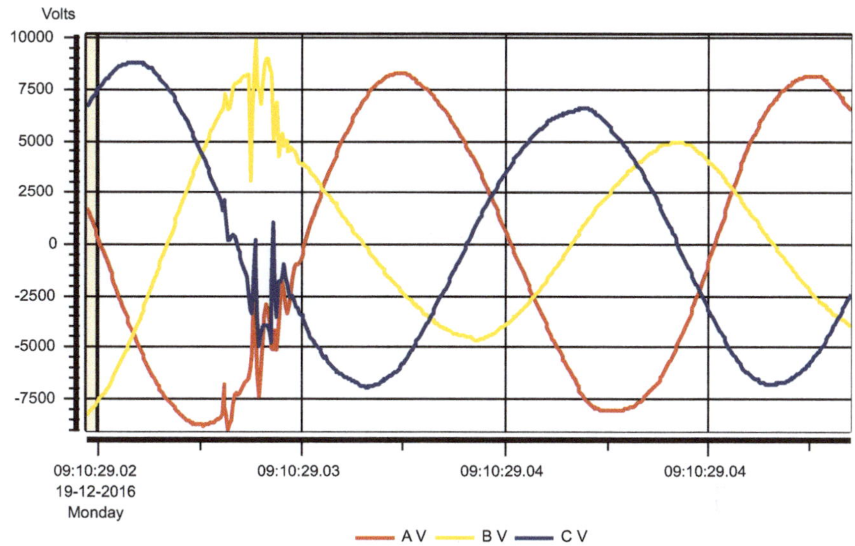

FIGURE 1.5 Oscillatory transients in all three-phase voltage wave shapes.

1.4.1 Voltage sag

Voltage sags are short duration reductions in voltage magnitude and duration lasting typically from a few cycles to a few seconds [3,13]. Definition as per IEEE 1159 is a decrease in RMS voltage between 0.1 and 0.9 pu and a time duration from 0.5 cycles to 1 min. The depth of voltage sag depends on

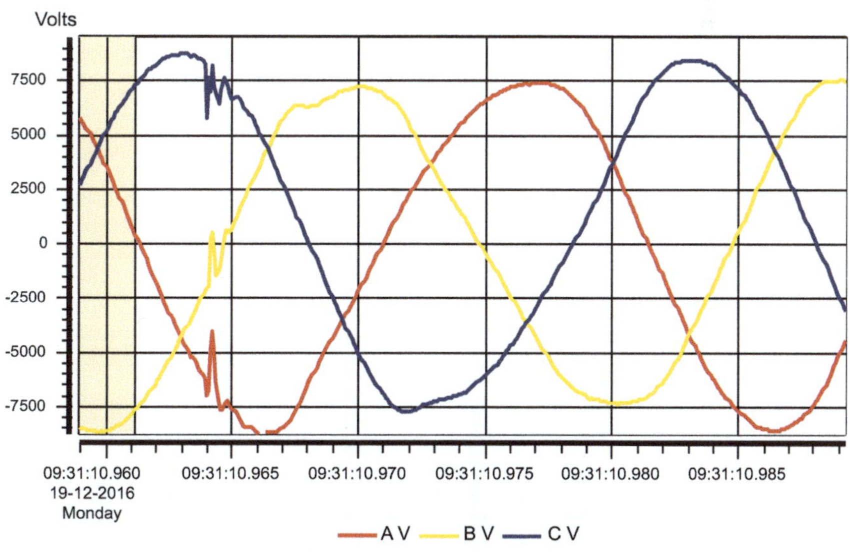

FIGURE 1.6 Oscillatory transients in all three-phase voltage wave shapes.

TABLE 1.3 Short duration voltage variation.

	Instantaneous		Momentary		Temporary	
Parameters	**Duration**	**Voltage magnitude**	**Duration**	**Voltage magnitude**	**Duration**	**Voltage magnitude**
Sag	0.5–30 cycles	0.1–0.9 pu	30 cycles–3 s	0.1–0.9 pu	>3 s–1 min	0.1–0.9 pu
Swell	0.5–30 cycles	1.1–1.8 pu	30 cycles–3 s	1.1–1.4 pu	>3 s–1 min	1.1–1.2 pu
Interruption	–	–	0.5 cycles–3 s	<0.1 pu	>3 s–1 min	<0.1 pu
Voltage imbalance	–	–	30 cycles–3 s	2%–15%	>3 s–1 min	2%–15%

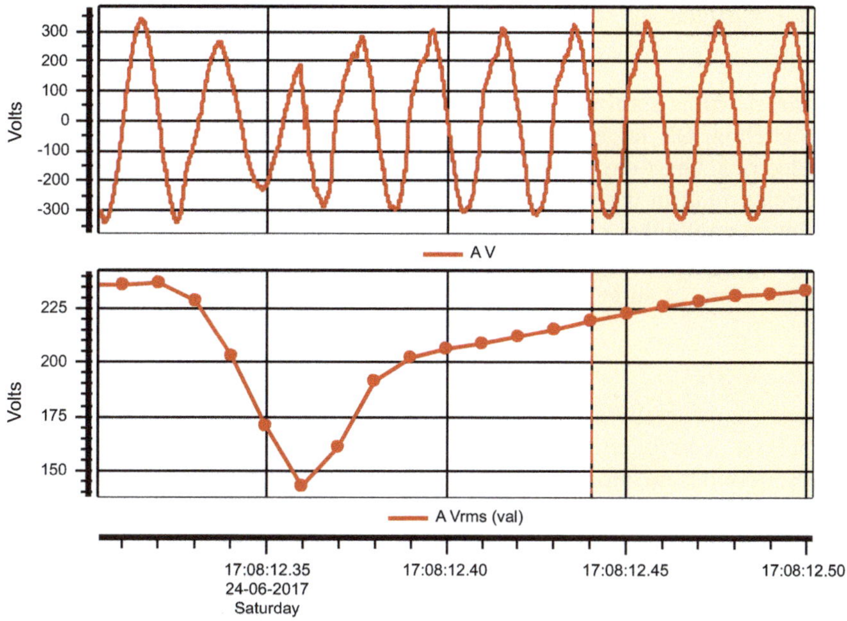

FIGURE 1.7 Wave shape of voltage sag.

multiple factors like network impedance, distance of the fault occurrence, voltage level at the fault location, connected loads at the time of fault, and voltage improvement due to reactive power components in the power system. The duration of voltage sag is determined by the dynamics of rotating loads, system impedance, and fault clearing time.

Fig. 1.7 top shows the wave shape of the voltage sag; the bottom image captures the disturbance in the RMS voltage trend. Table 1.4 describes the peak and RMS voltage recorded just before the sag, during the sag, and after recovering from the sag.

TABLE 1.4 Peak and root mean square (RMS) voltage variations during sag event.

Parameter	Before sag event at 17:08:12.32	During sag event	After sag recovery at 17:08:12.47
Voltage wave shape (peak) in V	340	178	333
RMS voltage in V	235	142	232

1.4.1.1 Causes

Voltage sag can be caused by many conditions, a few of which are:

1. Direct online starting of large motors
2. Switching ON large loads in one instance
3. Electrical short circuit
4. Arcing fault in the system
5. Energizing higher-capacity power transformer
6. Switching OFF large reactive power source
7. Operations of surge suppressors in the power system

1.4.1.2 Impact

Voltage sag generally affects the end-user equipment in terms of malfunction of process controllers, programmable logic controllers, adjustable speed drives, robotics in modem industries, reduction in the performance of the motor, flickering in illumination devices, etc.

1.4.1.3 Example

Voltage sag due to a fault in a remote location at an 11 kV feeder by practical measurement

The grid supply voltage of 11 kV is stepped down to 433 V by a 2.5 MVA step-down transformer [11]. An SLD of electrical distribution is shown in Fig. 1.8.

The power quality analyzer configured in 3ϕ 4 W Y mode is connected at the metering PTs and CTs of an 11 kV supply voltage between 19:35:06 h on December 18, 2016 and 11:26:10 h on December 19, 2016. During this period,

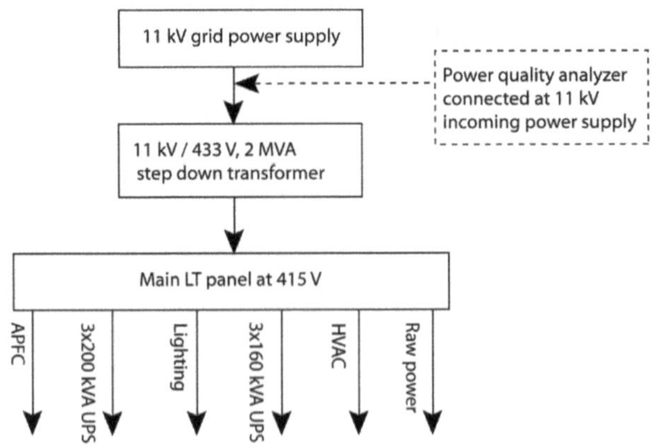

FIGURE 1.8 Schematic diagram of power distribution. *APFC*, Automatic power factor controller; *HVAC*, heating, ventilation, and air conditioning; *LT*, low tension; *UPS*, uninterruptible power supply.

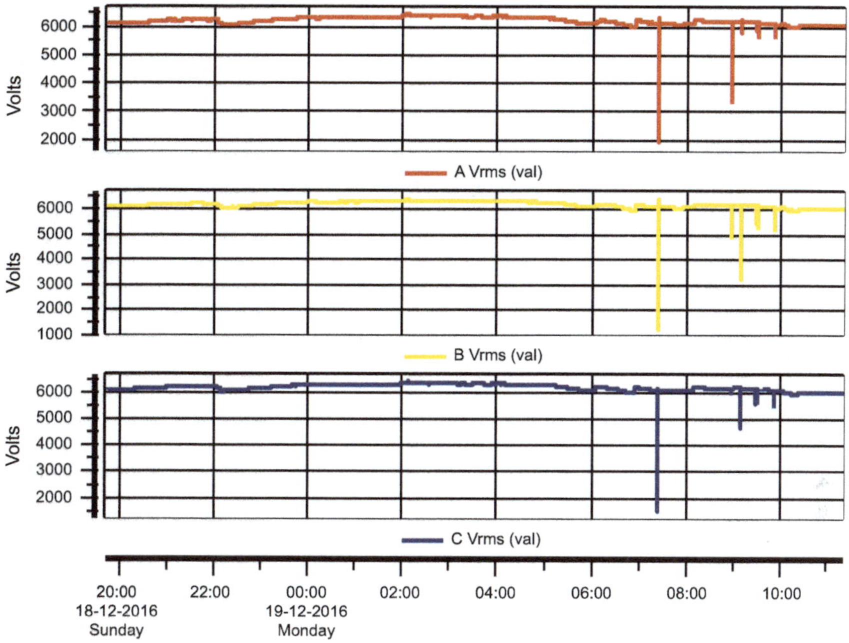

FIGURE 1.9 Root mean square voltage trend.

voltage sag occurred multiple times (five times) in all three phases. The RMS voltage trend is shown in Fig. 1.9.

Event 1: @07:25:19:93 h, Fig. 1.10 shows the voltage sag in the R phase voltage wave shape and RMS voltage trend. Peak and RMS voltage variations during sag event is listed in Table 1.5.

The duration of the voltage sag is 223 ms.

Event 2: @08:57:34:80 h, the power quality analyzer captured another voltage sag event in the same circuit. Three-phase voltage wave shape and RMS voltage trend are shown in Fig. 1.11. Peak and RMS voltage variations during sag event is listed in Table 1.6.

The duration of the voltage sag is 59.9 ms.

Event 3: @09:10:29:03 h, Fig. 1.12 shows the voltage sag in all three-phase voltage wave shapes and RMS voltage trend. Peak and RMS voltage variations during sag event is listed in Table 1.7.

The duration of the voltage sag is 56 ms.

Event 4: @09:30:09:76 h, the power quality analyzer captured another voltage sag event in the same circuit. Three-phase voltage wave shape and RMS voltage trend are shown in Fig. 1.13. Peak and RMS voltage variations during sag event is listed in Table 1.8.

The duration of the voltage sag is 51 ms.

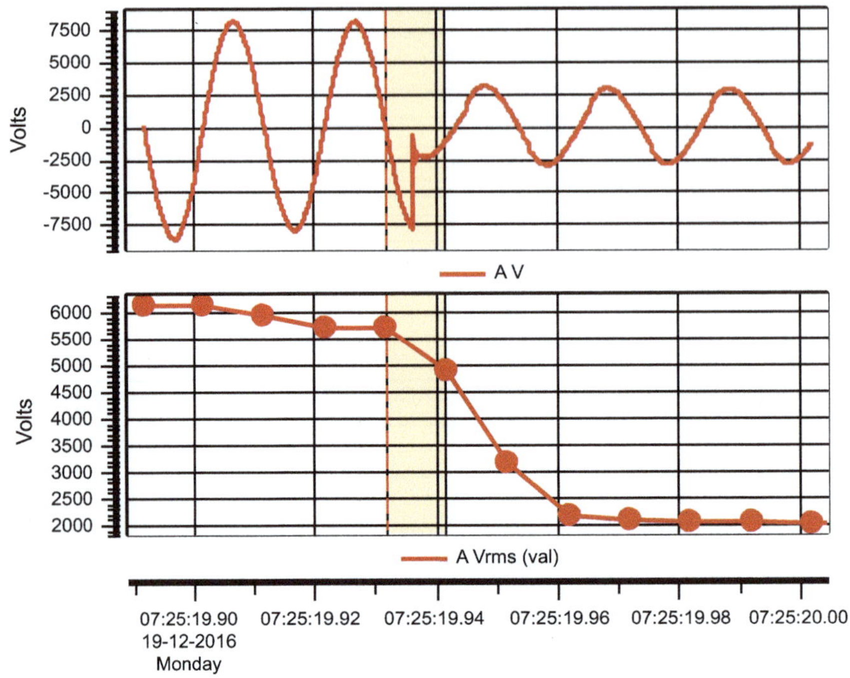

FIGURE 1.10 Wave shape and root mean square voltage trend during voltage sag event.

Event 5: @09:51:58:66 h, the power quality analyzer captured another voltage sag event in the same circuit. Three-phase voltage wave shape and RMS voltage trend are shown in Fig. 1.14. Peak and RMS voltage variations during sag event is listed in Table 1.9.

The duration of the voltage sag is 80 ms.

1.4.2 Voltage swell

Voltage swells are short duration increases in voltage magnitude lasting typically from a few cycles to a few seconds. Definition as per IEEE 1159 is an

TABLE 1.5 Peak and root mean square (RMS) voltage variations during sag event.

Parameter	Before sag event	During sag event
R ϕ voltage wave shape (peak) in kV	8156	2965
RMS voltage of R ϕ in kV	6299	1924

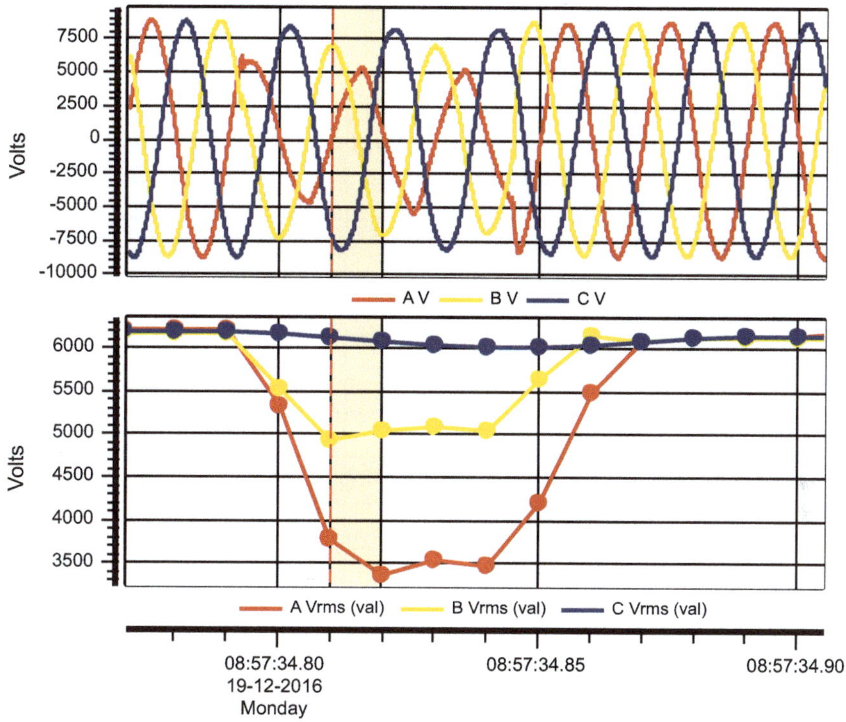

FIGURE 1.11 Wave shape and root mean square voltage trend during voltage sag event.

increase in RMS voltage between 1.1 and 1.8 pu and a time duration from 0.5 cycles to 1 min [2]. The typical voltage magnitude is between 1.1 and 1.2 pu.

Fig. 1.15 top shows the wave shape of voltage sag; the bottom image captures the disturbance in the RMS voltage trend. Table 1.10 describes the peak and RMS voltage recorded just before the swell, during the swell, and after recovering from the swell.

TABLE 1.6 Peak and root mean square (RMS) voltage variations during sag event.

Parameter	Before sag event	During sag event	After sag event
R ϕ voltage wave shape (peak) in kV	8806	5535	8727
RMS voltage of R ϕ in kV	6211	1924	6183

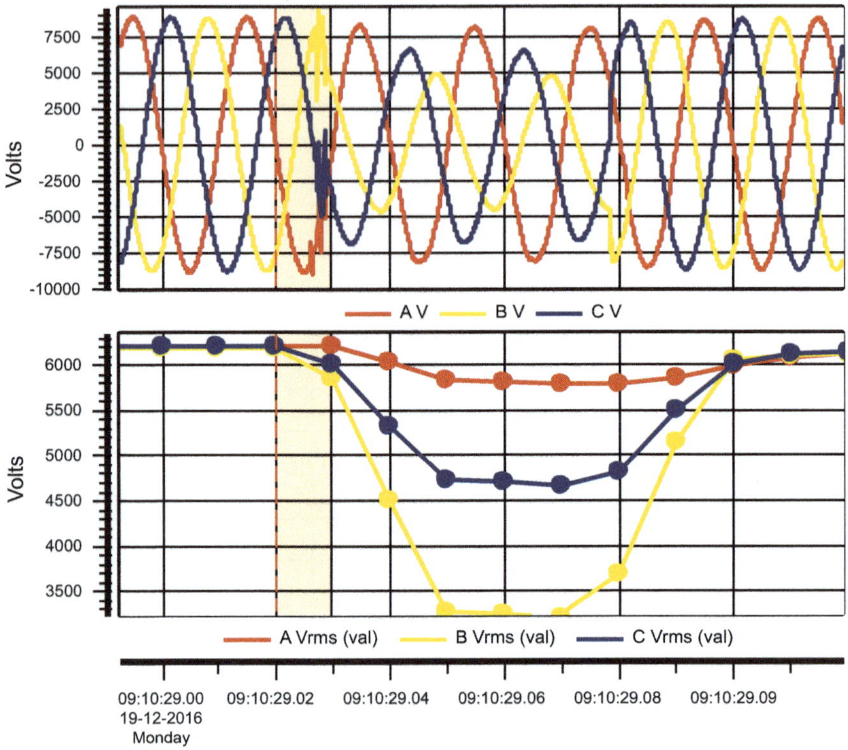

FIGURE 1.12 Wave shape and root mean square voltage trend during voltage sag event.

1.4.2.1 Cause

Voltage swell can be caused by many conditions, a few of which are:

1. Switching OFF large motors
2. Switching OFF large loads in one instance
3. Electrical short circuit (L-G fault)
4. Switching ON a large reactive power source

TABLE 1.7 Peak and root mean square (RMS) voltage variations during sag event.

Parameter	Before sag event	During sag event	After sag event
Y ϕ voltage wave shape (peak) in kV	8806	4716	8662
RMS voltage of Y ϕ in kV	6201	3200	6119

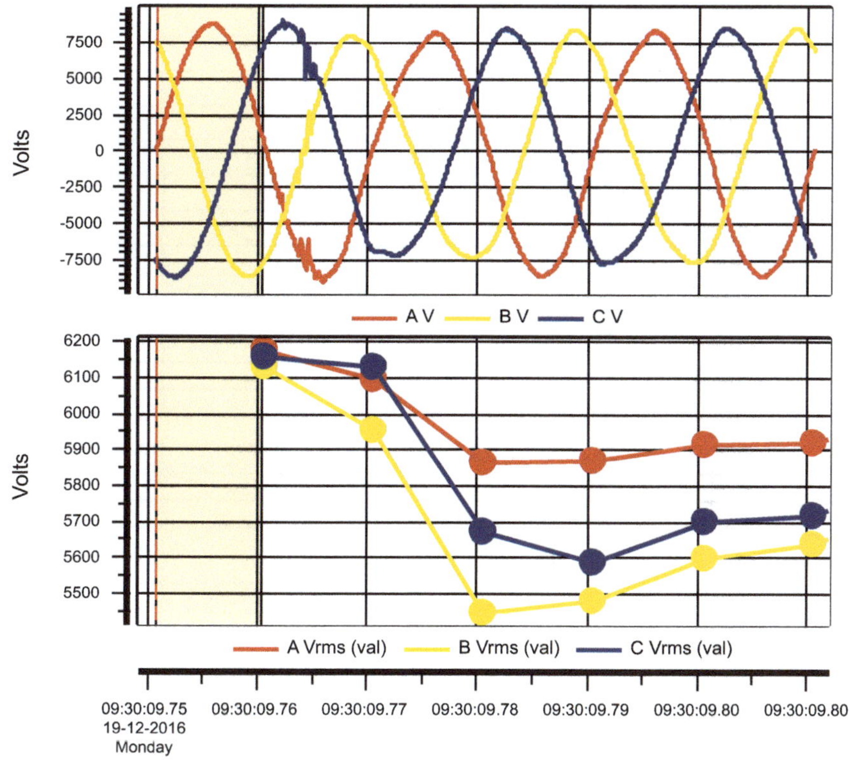

FIGURE 1.13 Wave shape and root mean square voltage trend during voltage sag event.

1.4.2.2 Impact

Voltage swell generally affects the end-user equipment in terms of malfunction of process controllers, programmable logic controllers, adjustable speed drives, robotics in modem industries, reduction in the performance of the motor, flickering in illumination, etc.

TABLE 1.8 Peak and root mean square (RMS) voltage variations during sag event.

Parameter	Before sag event	During sag event	After sag event
Y ϕ voltage wave shape (peak) in kV	8490	7236	8682
RMS voltage of Y ϕ in kV	6177	5444	6121

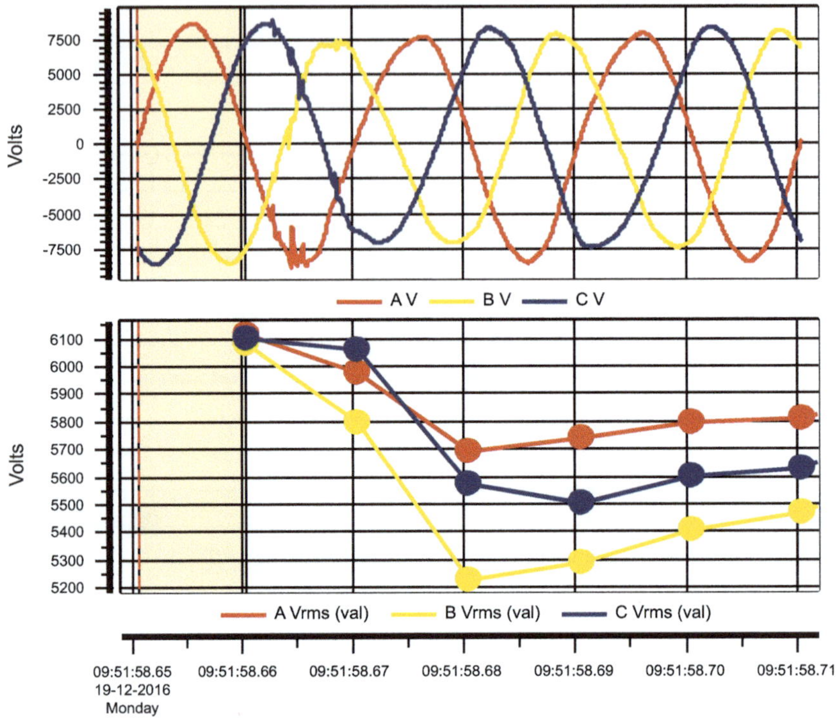

FIGURE 1.14 Wave shape and root mean square voltage trend during voltage sag event.

1.4.2.3 Example

Voltage swell due to single line-to-ground fault in the ungrounded system through PSCAD simulation

Fig. 1.16 shows a transformer of 2.5 MVA stepping down the voltage from 11 kV to 433 V to power a large induction motor of 500 HP and 560 kVA, with a 0.9 PF lumped load. All the respective breakers, namely HVBRK, LVBRK1, LVBRK2 and LVBRK3, are enclosed and the system is operating under normal conditions. The transformer low-voltage side (secondary) is not

TABLE 1.9 Peak and root mean square (RMS) voltage variations during sag event.

Parameter	Before sag event	During sag event	After sag event
Y ϕ voltage wave shape (peak) in kV	8655	7354	8638
RMS voltage of Y ϕ in kV	6081	5224	6067

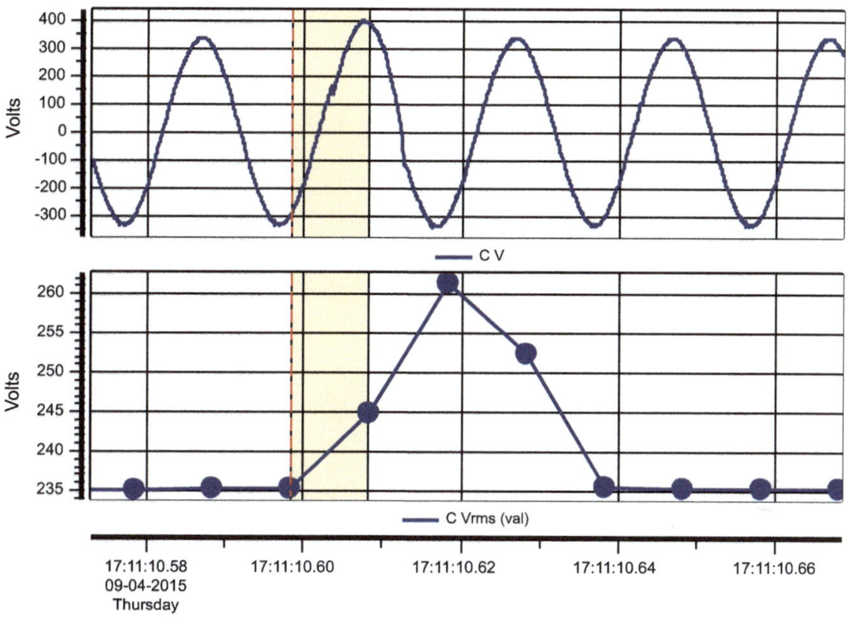

FIGURE 1.15 Wave shape of voltage swell.

grounded. A single line-to-ground fault (R phase to ground) is injected at the terminals of a lumped load at 0.5 s. The LVBRK3 breaker clears the fault after 40 ms.

Figs. 1.17 and 1.18 show the voltage oscillogram and RMS voltage trend at the lumped load feeder terminals, respectively, while Figs. 1.19 and 1.20 show the voltage oscillogram and RMS voltage trend at the motor feeder terminals. The fault is cleared by the LVBKR3 breaker and the voltage falls to zero in Figs. 1.17 and 1.18. As the fault is cleared by LVBKR3, the voltage restores to normal at the motor terminals as shown in Figs. 1.19 and 1.20. Table 1.11 describes the peak and RMS voltage recorded just before the swell, during the swell, and after recovering from the swell at the motor load feeder.

TABLE 1.10 Peak and root mean square (RMS) voltage variations during a swell event.

Parameter	Before sag event at 17:08:12.32	During sag event	After sag recovery at 17:08:12.47
Voltage wave shape (peak) in V	335.7	396.6	335.6
RMS voltage in V	235.2	261.2	235.4

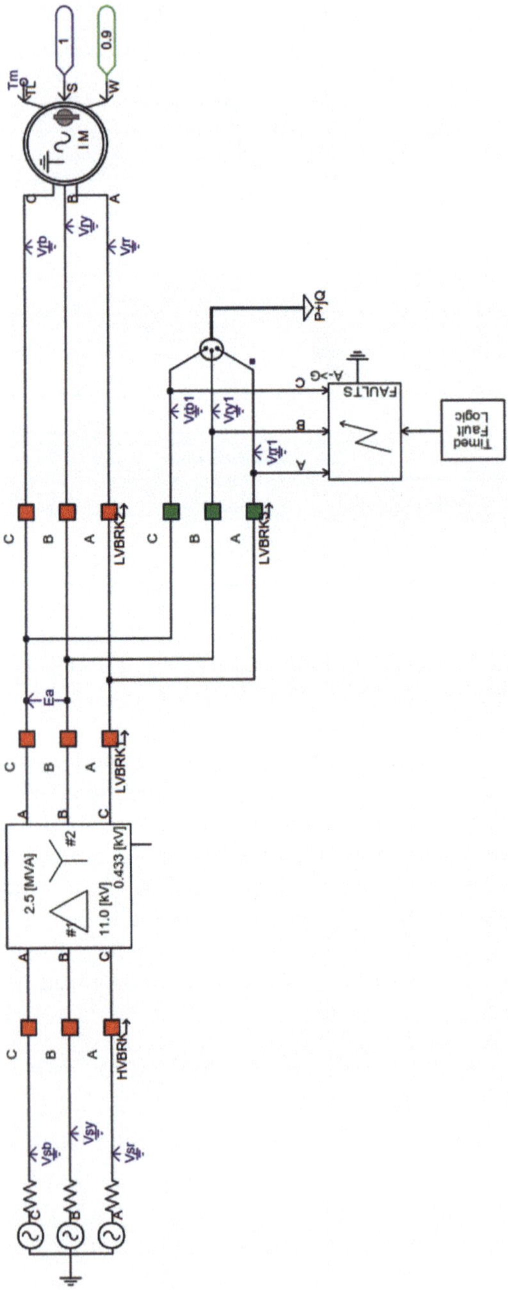

FIGURE 1.16 Schematic diagram of the circuit.

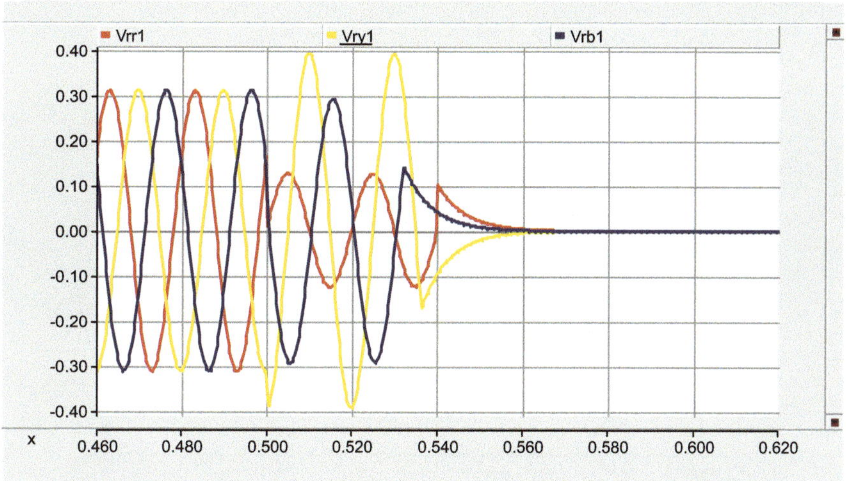

FIGURE 1.17 Voltage oscillogram at lumped load terminals at lumped load.

A swell event lasting for 40 ms was captured wherein voltage increased to 127% of the nominal value at the motor terminals.

1.4.3 Interruption

Momentary interruptions are short duration reductions in voltage magnitude of less than 0.1 pu for less than 1 min time duration. The duration of the momentary interruption is determined by the fault clearing time and breaker closing time. Fig. 1.21 shows momentary interruption in the RMS voltage

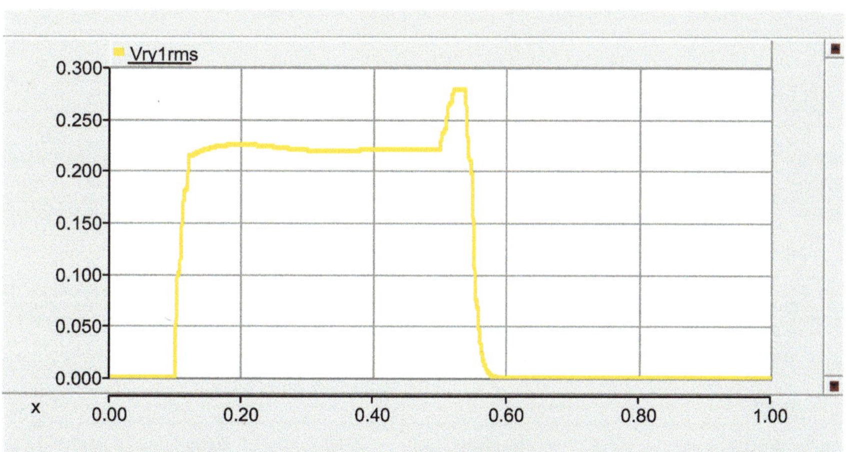

FIGURE 1.18 Root mean square voltage trend at lumped load terminals at lumped load.

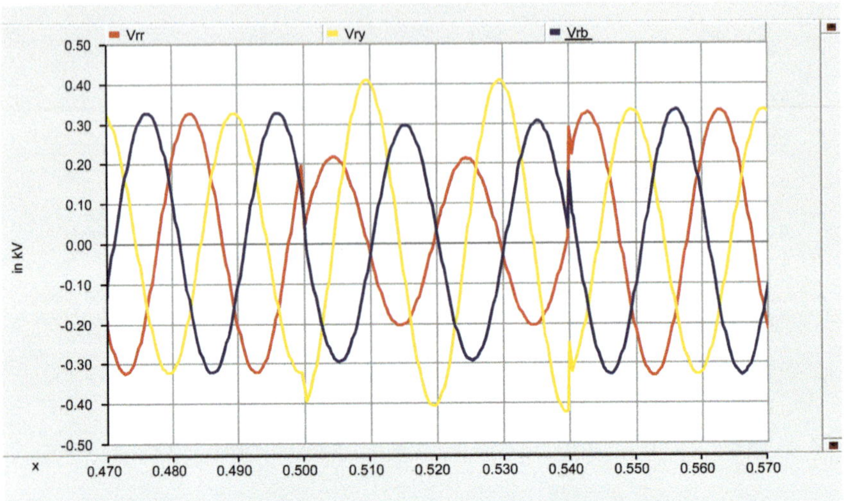

FIGURE 1.19 Voltage oscillogram at lumped load terminals at motor load.

trend. Table 1.12 describes the RMS voltage recorded just before, during, and after the momentary interruption.

The duration of the momentary interruption is 19 s.

1.4.3.1 Cause

Interruption can be caused by many conditions, a few of which are:

1. Power system faults (electrical faults)
2. Equipment failures
3. Control equipment malfunctions

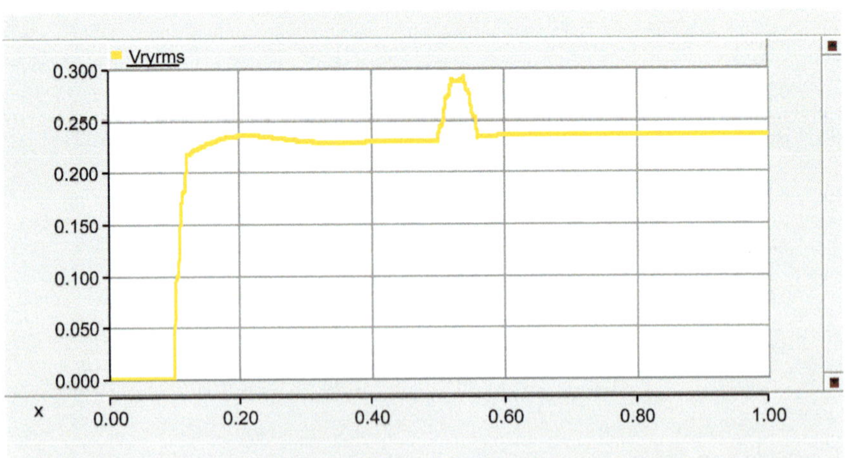

FIGURE 1.20 Root mean square voltage trend at lumped load terminals at motor load.

TABLE 1.11 Peak and root mean square (RMS) voltage variations during a swell event.

Parameter	Before swell event at 500 ms	During swell event	After swell recovery at 530 ms
Y ϕ voltage wave shape (peak) in kV	0.32	0.4	0.32
RMS voltage of Y ϕ in kV	0.22	0.28	0.23

1.4.3.2 Impact

Production stopped due to interruption, which leads to production losses in manufacturing.

1.4.3.3 Example

Interruption through field measurements

The grid supply voltage of 22 kV is stepped down to 433 V by a 2.5 MVA stepdown transformer. An SLD of electrical distribution is shown in Fig. 1.22.

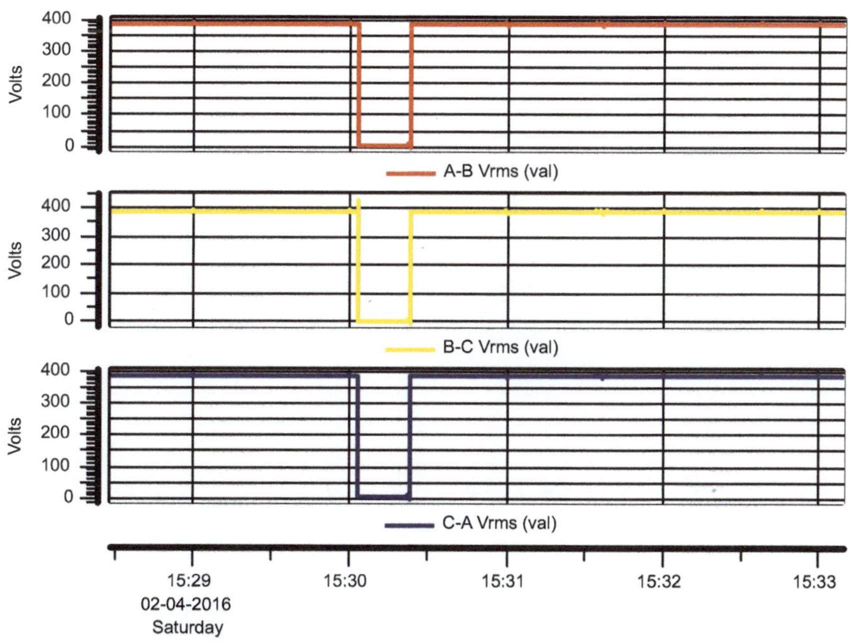

FIGURE 1.21 Root mean square voltage trend.

TABLE 1.12 Root mean square (RMS) voltage variations during the momentary interruption.

Parameter	Before the interruption event	During the interruption event	After interruption recovery
RMS voltage (R–Y) in V	400	4.5	400
RMS voltage (Y–B) in V	402	2.8	401
RMS voltage (R–B) in V	401	6	401

The power quality analyzer configured in 3ϕ 4 W Y mode is connected directly at the 433 V incoming side of the main low-tension panel as shown in Fig. 1.22 from 15.28 h to 15.32 h captured voltage interruption in all three-phase wave shapes. The voltage wave shapes and RMS voltage trend are shown in Figs. 1.23 and 1.24.

Interruption occurs between 15:30:03 and 15:30:22 h. The duration of the interruption is 19 s or 190 ms observed at the main incoming supply.

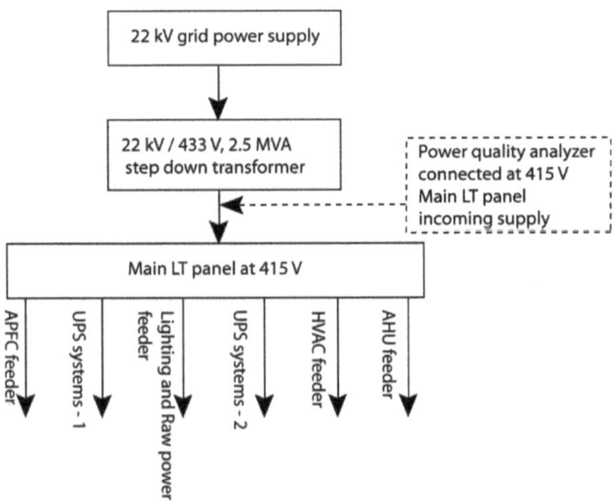

FIGURE 1.22 Single-line diagram of electrical distribution. *AHU*, Air-handling unit; *APFC*, automatic power factor controller; *HVAC*, heating, ventilation, and air conditioning; *LT*, low tension; *UPS*, uninterruptible power supply.

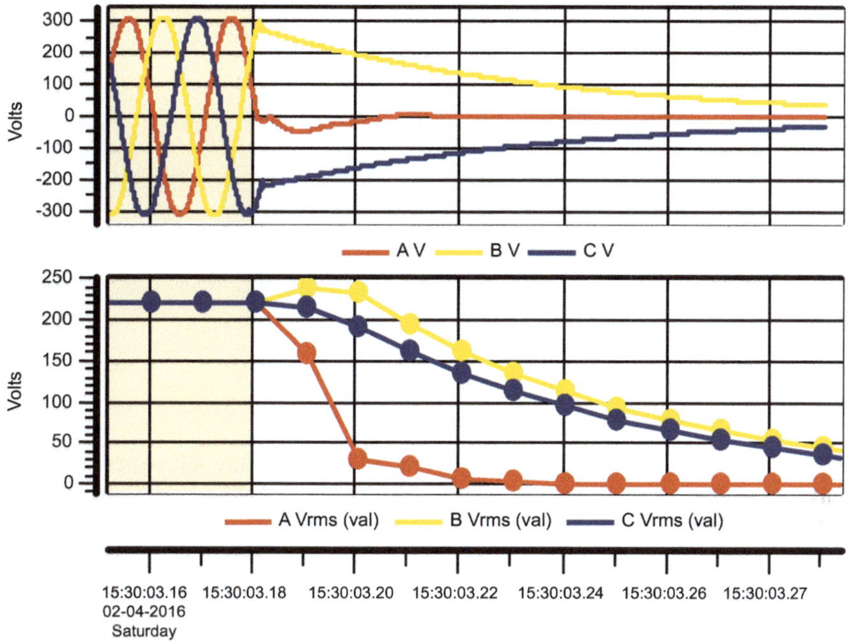

FIGURE 1.23 Voltage wave shape during an interruption.

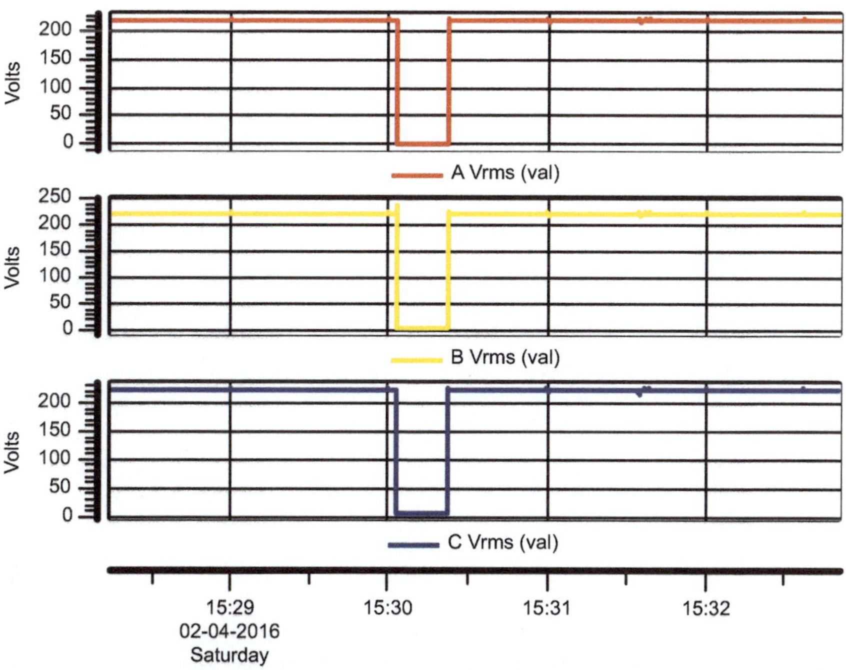

FIGURE 1.24 Root mean square voltage trend at main low-tension panel of incoming supply.

TABLE 1.13 Long duration voltage variation.

Parameters	Duration	Voltage magnitude
Undervoltage	>1 min	0.8–0.9 pu
Overvoltage	>1 min	1.1–1.2 pu
Sustained interruption	>1 min	0.0 pu
Current overload	>1 min	–

1.5 Long duration voltage variation

Variations in RMS voltage lasting more than 1 min are classified as long duration voltage variations as per IEEE 1159–2019 [2]. Accordingly, reduction in voltage is termed undervoltage, while an increase in the voltage is termed overvoltage. The absence of voltage is termed a sustained interruption. The different classifications of long duration voltage variations are listed in Table 1.13.

1.5.1 Undervoltage

Undervoltages are long duration reductions in voltage magnitude and a duration for more than 1 min. Definition as per IEEE 1159 is a decrease in RMS voltage between 0.8 and 0.9 pu and a time duration for more than 1 min [2]. Fig. 1.25 shows the voltage reduction in the RMS voltage trend from 1 to less than 0.9 pu for the duration of 65 s.

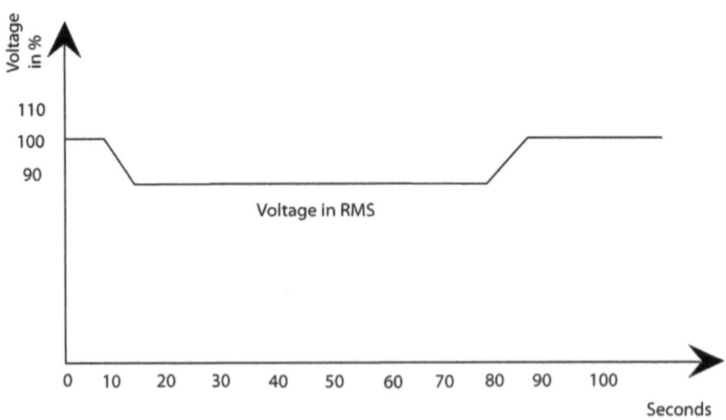

FIGURE 1.25 Root mean square (RMS) voltage trend.

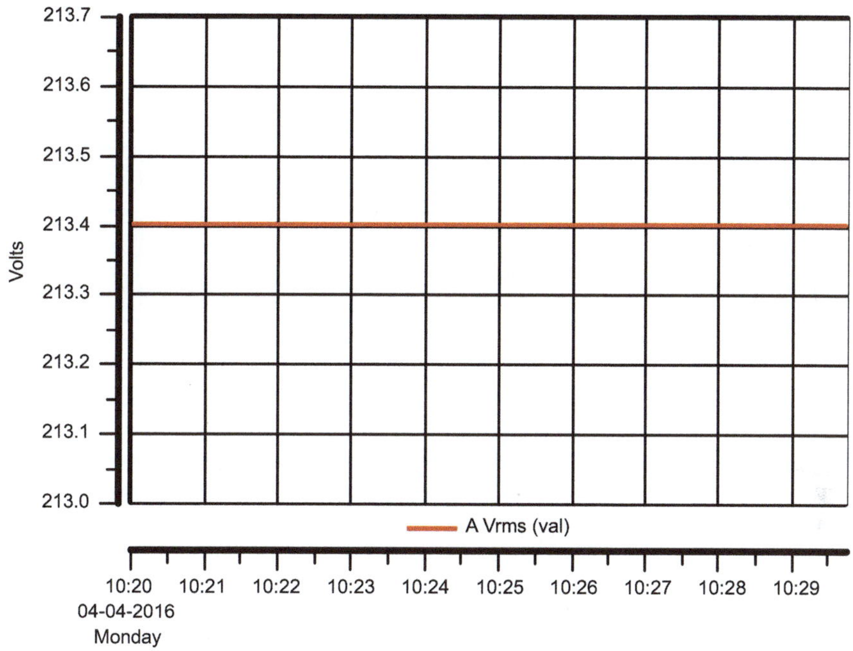

FIGURE 1.26 Root mean square voltage trend.

Fig. 1.26 shows the sustained voltage reduction in the RMS voltage trend between 10:20 and 10:30 h; the voltage recorded during this duration is 213.4 V in a nominal system voltage of 240 V. Reduction in voltage from the nominal voltage is 0.88 pu. Fig. 1.27 shows the voltage wave shape during the undervoltage.

1.5.1.1 Cause

Undervoltage can be caused by many conditions, a few of which are:

1. Overload in the circuit
2. Higher voltage drop due to undersized cables/conductors
3. Switching ON higher capacity loads continuously
4. Switching OFF large reactive power source
5. Improper tap selection of transformers
6. Unbalanced loading in three-phase, four-wire distribution systems

1.5.1.2 Impact

1. Constant power load can draw higher current, which leads to higher losses
2. Reduced performance in equipment like motors and incandescent lamps

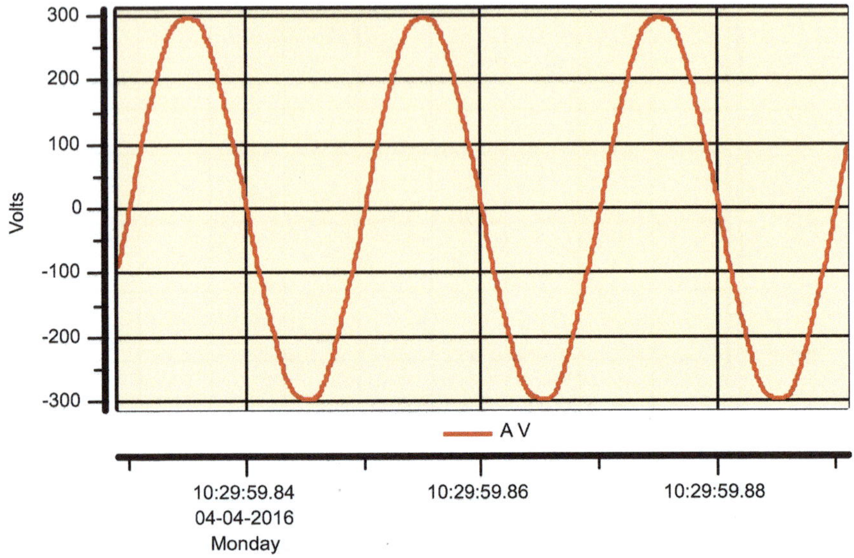

FIGURE 1.27 Voltage wave shape.

1.5.1.3 Example

Undervoltage due to improper tap selection through practical measurement.

Fig. 1.28 shows a transformer of 10 MVA stepping down the voltage from 66 to 11 kV to power a large institutional campus. The institution consists of various departments, administration offices, hostels, guest houses, laboratories, etc. Most of the electrical loads are connected in low voltage at 415 V. To supply to various departments, 11 kV voltage is further stepped down to 433 V through various ratings of the distribution transformer. There is a problem with the tap changer in the 10 MVA 66 kV/11 kV main receiving transformer. The output voltage is not controlled by tap changers in the primary. During peak hours, the secondary of the 66 kV/11 kV step transformer is not maintained at the 11 kV supply voltage. A 600 kVA 11 kV/433 V stepdown transformer has an offline tap changer in the high-voltage side powering department A. Due to undervoltage, three-phase induction motors are not receiving the desired voltage level to start and the three-phase uninterruptible power supply (UPS) system goes into battery mode because input voltage to the UPS is less than the UPS voltage requirements. The nominal system voltage is 415 V.

The power quality analyzer configured in 3ϕ 3 W delta mode is connected directly at the 433 V incoming side of the main low-tension panel from April 4, 2016 18.28 h to May 5, 2016 17.20 h. The RMS voltage trend for all three-phase wave shapes is shown in Fig. 1.29.

The RMS voltage trend in the zoomed view between April 5, 2016 09:30 h and April 5, 2016 11:45 h is shown in Fig. 1.30.

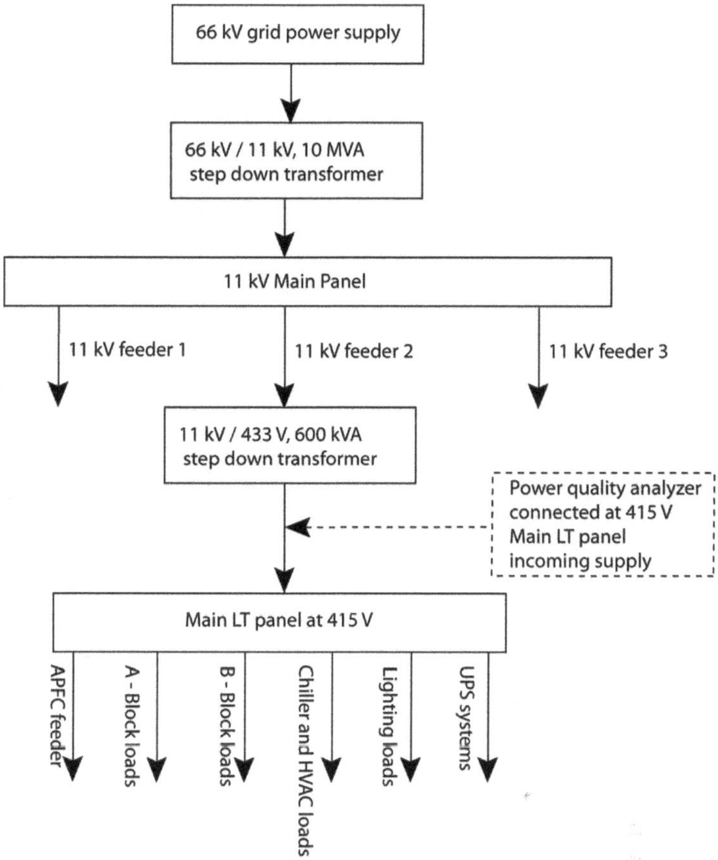

FIGURE 1.28 Schematic of power distribution. *APFC*, Automatic power factor controller; *HVAC*, heating, ventilation, and air conditioning; *LT*, low tension; *UPS*, uninterruptible power supply.

1.5.2 Overvoltage

Overvoltages are long duration increases in voltage magnitude and a duration for more than 1 min. Definition as per IEEE 1159 is an increase in RMS voltage above 1.1 pu or 110% of its nominal and a time duration for more than 1 min [2]. Fig. 1.31 shows the voltage increase in the RMS voltage trend above 110% for the duration of 65 s.

1.5.2.1 Cause

Overvoltage can be caused by many conditions, a few of which are:

1. Switching OFF large loads
2. Variations in reactive power compensation

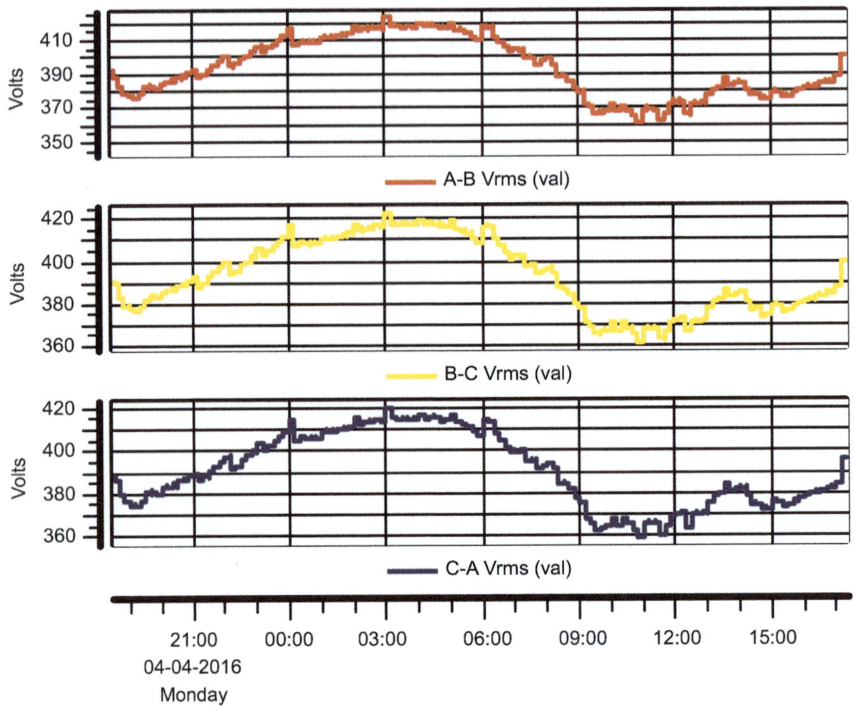

FIGURE 1.29 Root mean square voltage trend.

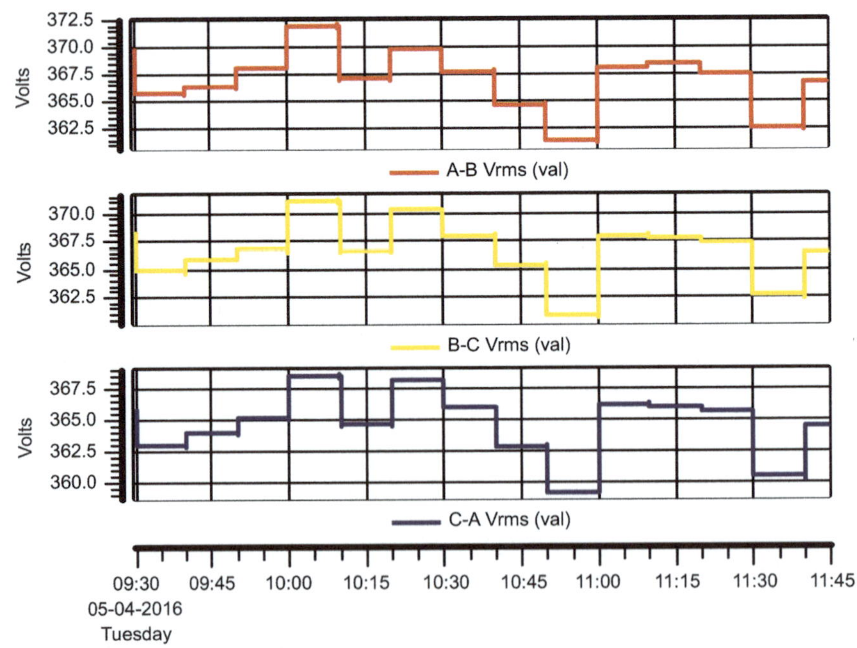

FIGURE 1.30 Root mean square voltage trend.

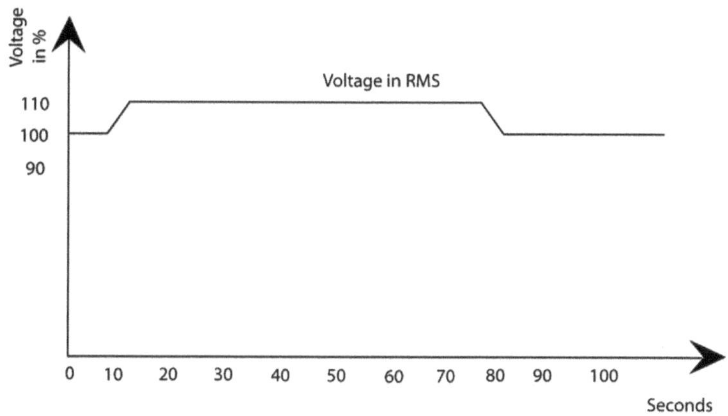

FIGURE 1.31 Root mean square (RMS) voltage trend.

3. Incorrect tap settings on transformers
4. Switching OFF higher capacity loads continuously
5. Unbalanced loading in three-phase, four-wire distribution systems

1.5.2.2 Impact

1. Prolonged overvoltage leads to insulation failure
2. Reduced performance of equipment

1.5.3 Sustained interruption

Sustained interruptions are long duration reductions in voltage magnitude of less than 0.1 pu and a duration for more than 1 min. Definition as per IEEE 1159 is RMS voltage is reduced to less than 0.1 pu or 10% of its nominal voltage and a time duration for more than 1 min [2]. Fig. 1.32 shows the voltage reduction in the RMS voltage trend to less than 0.1 pu for the duration of 65 s.

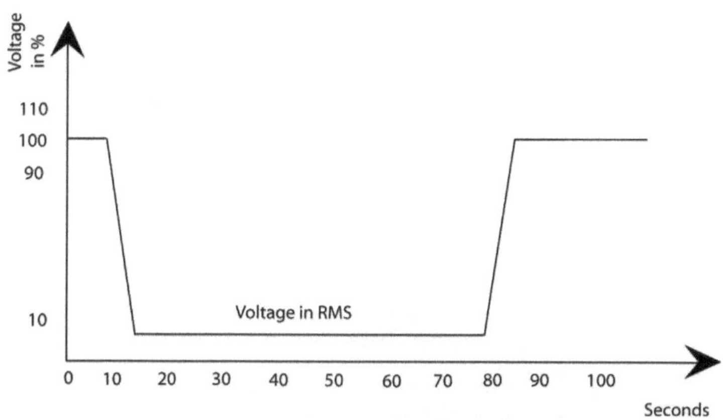

FIGURE 1.32 Root mean square (RMS) voltage trend.

1.5.3.1 Cause

Sustained interruption can be caused by many conditions, a few of which are:

1. A permanent fault that requires manual fault clearance and supply restoration
2. Breaker operation due to overload

1.5.3.2 Impact

1. Production stopped due to supply interruption
2. Motors and transformers drawing inrush current during every switch-ON

1.5.3.3 Example

Interruption through field measurement

The grid supply voltage of 22 kV is stepped down to 433 V by a 2.5 MVA stepdown transformer. An SLD of electrical distribution is shown in Fig. 1.33.

The power quality analyzer configured in 3ϕ 4 W Y mode is connected directly as shown in Fig. 1.33 from 15.10 h to 15.30 h captured voltage interruption in all three-phase wave shapes. The voltage wave shapes and RMS voltage trend are shown in Figs. 1.34 and 1.35.

Interruption occurs between 15:20:55 and 15:27:43 h and total interruption time is 6 min and 13 s.

The voltage wave shapes and RMS voltage trend resuming after the interruption are shown in Fig. 1.36.

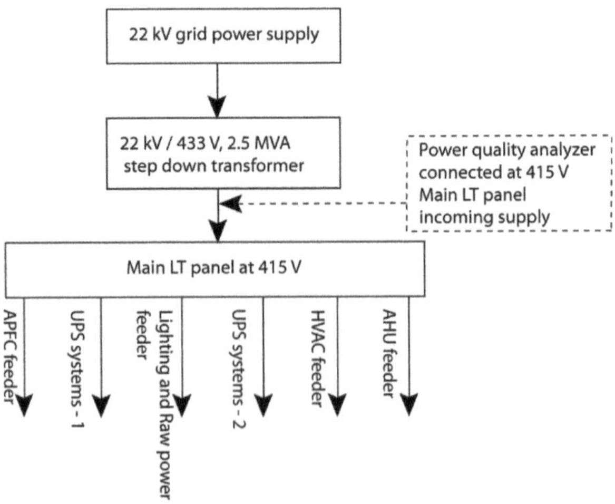

FIGURE 1.33 Single-line diagram of electrical distribution. *AHU*, Air-handling unit; *APFC*, automatic power factor controller; *HVAC*, heating, ventilation, and air conditioning; *LT*, low tension; *UPS*, uninterruptible power supply.

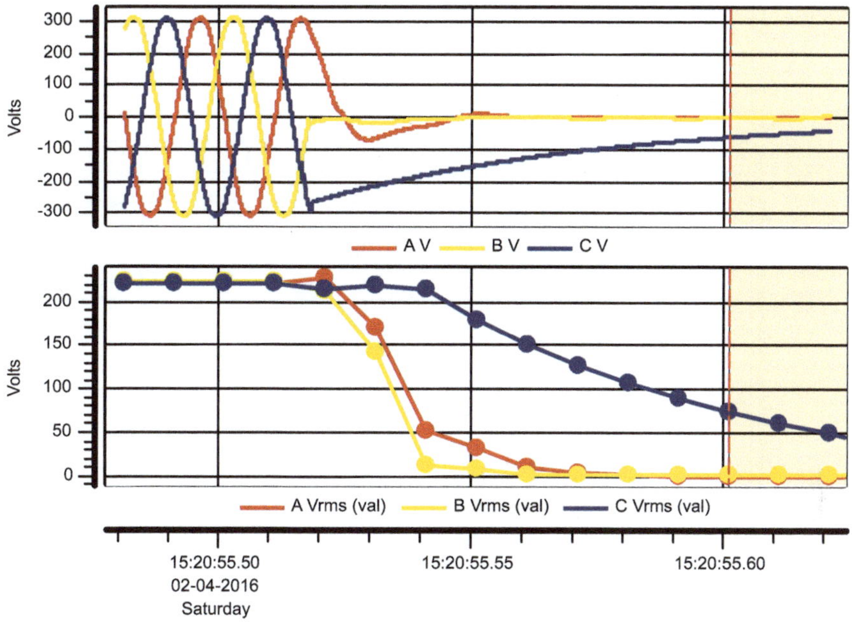

FIGURE 1.34 Voltage wave shape during an interruption.

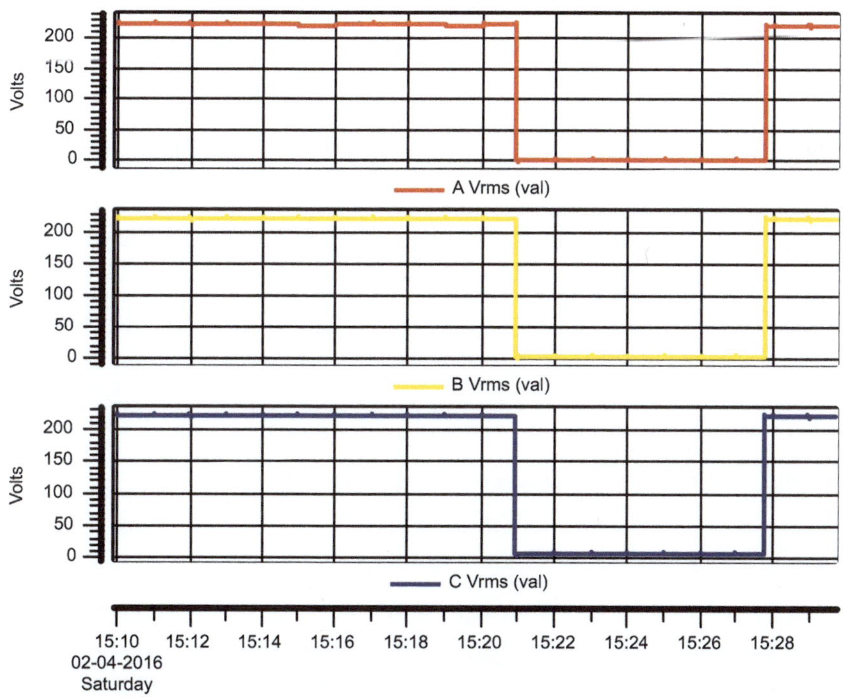

FIGURE 1.35 Root mean square voltage trend at the main low-tension panel incoming supply.

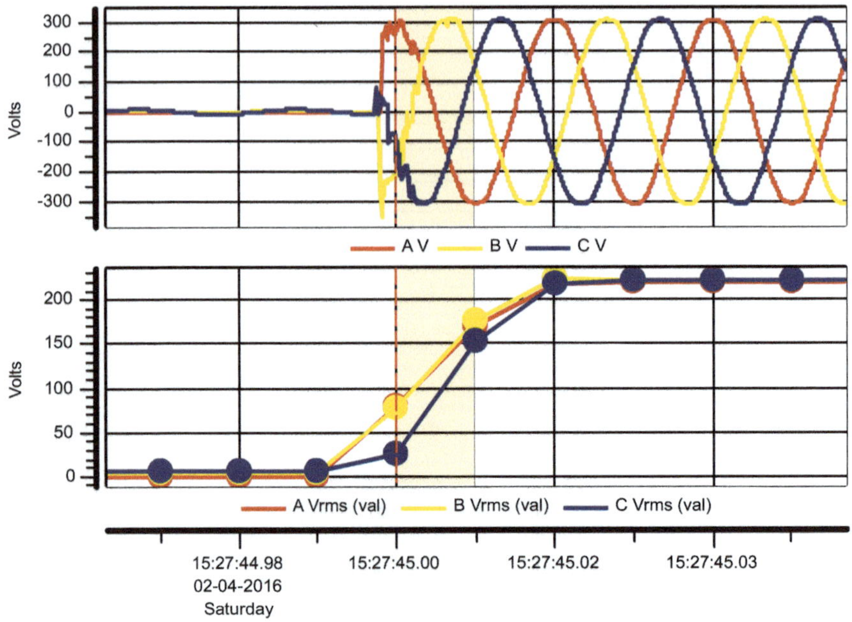

FIGURE 1.36 Voltage wave shape resuming from interruption.

1.6 Unbalance or imbalance

IEEE 1159–2019 defines unbalance or imbalance in a three-phase system as the ratio of the magnitude of the negative sequence component to the magnitude of the positive sequence component expressed in percentage [2]. Unbalance is classified into voltage unbalance and current unbalance (Tables 1.14 and 1.15).

Voltage unbalance is the ratio of the magnitude of the negative sequence component to the magnitude of the positive sequence component expressed in percentage. The expression for voltage unbalance is given in Eq. (1.1):

$$V_{unb}(\%) = \frac{V_{neg}}{V_{pos}} \times 100\% \tag{1.1}$$

TABLE 1.14 Root mean square (RMS) voltage.

Parameter	Minimum voltage (V)	Voltage in pu
RMS voltage (R–Y)	361.2	0.87
RMS voltage (Y–B)	360.6	0.87
RMS voltage (R–B)	359.2	0.86

TABLE 1.15 Peak and root mean square (RMS) voltage variations during an interruption event.

Parameter	Before interruption	During interruption	After interruption
R ϕ voltage wave shape (peak) in V	307.8	0.0	307.2
RMS voltage of R ϕ in V	227.3	0.0	220.9

The current unbalance is the ratio of the magnitude of the negative sequence component to the magnitude of the positive sequence component expressed in percentage. The expression for current unbalance is given in Eq. (1.2):

$$I_{unb}(\%) = \frac{I_{neg}}{I_{pos}} \times 100\% \tag{1.2}$$

1.6.1 Cause

Unbalance is caused by many conditions, a few of which are [8]:

1. Unbalanced loading among all three phases
2. Higher deployment of single-phase loads in three-phase, four-wire distribution systems
3. Unequal distribution impedance among three phases
4. Improper sizing and laying of single cables among three phases

1.6.2 Impact

An unbalanced voltage or current affects the end-user equipment in terms of performance. For example:

1. Increased heat in motor windings
2. Reduced performance of motors and transformers
3. Higher neutral current in three-phase, four-wire distribution systems
4. Higher voltage total harmonic distortion due to higher impedance in the phase

1.6.3 Example

Unbalance through practical measurement

The grid supply voltage of 11 kV is stepped down to 433 V by a 2.5 MVA stepdown transformer. An SLD of electrical distribution is shown in Fig. 1.37. The power quality analyzer configured in 3ϕ 4 W Y mode is directly connected at the 433 V supply voltage between June 25, 2017 03:31 h and June 26, 2017 11:10 h.

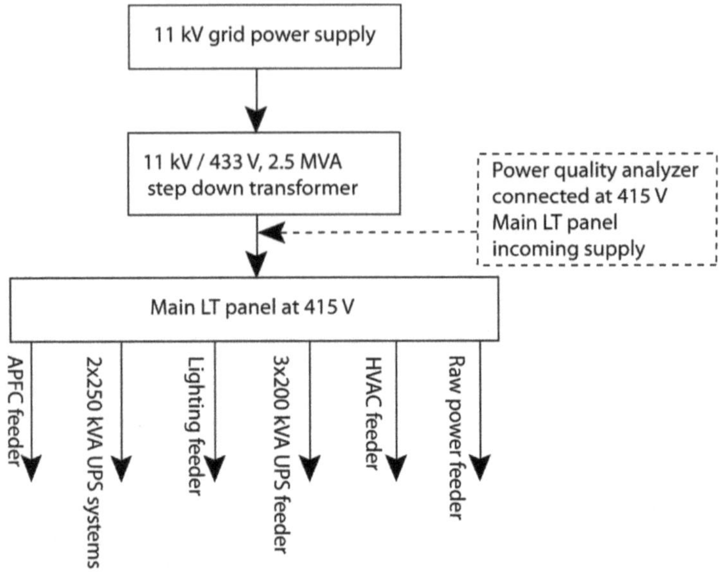

FIGURE 1.37 Schematic diagram of power distribution. *APFC*, Automatic power factor controller; *HVAC*, heating, ventilation, and air conditioning; *LT*, low tension; *UPS*, uninterruptible power supply.

Figs. 1.38 and 1.39 show the voltage positive sequence trend and voltage negative sequence trend. Table 1.16 describes the average voltage positive sequence and average voltage negative sequence.

From Eq. (1.1), the percentage of voltage unbalance is 0.31%.

Figs. 1.40 and 1.41 show the current positive sequence trend and current negative sequence trend. Table 1.17 describes the average current positive sequence and average current negative sequence.

From Eq. (1.2), the percentage of current unbalance is 6.87%.

1.7 Voltage fluctuations

Voltage fluctuation is a continuous change in the instantaneous voltage (cycle to cycle) due to variation in load resistance in every cycle. Continuous voltage changes are called voltage fluctuation. The definition of voltage fluctuation by IEC 61000-3-3 is a series of changes of RMS voltage evaluated as a single value for each successive half period between zero crossings of the source voltage [6]. The major cause of voltage fluctuation is arc furnaces in industrial plant. Voltage fluctuations have an impact on illumination intensity from the lamp. That is, continuous variation in voltage has an impact on illumination density resulting in a noticeable change in illumination by the normal human eye. This phenomenon is called flicker or voltage flicker.

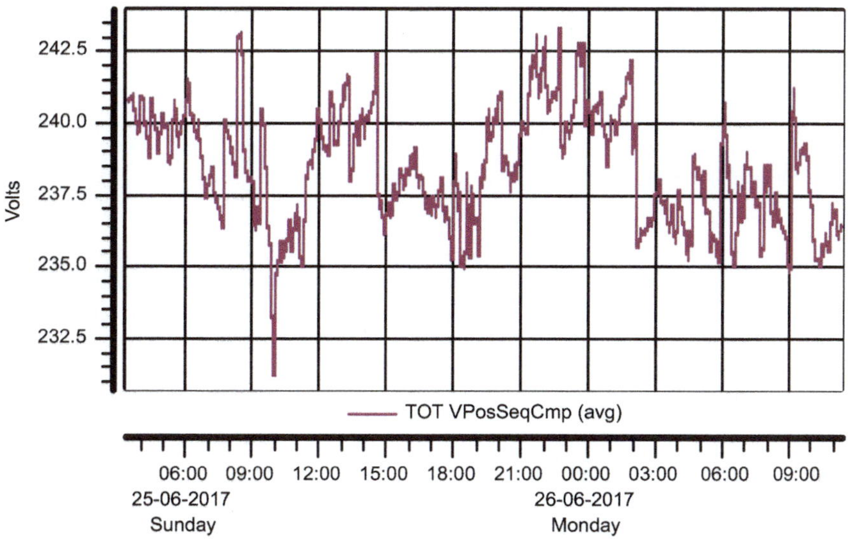

FIGURE 1.38 Voltage positive sequence.

The definition of flicker by IEC 61000-3-3 is the impression of unsteadiness of visual sensation induced by a light stimulus whose luminance or spectral distribution fluctuates with time.

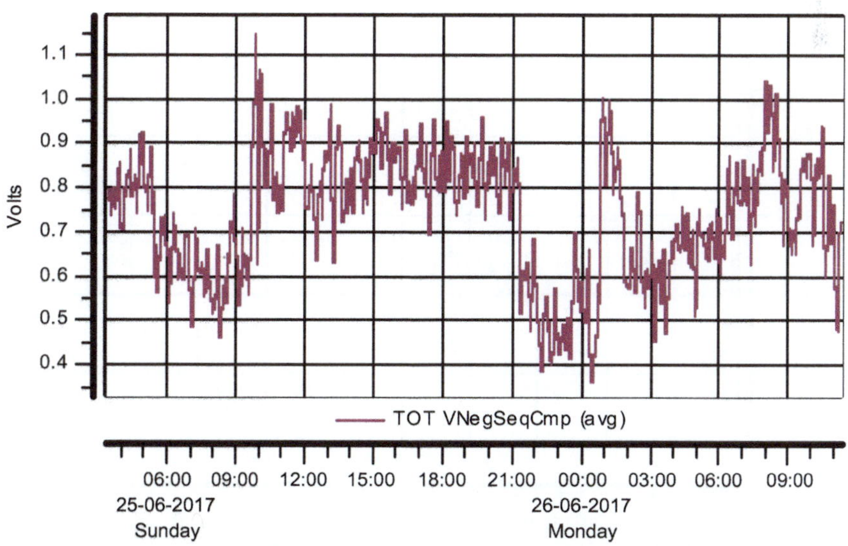

FIGURE 1.39 Voltage negative sequence.

TABLE 1.16 Voltage positive and negative sequence.

Parameter	Voltage in V
Positive sequence	238.5
Negative sequence	0.74

Flicker is classified into two categories: short-term flicker (P_{st}) and long-term flicker (P_{lt}). Short-term flicker is measured and evaluated over a short time period in minutes. Long-term flicker is measured and evaluated over a long time period in hours [7]. As per IEC 61000-3-3, the limits of short-term flicker shall not be more than 1.0 and long-term flicker shall not be more than 0.65. The typical time duration for analyzing short-term flicker is 10 min and long-term flicker is 2 h. These limits shall not apply to voltage fluctuations due to emergency switching operations and manual switching operations. The expression for short-term flicker is given in Eq. (1.3):

$$P_{st} = \left(\frac{SL}{STR}\right)^{\frac{1}{3}} \tag{1.3}$$

The expression for long-term flicker is given in Eq. (1.4):

$$P_{lt} = 0.65\left(\frac{SL}{STR}\right)^{\frac{1}{3}} \tag{1.4}$$

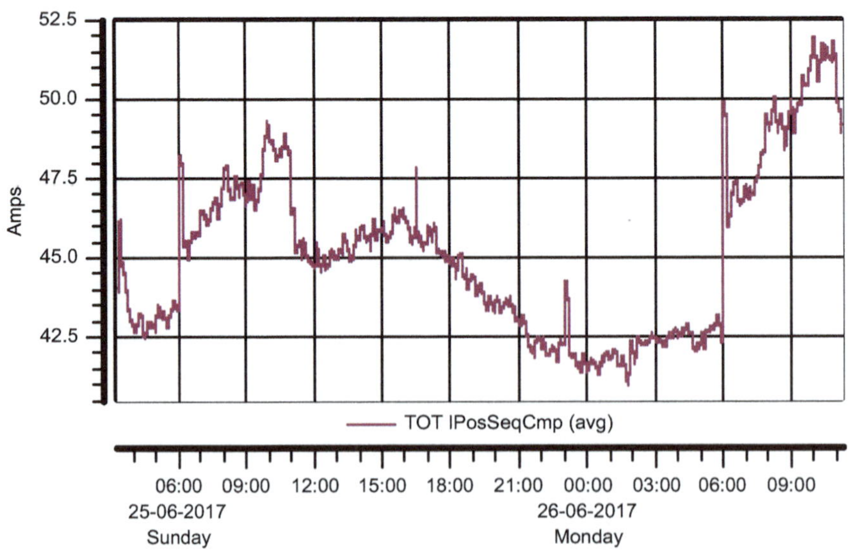

FIGURE 1.40 Current positive sequence.

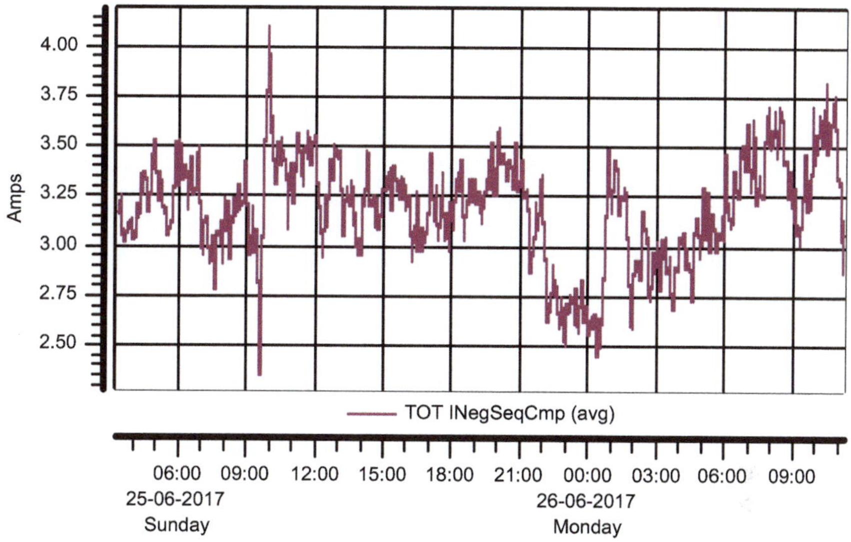

FIGURE 1.41 Current negative sequence.

where:

S_L is rated apparent power of the connected load
S_{TR} is rated apparent power of the feeding transformer

1.7.1 Cause

Voltage fluctuations can be caused by:

1. Arc furnace
2. Arc welding
3. Arcing load

1.7.2 Impact

1. Reduced performance of the equipment connected to the same bus
2. Flickering in an incandescent lamp

TABLE 1.17 Current positive and negative sequence.

Parameter	Current in A
Positive sequence	45.1
Negative sequence	3.1

1.7.3 Example

Flicker through practical measurement

The grid supply voltage of 11 kV is stepped down to 433 V by a 2.5 MVA stepdown transformer. An SLD of electrical distribution is shown in Fig. 1.42.

The power quality analyzer configured in 3ϕ 4 W Y mode is connected directly at the 415 V outgoing of the lighting feeder from June 17, 2015 10:43 h to June 18, 2015 10:55 h. The short-term flicker trend is shown in Fig. 1.43. The minimum and maximum values of short-term flicker are listed in Table 1.18.

Short-term flicker in the entire monitoring duration is listed in Annexure 1.1.

The long-term flicker trend is shown in Fig. 1.44. The minimum and maximum values of long-term flicker are listed in Table 1.19.

Long-term flicker in the entire monitoring duration is listed in Annexure 1.2.

1.8 Waveform distortion

IEEE 1159–2019 defines waveform distortion as a steady-state deviation from ideal sinusoidal characteristics [2,4]. Waveform distortion is classified into the following five types:

1. Harmonics
2. Interharmonics
3. DC offset
4. Notching
5. Noise

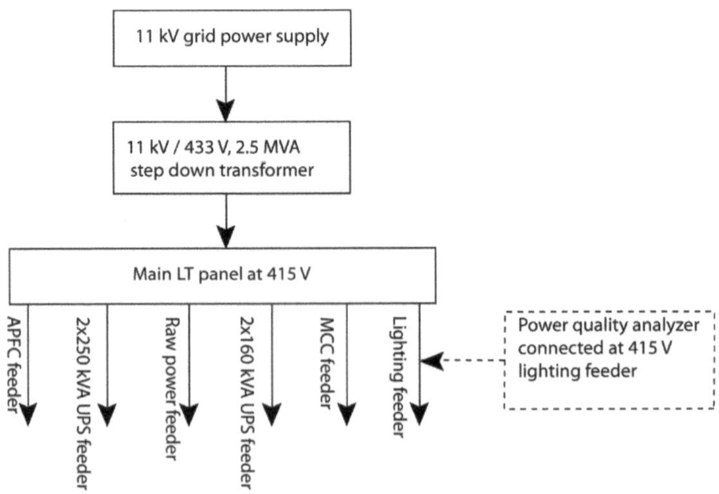

FIGURE 1.42 Single-line diagram of electrical distribution. *APFC*, Automatic power factor controller; *MCC*, motor control center; *LT*, low tension; *UPS*, uninterruptible power supply.

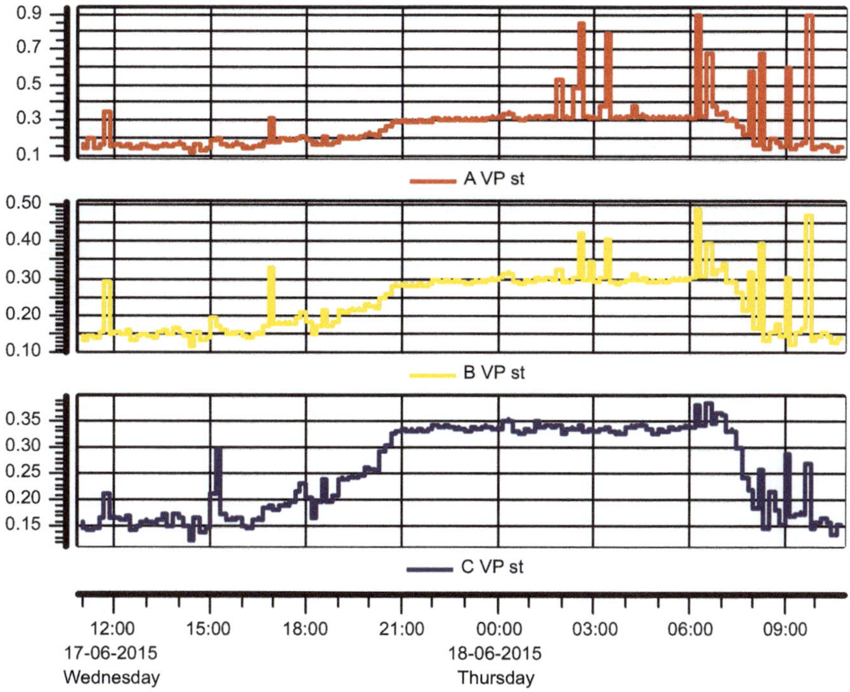

FIGURE 1.43 Short-term flicker trend.

1.8.1 Harmonics

Harmonics are the presence of multiple frequencies superimposed on the fundamental frequency of voltage or current or both. The definition as per IEEE 1159 is harmonics are sinusoidal voltages or currents having frequencies that are integer multiples of the frequency at which the supply system is designed to operate [1]. The fundamental system frequency is either 50 or 60 Hz. Due to the presence of harmonics, a system fundamental frequency of 50 or 60 Hz is added with multiple frequency ranges of 150, 250, 350, etc. [12].

TABLE 1.18 Minimum and maximum values of short-term flicker.

Parameters	R phase	Y phase	B phase
Maximum value	0.89	0.49	0.38
Minimum value	0.12	0.12	0.13

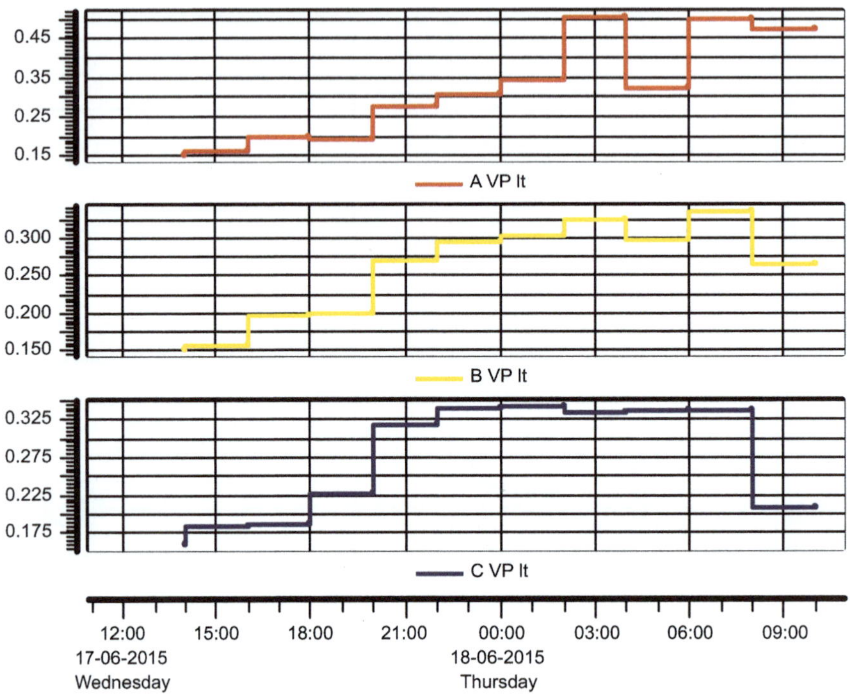

FIGURE 1.44 Long-term flicker trend.

Due to multiple frequencies, the fundamental frequency deviates from its ideal sinusoidal characteristics [10]. The instantaneous phase-to-neutral voltage is shown in Fig. 1.45.

From Fig. 1.45, it is clearly visible that due to the presence of multiple frequencies in voltage, the fundamental frequency is deviated from its ideal sinusoidal characteristics. This is called voltage harmonic distortion or simply voltage distortion.

The instantaneous phase current is shown in Fig. 1.46.

From Fig. 1.46, it is clearly visible that duc to thc presence of multiple frequencies in current, the fundamental frequency is deviated from its ideal sinusoidal characteristics. This is called current harmonic distortion or simply current distortion.

TABLE 1.19 Minimum and maximum values of long-term flicker.

Parameters	R phase	Y phase	B phase
Maximum value	0.5	0.33	0.34
Minimum value	0.19	0.15	0.16

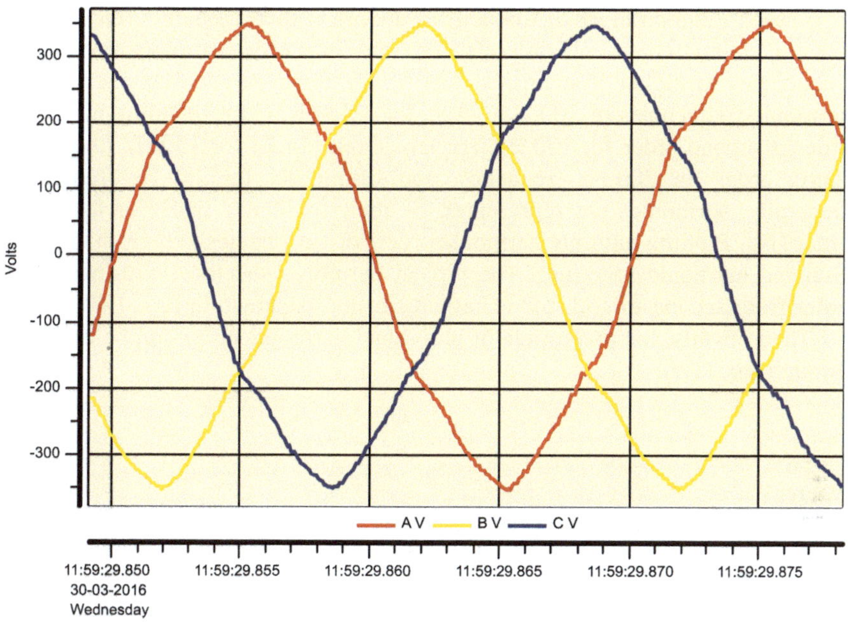

FIGURE 1.45 Phase-to-neutral instantaneous voltage.

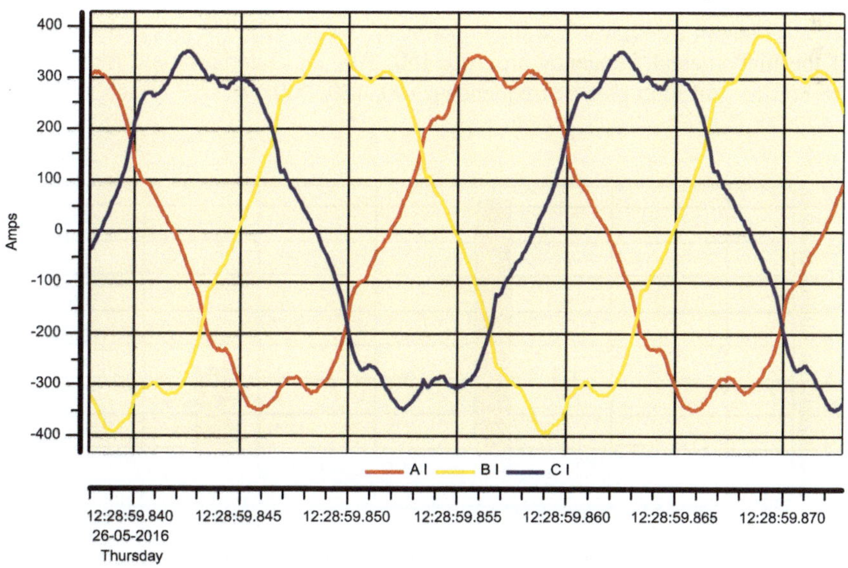

FIGURE 1.46 Instantaneous phase current.

These multiple ranges of frequencies are also called harmonic frequencies, and are related to harmonic order [9]. The harmonic order increases from lower frequencies to higher frequencies. The 50 Hz fundamental frequencies have harmonic frequencies of 100, 150, 200, and 250 Hz. Corresponding harmonic order is second order for 100 Hz frequency, third order for 150 Hz frequency, fourth order for 200 Hz frequency, and fifth order for 250 Hz frequency. Generally, harmonics are represented by their frequency in Hz or harmonic order [14]. Plotting multiple harmonic orders or frequencies to their magnitude is called harmonic spectrum. The typical harmonic spectrum with harmonic order from second to 15th for voltage distortion is shown in Fig. 1.47.

Theoretically, the magnitude of individual harmonic order can be obtained through Eq. (1.5):

$$I_h = \frac{I_f}{h} \tag{1.5}$$

where:

I_h is the magnitude of h order harmonic current
I_f is the magnitude of the fundamental current
h is harmonic order

The waveforms for odd harmonics, even harmonics, and tripplen harmonics are given in Annexure 1.3.

1.8.2 Interharmonics

Interharmonics are not integral multiples of a fundamental frequency. Multiple harmonic frequencies are added with a fundamental frequency whose frequencies are not integral multiples of a fundamental frequency [2]. The integral multiples of the fundamental frequency are 100, 150, 200, etc. Interharmonic frequencies are not integral multiples of frequencies 100, 150, 200, etc.

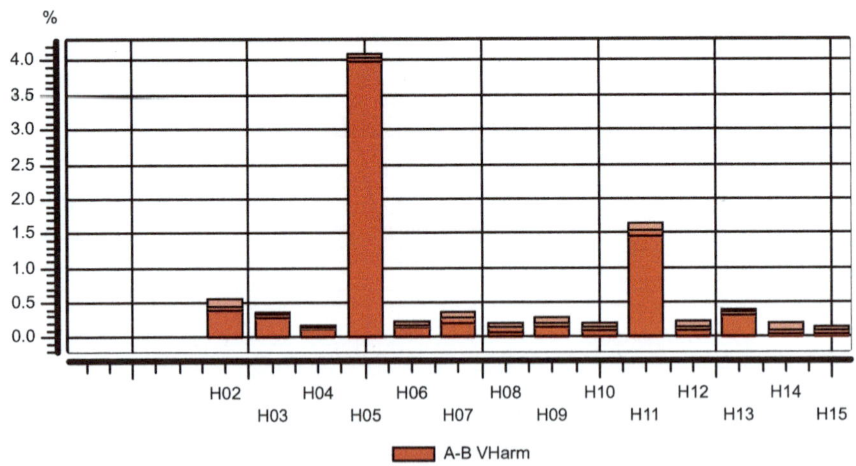

FIGURE 1.47 Harmonic spectrum for voltage distortion.

1.8.2.1 Cause

Interharmonics can be caused by:

1. Cycloconverters
2. Induction furnaces
3. Arcing loads

1.8.2.2 Impacts

- Transformer saturation
- Insulation stress

1.8.3 DC offset

DC offset is the presence of DC current or voltage in an AC power system [2].

1.8.3.1 Cause

DC offset can be caused by:

- Half-wave rectification

1.8.3.2 Impacts

- Transformer saturation
- Insulation stress

1.8.4 Notching

Periodic voltage disturbance caused by the normal operation of power electronics devices is called notching [2]. Notching occurs by commutation between one phase and another phase. Notching in a three-phase wave shape is shown in Fig. 1.48.

During commutation there is a momentary short circuit between two phases. The average depth of line voltage from the sinusoidal wave of voltage is called notch depth and the width of notch depth is based on the commutation angle. The notch depth in a voltage wave shape is shown in Fig. 1.49.

From Fig. 1.49, notch depth in a voltage wave shape is 71.037 V.

Notch width in a voltage wave shape is shown in Fig. 1.50.

From Fig. 1.50, notch width in a voltage wave shape is 304.937 μs.

At any point in the system, the notch depth can be determined by the magnitude of current flow, source inductance, and isolating inductance between the converter and monitoring point.

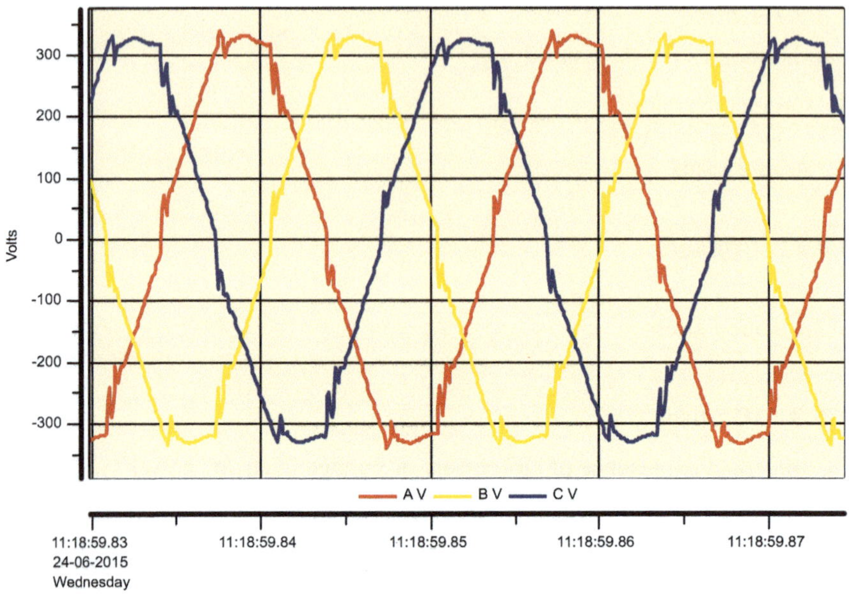

FIGURE 1.48 Three-phase voltage wave shape.

1.8.4.1 Cause

Notching can be caused by:

- Commutation of power electronic devices

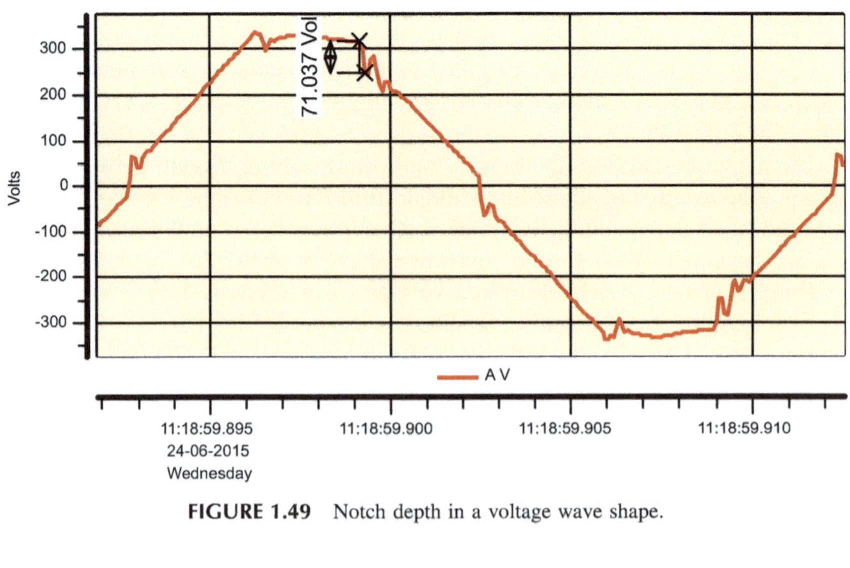

FIGURE 1.49 Notch depth in a voltage wave shape.

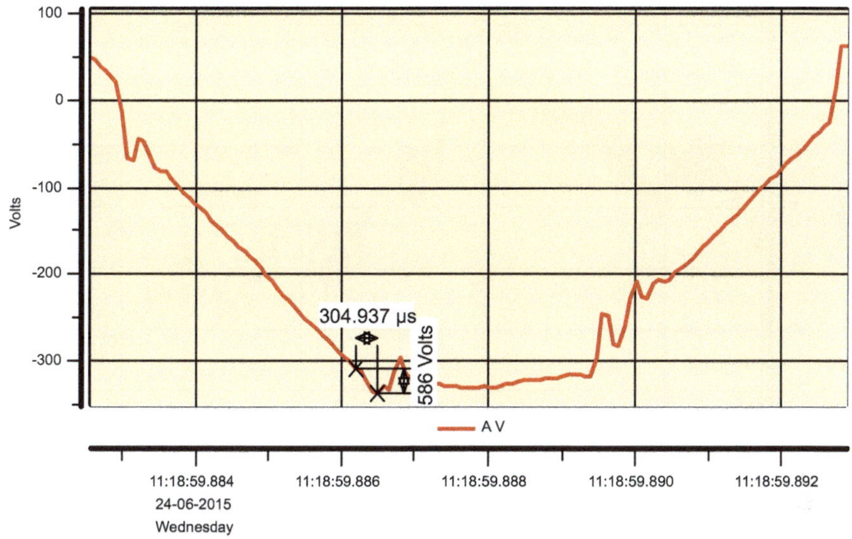

FIGURE 1.50 Notch width in a voltage wave shape.

1.8.4.2 Impacts

- Frequency errors
- Timing errors

1.8.5 Noise

An unwanted signal added to the fundamental frequency of voltage or current in the phase or neutral conductor is called noise [2]. Noise in the current wave shape is shown in Fig. 1.51 and the neutral conductor is shown in Fig. 1.52.

Basically, noise cannot be classified as transients or harmonic distortion. This problem is frequently increased by improper or poor grounding in the system. Noise magnitude and frequency range generally depend on the source and system characteristics. Typically, magnitude is less than 1% of voltage magnitude and spectral content is less than 200 kHz.

1.8.5.1 Cause

Noise can be caused by any of the following equipment:

- Power electronics devices
- Control circuits
- Arcing equipment
- Switched-mode power supply
- Solid-state rectifiers

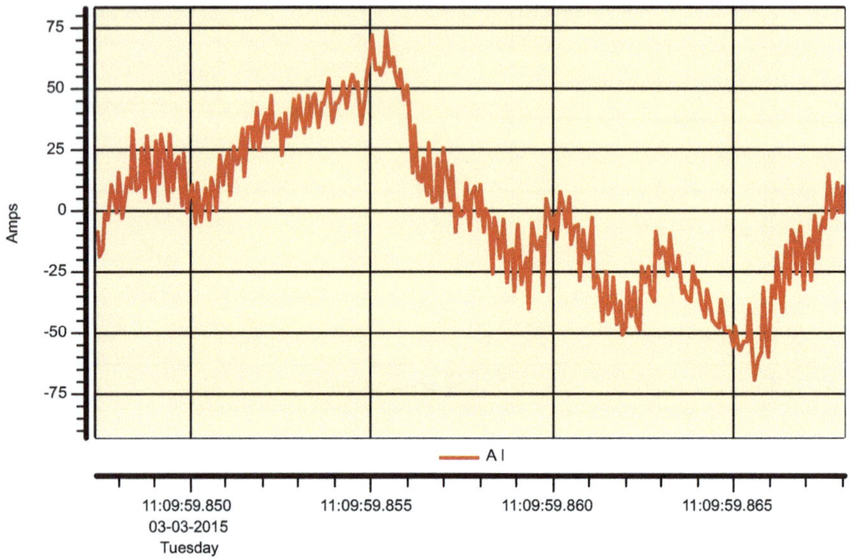

FIGURE 1.51 Phase current wave shape.

1.8.5.2 Impacts

- Electronic equipment
- Programmable logic controllers
- Microcomputers

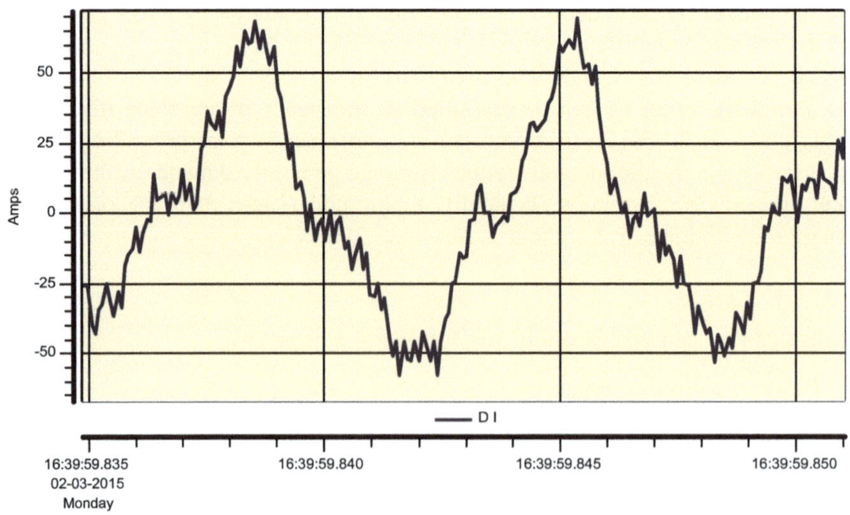

FIGURE 1.52 Neutral current wave shape.

1.9 Annexure 1.1

Time	P_{st} value in R phase	P_{st} value in Y phase	P_{st} value in B phase
11:00:00	0.14806	0.14233	0.15478
11:10:00	0.14061	0.13506	0.14731
11:20:00	0.20164	0.14501	0.1434
11:30:00	0.13913	0.1376	0.1462
11:40:00	0.15861	0.15509	0.16771
11:50:00	0.3492	0.29264	0.21165
12:00:00	0.15529	0.14854	0.16806
12:10:00	0.15803	0.15449	0.16547
12:20:00	0.15738	0.14886	0.16425
12:30:00	0.16012	0.1611	0.16895
12:40:00	0.13615	0.13675	0.1442
12:50:00	0.14092	0.14334	0.15143
13:00:00	0.15758	0.15259	0.15728
13:10:00	0.15183	0.14678	0.1584
13:20:00	0.14491	0.14136	0.15458
13:30:00	0.15284	0.15484	0.16274
13:40:00	0.15789	0.1622	0.17252
13:50:00	0.15205	0.14822	0.15314
14:00:00	0.16523	0.16574	0.17331
14:10:00	0.16246	0.1552	0.16765
14:20:00	0.1446	0.14327	0.15208
14:30:00	0.12158	0.12036	0.1259
14:40:00	0.16051	0.15752	0.16708
14:50:00	0.12945	0.13458	0.1399
15:00:00	0.1407	0.14057	0.14705
15:10:00	0.1897	0.19476	0.21418
15:20:00	0.19695	0.16994	0.29904
15:30:00	0.16397	0.1602	0.17533
15:40:00	0.15496	0.15168	0.16107
15:50:00	0.15987	0.15104	0.16122
16:00:00	0.15766	0.15667	0.16651
16:10:00	0.14432	0.1465	0.15328
16:20:00	0.13766	0.13932	0.14732
16:30:00	0.15378	0.14894	0.16249
16:40:00	0.15415	0.15217	0.16434
16:50:00	0.17351	0.17059	0.18655
17:00:00	0.31343	0.3323	0.1909
17:10:00	0.17907	0.1763	0.1838
17:20:00	0.19868	0.17885	0.18898
17:30:00	0.18133	0.17705	0.19075
17:40:00	0.18445	0.17811	0.19717
17:50:00	0.19374	0.1938	0.21716
18:00:00	0.20385	0.21136	0.23257

18:10:00	0.18289	0.18492	0.20654
18:20:00	0.16444	0.15277	0.16822
18:30:00	0.16768	0.17094	0.19796
18:40:00	0.20362	0.21401	0.24094
18:50:00	0.16542	0.17214	0.19833
19:00:00	0.17499	0.18179	0.20774
19:10:00	0.20516	0.21388	0.24037
19:20:00	0.202	0.21104	0.23954
19:30:00	0.20044	0.21333	0.24411
19:40:00	0.20276	0.21454	0.24452
19:50:00	0.20401	0.21661	0.24638
20:00:00	0.22126	0.23057	0.26385
20:10:00	0.21763	0.22833	0.26033
20:20:00	0.20965	0.21905	0.25491
20:30:00	0.24794	0.24829	0.29226
20:40:00	0.26161	0.25718	0.30502
20:50:00	0.28915	0.27883	0.3295
21:00:00	0.29409	0.27986	0.3335
21:10:00	0.29805	0.28453	0.3356
21:20:00	0.29395	0.28185	0.3312
21:30:00	0.29303	0.2833	0.3313
21:40:00	0.29549	0.28658	0.3347
21:50:00	0.29152	0.2823	0.331
22:00:00	0.29507	0.28675	0.3369
22:10:00	0.30869	0.29753	0.3427
22:20:00	0.30283	0.29185	0.3397
22:30:00	0.30347	0.2949	0.3434
22:40:00	0.3031	0.29431	0.339
22:50:00	0.30857	0.29521	0.336
23:00:00	0.30503	0.29242	0.3354
23:10:00	0.29814	0.28913	0.331
23:20:00	0.30121	0.29035	0.339
23:30:00	0.29769	0.29046	0.3372
23:40:00	0.30336	0.29187	0.34
23:50:00	0.31312	0.30001	0.3397
00:00:00	0.31365	0.29733	0.3363
00:10:00	0.31396	0.30113	0.3369
00:20:00	0.3322	0.31446	0.35
00:30:00	0.3316	0.31329	0.3514
00:40:00	0.31044	0.29237	0.3322
00:50:00	0.30366	0.28773	0.3299
01:00:00	0.30835	0.29519	0.3344
01:10:00	0.3079	0.29439	0.3331
01:20:00	0.3186	0.30292	0.3506
01:30:00	0.31522	0.29865	0.3389
01:40:00	0.3202	0.30433	0.3449
01:50:00	0.31332	0.29534	0.3398
02:00:00	0.5298	0.3266	0.3442
02:10:00	0.31014	0.28967	0.3298

02:20:00	0.3122	0.29414	0.3339
02:30:00	0.4842	0.30439	0.3361
02:40:00	0.8475	0.4221	0.3447
02:50:00	0.3217	0.29763	0.3325
03:00:00	0.31541	0.3467	0.3358
03:10:00	0.31541	0.29345	0.3318
03:20:00	0.3848	0.30163	0.3344
03:30:00	0.7923	0.4047	0.3401
03:40:00	0.30749	0.29424	0.3329
03:50:00	0.31365	0.28882	0.3266
04:00:00	0.31262	0.28922	0.3288
04:10:00	0.3198	0.29765	0.3393
04:20:00	0.3857	0.31601	0.3389
04:30:00	0.3181	0.29774	0.3433
04:40:00	0.3202	0.30273	0.3421
04:50:00	0.31519	0.2943	0.334
05:00:00	0.30855	0.29325	0.3297
05:10:00	0.31044	0.29499	0.3346
05:20:00	0.3115	0.29157	0.3302
05:30:00	0.3174	0.2979	0.3405
05:40:00	0.3172	0.29701	0.3356
05:50:00	0.31416	0.29933	0.3377
06:00:00	0.3193	0.29939	0.3409
06:10:00	0.31594	0.30364	0.3391
06:20:00	0.893	0.4852	0.3817
06:30:00	0.3186	0.30576	0.3444
06:40:00	0.6803	0.3971	0.3842
06:50:00	0.3826	0.31164	0.3463
07:00:00	0.3345	0.3251	0.3663
07:10:00	0.3419	0.3428	0.3622
07:20:00	0.29909	0.29273	0.333
07:30:00	0.29624	0.29158	0.3301
07:40:00	0.26764	0.26402	0.30249
07:50:00	0.21942	0.21466	0.24485
08:00:00	0.579	0.3178	0.21884
08:10:00	0.16352	0.16468	0.1856
08:20:00	0.6793	0.3947	0.25704
08:30:00	0.13855	0.13532	0.14839
08:40:00	0.19854	0.15891	0.2162
08:50:00	0.17778	0.17814	0.18068
09:00:00	0.15376	0.14595	0.15455
09:10:00	0.5947	0.30274	0.28772
09:20:00	0.13729	0.1255	0.17247
09:30:00	0.16048	0.15758	0.17495
09:40:00	0.17363	0.16492	0.17428
09:50:00	0.8926	0.4693	0.26936
10:00:00	0.14042	0.13672	0.14634
10:10:00	0.14957	0.14321	0.15727
10:20:00	0.15906	0.15497	0.1653

10:30:00	0.15334	0.1465	0.15937
10:40:00	0.12744	0.12675	0.13652
10:50:00	0.14834	0.14131	0.15501

1.10 Annexure 1.2

Time	P_{lt} value in R phase	P_{lt} value in Y phase	P_{lt} value in B phase
14:00:00	0.15333	0.15182	0.16097
16:00:00	0.15961	0.15505	0.18482
18:00:00	0.1961	0.19821	0.1877
20:00:00	0.19294	0.20067	0.22799
22:00:00	0.2772	0.26996	0.3172
00:00:00	0.30498	0.29382	0.3383
02:00:00	0.3459	0.30257	0.3407
04:00:00	0.5026	0.3255	0.334
06:00:00	0.3227	0.29861	0.3373
08:00:00	0.5001	0.3344	0.3362
10:00:00	0.4735	0.26575	0.2084

1.11 Annexure 1.3: harmonics

1.11.1 Odd harmonics

Odd harmonics are harmonics in which frequencies are odd numbers such as 150, 250, 350 Hz, etc. in the fundamental frequency of 50 Hz. The odd harmonics present in the system are listed in Table A.1.1.

Theoretical magnitude, wave shape of fundamental frequency, and third order harmonics are shown in Fig. A.1.1.

The resultant wave shape of third order harmonic frequency superimposed in fundamental frequency is shown in Fig. A.1.2.

Similarly, resultant wave shape of fifth order and seventh order harmonic frequency superimposed in fundamental frequency is shown in Figs. A.1.3 and A.1.4 respectively.

TABLE A.1.1 Odd harmonics.

Harmonic order	Harmonic frequency at 50 Hz	Harmonic frequency at 60 Hz
3rd	150	180
5th	250	300
7th	350	420
9th	450	540
11th	550	660
13th	650	780
15th	750	900
17th	850	1020
19th	950	1140
21st	1050	1260
23rd	1150	1380
25th	1250	1500
27th	1350	1620
29th	1450	1740
31st	1550	1860
33rd	1650	1980
35th	1750	2100
37th	1850	2220
39th	1950	2340
41st	2050	2460
43rd	2150	2580
45th	2250	2700
47th	2350	2820
49th	2450	2940

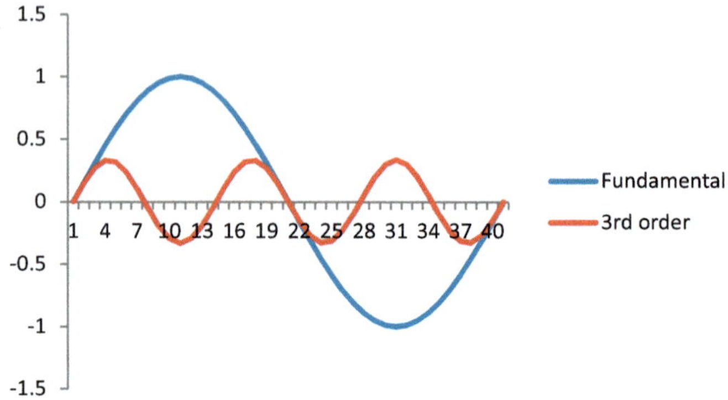

FIGURE A.1.1 Magnitude, wave shape of fundamental frequency, and third order harmonics.

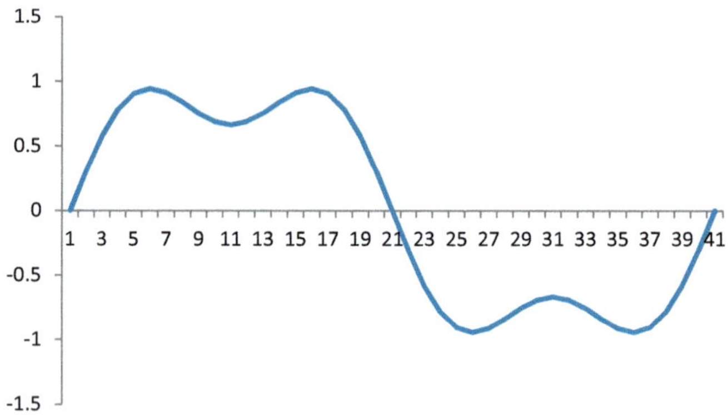

FIGURE A.1.2 Resultant wave shape.

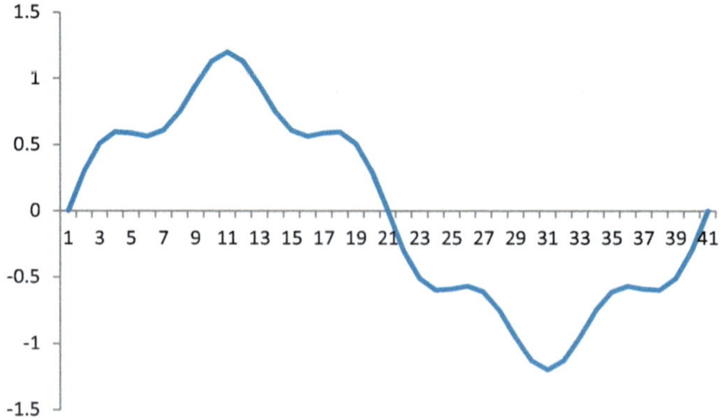

FIGURE A.1.3 Resultant wave shape.

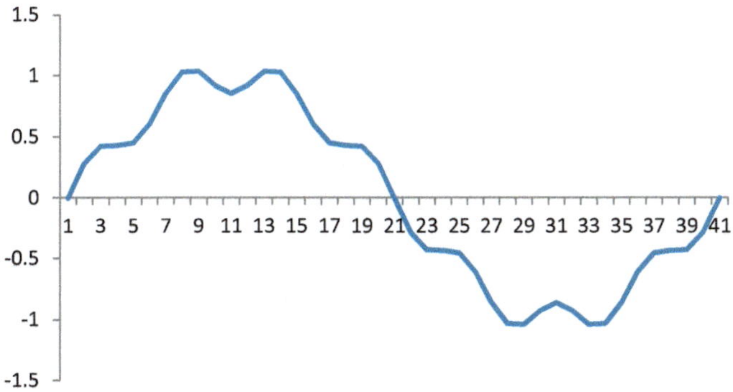

FIGURE A.1.4 Resultant wave shape.

1.11.2 Even harmonics

Even harmonics are harmonics whose frequencies are even numbers such as 100 Hz, 200 Hz, 300 Hz, etc. in the fundamental frequency of 50 Hz.

The most predominant even harmonics present in the system are listed in Table A.1.2.

Theoretical magnitude, wave shape of fundamental frequency, and third order harmonics are shown in Fig. A.1.5.

The resultant wave shape of second order harmonic frequency superimposed in fundamental frequency is shown in Fig. A.1.6.

Generally, power systems are designed free from even harmonics. In recent years, usage of power electronics-based devices in power systems introduced even order harmonics. When power converters are working normally the even order harmonics are very minimal and whenever there is any problem in firing pulse to power electronic components, it generates even order harmonics into the system.

Example

Even order harmonics produced by improper firing in a UPS system.

Fig. A.1.7 shows an SLD of electric distribution and location of power quality monitoring for a UPS system.

The current wave shape drawn by a UPS system is shown Fig. A.1.8.

For better understanding, the current wave shape (R phase) alone is shown Fig. A.1.9.

From Fig. A.1.9, a UPS system is drawing different waveform patterns in both positive and negative half cycles. That is, the positive cycle current wave shape does not match the negative cycle current wave shape. This is due to improper firing of the rectifier circuit in the UPS system and results in the

TABLE A.1.2 Even harmonics.

Harmonic order	Harmonic frequency at 50 Hz	Harmonic frequency at 60 Hz
2nd	100	120
4th	200	120
6th	300	360
8th	400	480
10th	500	600
12th	600	720
14th	700	840
16th	800	960
18th	900	1080
20th	1000	1200
22nd	1100	1320
24th	1200	1440
26th	1300	1560
28th	1400	1680
30th	1500	1800
32nd	1600	1920
34th	1700	2040
36th	1800	2160
38th	1900	2280
40th	2000	2400
42nd	2100	2520
44th	2200	2640
46th	2300	2760
48th	2400	2880

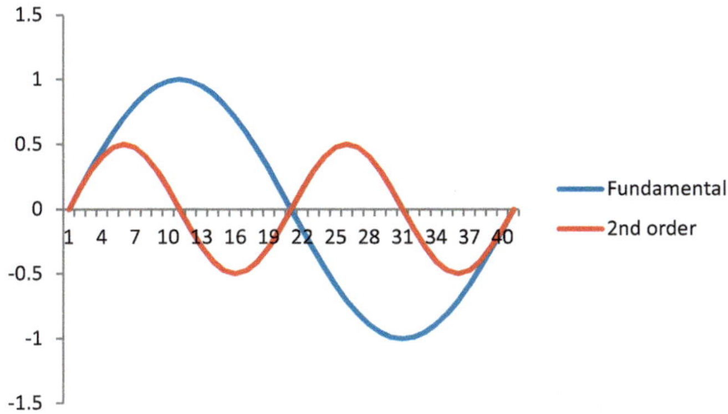

FIGURE A.1.5 Magnitude, wave shape of fundamental frequency, and second order harmonics.

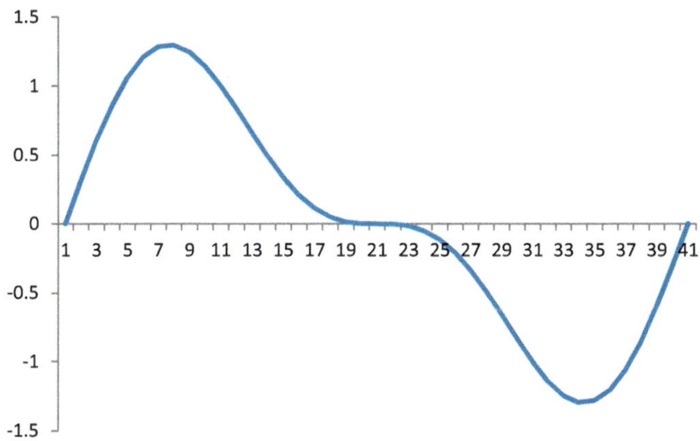

FIGURE A.1.6 Resultant wave shape.

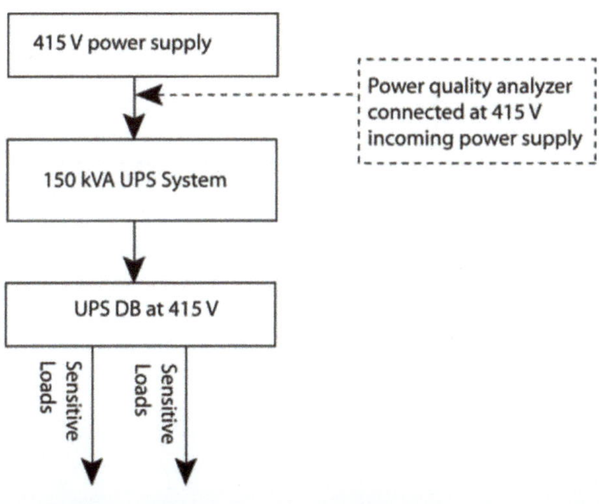

FIGURE A.1.7 Single-line diagram of a UPS system. *DB*, distribution board; *UPS*, uninterruptible power supply.

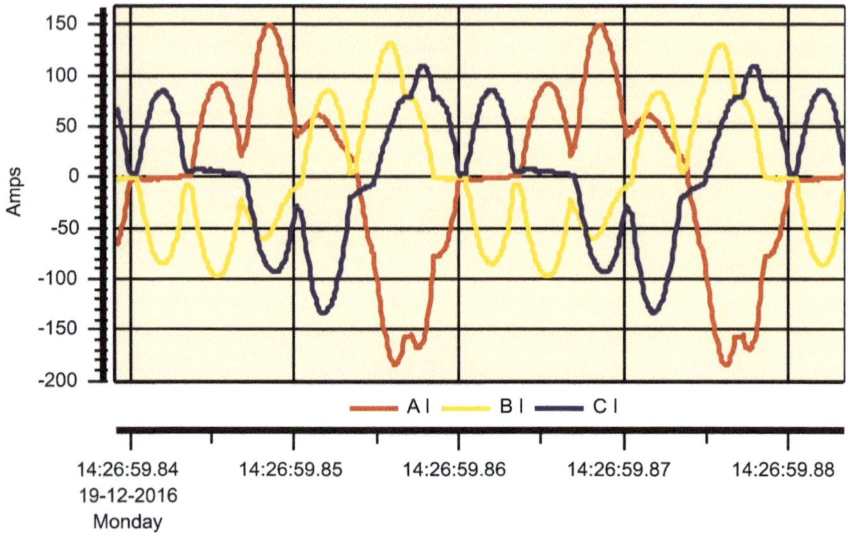

FIGURE A.1.8 Current wave shape drawn by an uninterruptible power supply system.

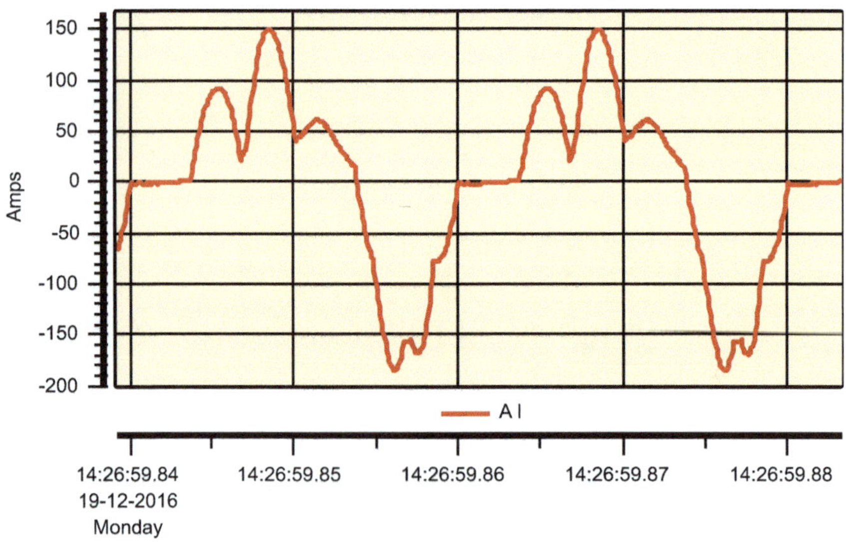

FIGURE A.1.9 Current wave shape of R phase.

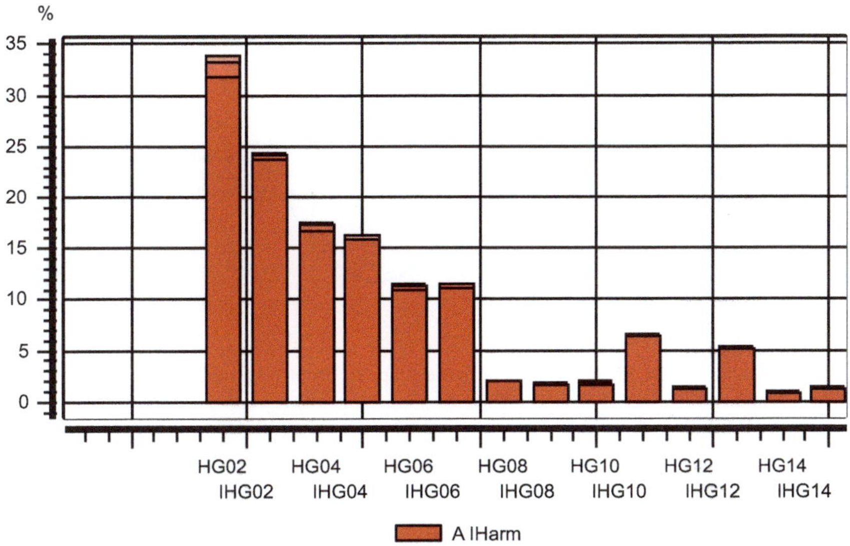

FIGURE A.1.10 Harmonic spectrum of R phase.

presence of even order harmonics in the system. The harmonic spectrum for Fig. A.1.9 is shown in Fig. A.1.10.

In Fig. A.1.10, even order harmonics such as second, fourth, and sixth orders are injected by the UPS system.

1.11.3 Tripplen harmonics

Tripplen harmonics are multiples of third order harmonic frequencies such as 150, 300, 450, 600 Hz, etc. in the fundamental frequency of 50 Hz. Tripplen harmonics of third order are three cycles compared to fundamental, and sixth order are six cycles compared to fundamental. These tripplen harmonics are created by nonlinear single-phase loads connected to the distribution system and will flow only in neutral conductors. They increase the current flow in the neutral conductor. When a three-phase, 4 W distribution system is operating in balanced loading condition, a 3.5 core cable is sufficient for current carrying. But due to the presence of tripplen harmonics, higher amount of current flowing through neutral conductor and its required higher neutral conductor size. i.e. double the size of phase conductor.

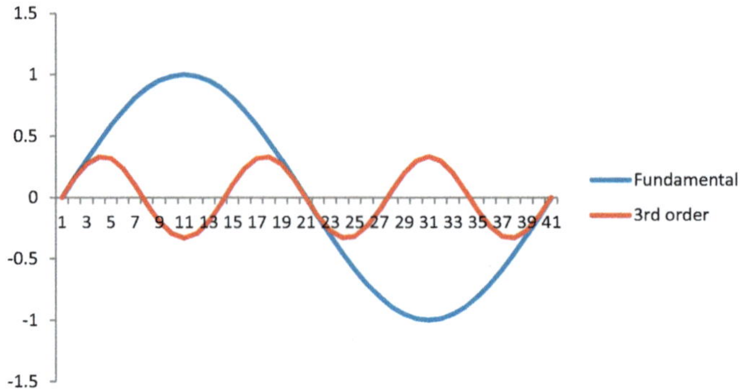

FIGURE A.1.11 Ideal wave shape of tripplen harmonics.

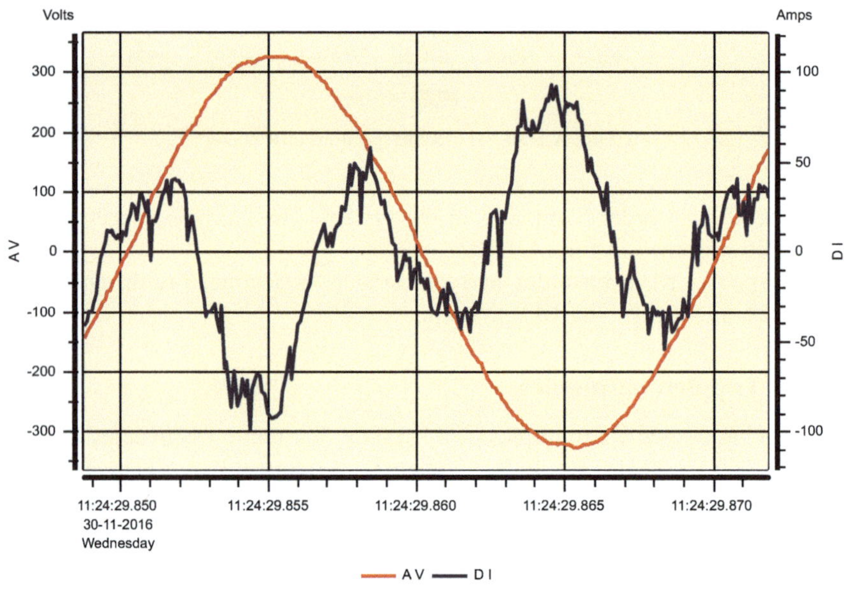

FIGURE A.1.12 Practical waveform of tripplen harmonics.

The wave shape of ideal tripplen harmonics is shown in Fig. A.1.11. In practical systems, actual tripplen harmonics flowing in the system may not be an ideal wave shape and are shown in Fig. A.1.12.

In Fig. A.1.12, red color waveform is fundamental voltage and black color is neutral current. For one cycle of voltage wave shape, the neutral wave shape completes three cycles.

The harmonic orders and frequencies of tripplen harmonics are listed in Table A.1.3.

TABLE A.1.3 Tripplen harmonics.

Harmonic order	Harmonic frequency at 50 Hz	Harmonic frequency at 60 Hz
3rd	150	180
6th	300	360
9th	450	540
12th	600	720
15th	750	900
18th	900	1080
21st	1050	1260
24th	1200	1440
27th	1350	1620
30th	1500	1800
33rd	1650	1980
36th	1800	2160
39th	1950	2340
42nd	2100	2520
45th	2250	2700
47th	2400	2820
50th	2500	3000

References

[1] IEEE Std 519 – 2014 IEEE Recommended Practice and Requirements for Harmonic Control in Electric Power Systems.

[2] IEEE Std 1159 – 2019 IEEE Recommended Practice for Monitoring Electric Power Quality.

[3] IEEE Std 1409 – 2012 IEEE Guide for Application of Power Electronics for Power Quality Improvement on Distribution Systems Rated 1 kV through 38 kV.

[4] F.C. De La Rosa, Harmonics and Power Systems, CRC press.

[5] K. Alexander, M.T. Thompson, Power Quality in Electrical Systems, Mc Graw hill.

[6] IEC 61000 – 3 – 3: Limitation of Voltage Changes, Voltage Fluctuations and Flicker in Public Low Voltage Power Supply Systems for Equipment with Rated Current ≤16 A Per Phase and Not Subject to Conditional Connection.

[7] IEC 61000 – 3 – 5: Limitation of Voltage Fluctuations and Flicker in Low Voltage Power Supply Systems for Equipment with Rated Current Greater than 75A.

[8] P. Sivaraman, C. Sharmeela, D.P. Kothari, Enhancing the Voltage Profile in Distribution System 40 GW of Solar PV Rooftop in Indian Grid by 2022: A Review, 1st International Conference on Large Scale Integration of Renewable Energy in India, 6 -8 September, 2017 (New Delhi, India).
[9] E. Ali, A. Nasiri, B. Stoyan, Bekiarov, Uninterruptible Power Supplies and Active Filters, CRC press, 2004.
[10] C. Sharmeela, B. Surendar, P. Sivaraman, R. Mahendran, Harmonic reduction and voltage flicker mitigation by using static VAR compensator and shunt passive filter for steel plant, GRIDTECH (2015) 524–528. April 8-10, 2015.
[11] P. Sivaraman, C. Sharmeela, Introduction to electric distribution system, in: B. Kahn, H. Haes Alhelou, Ghassan (Eds.), Handbook of Research on New Solutions and Technologies in Electrical Distribution Networks, IGI Global, Hershey, PA, 2020, pp. 1–31.
[12] Power Factor Correction and Harmonic Filtering in Electrical Plants, Technical Application Papers, ABB, July 2018.
[13] P. Sivaraman, C. Sharmeela, Existing issues associated with electric distribution system, in: B. Kahn, H. Haes Alhelou, Ghassan (Eds.), Handbook of Research on New Solutions and Technologies in Electrical Distribution Networks, IGI Global, Hershey, PA, 2020, pp. 1–31.
[14] R. Mahendran, P. Sivaraman, C. Sharmeela, Three Phase Grid Interfaced Renewable Energy Source using Active Power Filter, GRIDTECH (2015) 77–85.

Further reading

[1] D.P. Kothari, I.J. Nagrath, Modern Power System Analysis, fourth ed., Tata McGraw Hill, New Delhi, 2011.
[2] Dran-view 6 User's Guide AC Power Analysis Software.
[3] O. AArdakanian, S. Keshav, C. Rosenberg, Integration of Renewable Generation and Elastic Loads into Distribution Grids, Springer, 2016.
[4] H. William, Kersting, Distribution System Modelling and Analysis, fourth ed., CRC press, 2018.
[5] T. Gonen, Electric Power Distribution Engineering, third ed., CRC press, 2014.
[6] F. Saccomanno, Electric Power Systems: Analysis and Control, first ed., IEEE press, 2003.

Chapter 2

Power system harmonics

P. Sivaraman[1], C. Sharmeela[2]

[1]*Leading Engineering Organisation, Chennai, Tamil Nadu, India;* [2]*Department of Electrical and Electronics Engineering, College of Engineering Guindy, Anna University, Chennai, Tamil Nadu, India*

Chapter outline

2.1 Introduction **61**
2.2 Types of harmonic distortion **62**
2.2.1 Voltage harmonics 62
2.2.2 Current harmonics 62
2.3 Harmonic-producing loads **63**
2.3.1 Single-phase rectifiers 63
2.3.2 Three-phase rectifiers 65
2.4 Point of common coupling **73**
2.4.1 Source-side harmonics 76
2.4.2 Load-side harmonics 76
2.5 Impact of harmonics on connected loads sharing the same feeder **78**
2.6 Impact of harmonics in multitenanted premises **79**
2.7 Evaluation of harmonics in the system **80**
2.7.1 Total harmonic distortion 80
2.7.2 Total demand distortion 85
2.7.3 Total rated current distortion 87
2.8 Resonance **91**
2.8.1 Series resonance 91
2.8.2 Parallel resonance 92
2.8.3 How to prevent the system resonance condition 92
2.9 Annexure 2.1 **93**
References **103**
Further reading **103**

2.1 Introduction

Harmonics are the presence of multiple frequencies in the fundamental frequency of voltage or current or both. Due to this, a fundamental frequency of 50 Hz is superimposed by multiple frequencies like 100 Hz, 150 Hz, 250 Hz etc. resulting in deviating from ideal sinusoidal characteristics. Harmonics are classified into two types: voltage harmonics and current harmonics. The generation of harmonics in voltage is called voltage harmonic distortion and the generation of harmonics in current is called current harmonic distortion.

Power Quality in Modern Power Systems. https://doi.org/10.1016/B978-0-12-823346-7.00002-5

Voltage harmonics are produced by the nonlinear load as in an arc furnace and current harmonics are mainly generated by power converters. Compared to voltage harmonic distortions, current harmonic distortions affect the power system due to the deployment of large numbers of power electronics-based static power converters.

2.2 Types of harmonic distortion

Based on the type of harmonic generation, harmonic distortions are classified into two types [1]:

1. Voltage harmonics (generation of voltage harmonics)
2. Current harmonics (generation of current harmonics)

2.2.1 Voltage harmonics

Loads draw nonlinear voltage and current characteristics when sinusoidal voltage is applied across the terminals. Due to nonlinear characteristics, voltage is changed from ideal sinusoidal and injecting the voltage harmonics into the supplied voltage. These loads are called voltage harmonic sources or loads. Harmonics injected by these loads are called voltage harmonics. The loads that generate voltage harmonics are:

- ➢ Arc furnace [2]
- ➢ Arc welding
- ➢ Arcing loads

2.2.2 Current harmonics

Loads draw nonlinear voltage and current characteristics when sinusoidal voltage is applied across the terminals. Due to nonlinear characteristics, current drawn by the loads from the voltage is not sinusoidal, which results in the supply voltage changing from sinusoidal and injecting the current harmonics into the system. These loads are called current harmonic sources or loads. Harmonics injected by these loads are called current harmonics. The major sources of generation of current harmonics are due to the employment of power electronics-based loads [3].

The loads generating current harmonics in industrial system applications are:

1. Solid-state rectifiers and battery chargers
2. Variable frequency drives (VFDs)
3. Uninterruptible power supply (UPS) systems
4. Frequency changers or cycloconverters
5. AC voltage regulators

etc.

The loads generating current harmonics in commercial system applications are:

1. Switched mode power supplies (SMPSs)
2. Light-emitting diodes (LEDs) and compact fluorescent lamps (CFLs)
3. Electronic ballasts
4. UPS systems
5. VFDs
6. Voltage stabilizers
7. Solid-state rectifiers and battery chargers

etc.

The loads generating current residential system applications are:

1. Laptops and mobile chargers
2. LEDs and CFLs
3. Light and fan dimmers
4. Digital watches
5. Electronic stabilizers

etc.

Compared to commercial and industrial systems, current harmonics are generated from residential systems that are smaller in magnitude and higher by number of devices connected to the system.

2.3 Harmonic-producing loads

Rectifier circuits are used to convert AC supply into DC supply. They have power electronic devices like a diode, insulated-gate bipolar transistor (IGBT), metal—oxide—semiconductor field-effect transistor (MOSFET), etc., for conversion. Based on the number of phases, rectifiers are classified into single-phase rectifiers and three-phase rectifiers [1].

2.3.1 Single-phase rectifiers

The single-phase rectifiers are used for various applications like UPS systems, SMPSs in computers, LEDs, battery chargers, etc. Either diode-based (uncontrolled) or IGBT/MOSFET-based (controlled) rectifiers are widely used in these applications.

Example 2.1: A 240 V, 5 kVA single-phase UPS system is used to provide uninterrupted power to office computers. Power quality analyzers are used to record the voltage and current characteristics of the UPS systems.

The recorded parameters are voltage 236.7 V and current 8.8 A, drawing 1 kW real power at 0.5 lag power factor. Apparent power consumption is 2.1 kVA. The instantaneous phase to neutral voltage wave shape is shown in Fig. 2.1 and the instantaneous current wave shape is shown in Fig. 2.2.

From Figs. 2.1 and 2.2, when a sinusoidal voltage is supplied to a nonlinear load (rectifier part of a UPS system), it draws a distorted current wave shape from sinusoidal supply voltage.

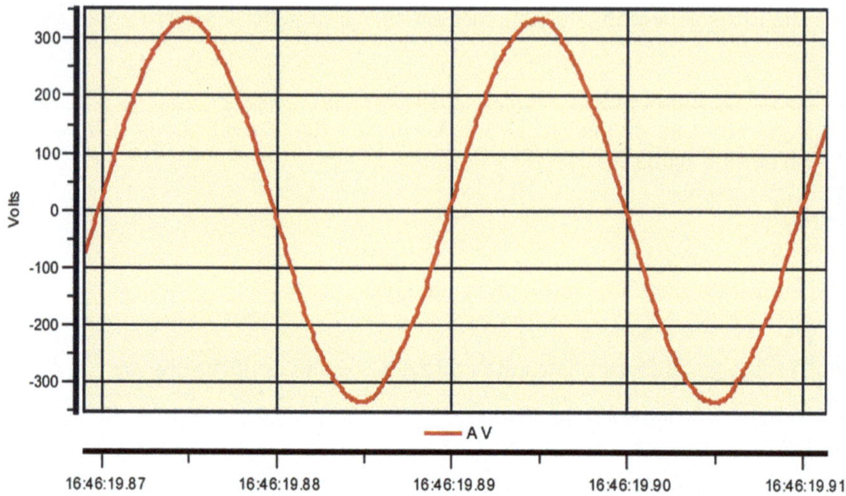

FIGURE 2.1 Instantaneous voltage wave shape.

The corresponding current harmonic spectrum in % and absolute ampere for Fig. 2.2 is shown in Figs. 2.3 and 2.4 respectively.

From Figs. 2.3 and 2.4, current total harmonic distortion (THD) is 124.5% and absolute current distortion is 7.1 A. The magnitudes of major harmonic orders are listed in Table 2.1.

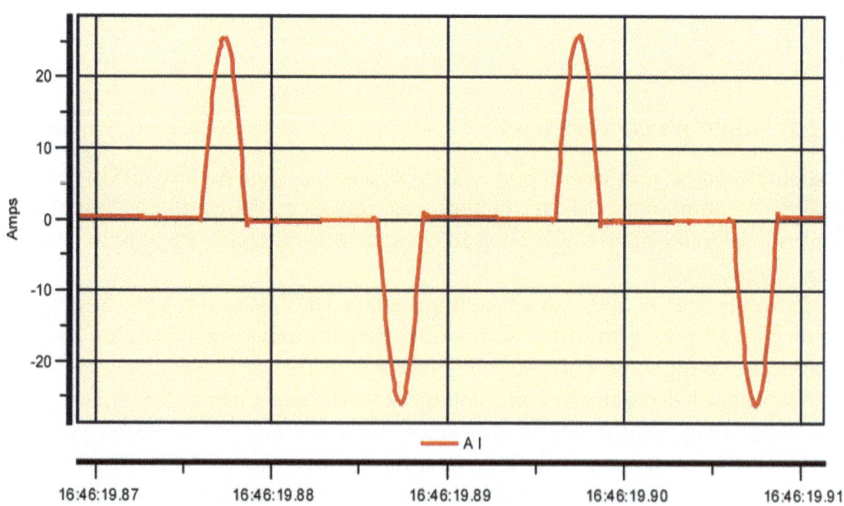

FIGURE 2.2 Instantaneous current wave shape.

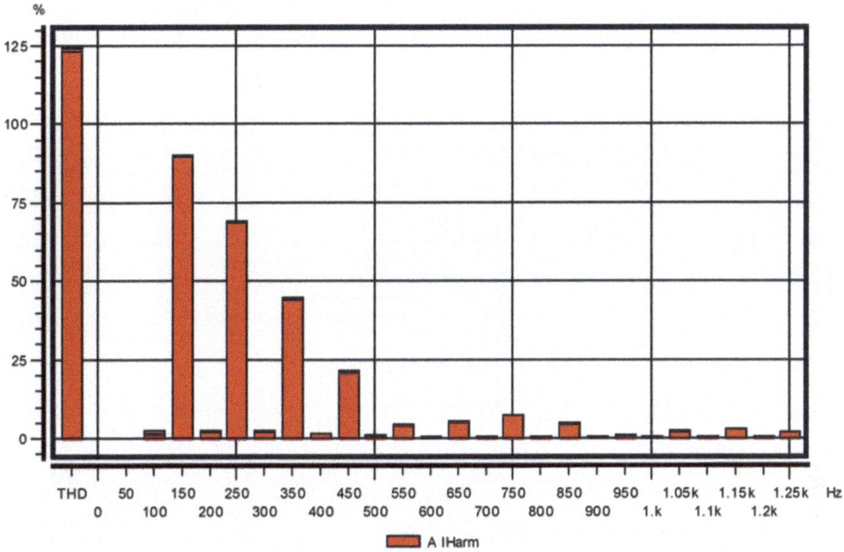

FIGURE 2.3 Current harmonic spectrum in %. *THD*, Total harmonic distortion.

2.3.2 Three-phase rectifiers

Three-phase rectifiers are used for various low-power and high-power applications like UPS systems, VFDs, cycloconverters, AC voltage regulators, etc. [4]. Most of these rectifiers are controlled by IGBTs, MOSFETs, or

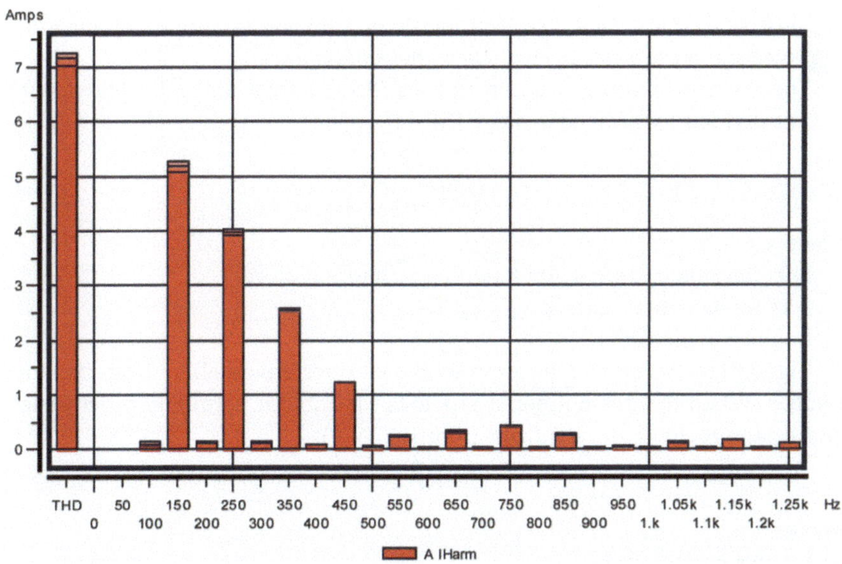

FIGURE 2.4 Current harmonic spectrum in A. *THD*, Total harmonic distortion.

TABLE 2.1 Current harmonics injected by a single-phase uninterruptible power supply system.

Harmonic order	Harmonics in %	Harmonics in A
Current total harmonic distortion	124.5	7.1
3rd	89.3	5.1
5th	68.8	3.9
7th	44.1	2.5
9th	20.8	1.2
11th	4	0.2
13th	5	0.3
15th	7.3	0.4
17th	4.5	0.2

silicon-controlled rectifiers (SCRs). The SCR-based rectifiers are classified based on the number of pulses used for conversion, which are:

- Six-pulse rectifier
- 12-pulse rectifier
- 18-pulse rectifier
- 24-pulse rectifier

Mostly, six-pulse and 12-pulse rectifiers are used in various applications due to economics, product size, and other constraints.

The current harmonic content for an ideal current wave is fundamental current divided by harmonic order [5]:

$$I_h = \frac{I_f}{h} \tag{2.1}$$

where:

I_h is harmonic current for a particular harmonic order

I_f is fundamental current

h is harmonic order.

The current harmonics injected by the rectifier are obtained from its pulse number. When the pulse number is known, the major harmonic order generated by the rectifier is calculated from Eq. (2.2):

$$h = NP \pm 1 \tag{2.2}$$

where:

h is harmonic order

N is a positive integer

P is pulse number.

Example 2.2: Obtain the harmonic order injected by a six-pulse rectifier.
Solution:
N is 1
P is 6
From Eq. (2.2), harmonic order is when it is "−1".
h = 5
From Eq. (2.2), harmonic order is when it is "+1".
h = 7

The major harmonic orders generated by a six-pulse converter are fifth and seventh orders.

Example 2.3: Obtain the harmonic order injected by a 12-pulse rectifier.
Solution:
N is 1
P is 12
From Eq. (2.2), harmonic order is when it is "−1".
h = 11
From Eq. (2.2), harmonic order is when it is "+1".
h = 13

The major harmonic orders generated by a 12-pulse converter are 11th and 13th orders.

Example 2.4: Calculate the fifth, seventh, 11th, 13th, and 19th harmonic current from the fundamental current of 100 A.

Solution:

The fundamental frequency is 100 A.

1. From Eq. (2.1), fifth order harmonic current $I_5 = \frac{100}{5} = 20$ A
2. From Eq. (2.1), seventh order harmonic current $I_7 = \frac{100}{7} = 14.3$ A
3. From Eq. (2.1), 11th order harmonic current $I_{11} = \frac{100}{11} = 9.1$ A
4. From Eq. (2.1), 13th order harmonic current $I_{13} = \frac{100}{13} = 7.7$ A
5. From Eq. (2.1), 17th order harmonic current $I_{17} = \frac{100}{17} = 5.9$ A
6. From Eq. (2.1), 19th order harmonic current $I_{19} = \frac{100}{19} = 5.3$ A

From the above calculations, fifth order harmonic current is 20 A, seventh order harmonic current is 14.3 A, 11th order harmonic current is 9.1 A, 13th order harmonic current is 7.7 A, 17th order harmonic current is 5.9 A, and 19th order harmonic current is 5.3 A. Harmonic current magnitude decreases with the increase of harmonic order. To reduce the harmonic current magnitude, a 12-pulse rectifier is used instead of a six-pulse rectifier. The advantage of a 12 pulse converter over a six-pulse converter is reduced harmonic current and reduced current THD, so a smaller size of harmonic filter is required for mitigation and the total current flowing in the circuit also reduces, leading to reduced power loss (I^2R).

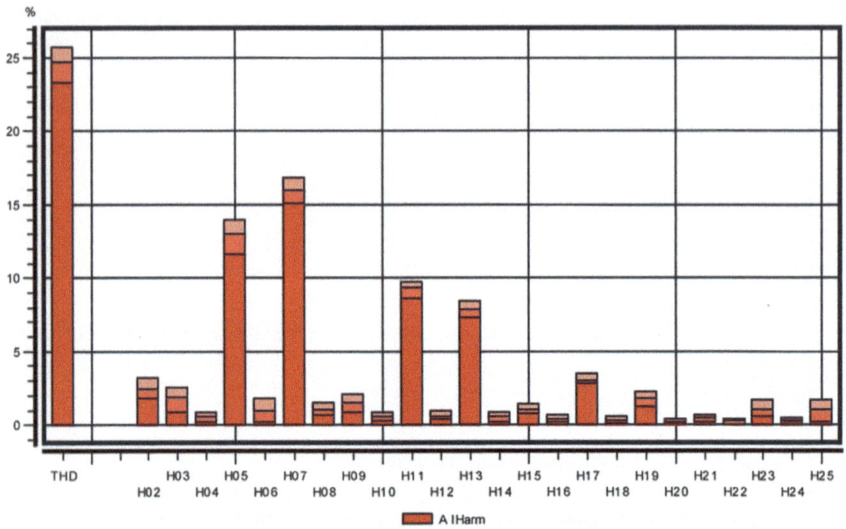

FIGURE 2.5 Current harmonic spectrum in % for a six-pulse rectifier. *THD*, Total harmonic distortion.

A. Six-pulse rectifier:

Six-pulse rectifiers are used in drives, UPS systems, cycloconverters, etc., for both lower and higher power applications. They generate fifth and seventh order harmonics in a higher magnitude. A typical six-pulse rectifier-generated current harmonic spectrum in % for the *R* phase is shown in Fig. 2.5.

From Fig. 2.5, the major harmonic orders are seventh, fifth, 11th, and 13th. Other harmonic orders are also present but have less percentages compared to these four orders.

Example 2.5: A three-phase, 415 V, 5 kW VFD is operating at 3.6 kW. The current drawn by the rectifier circuit in VFD is 5.7 A per phase, operating at a 0.9 lag power factor.

The instantaneous voltage of the *R* phase to neutral is shown in Fig. 2.6 and the instantaneous *R* phase current is shown in Fig. 2.7.

The corresponding current harmonic spectrum of the *R* phase in % for Fig. 2.7 is shown in Fig. 2.8 and the current harmonic spectrum in absolute ampere is shown in Fig. 2.9.

From Figs. 2.8 and 2.9, THD is 47% and absolute ampere is 2.5 A. The major individual harmonic order and its magnitudes are listed in Table 2.2.

Other practical examples are given in Annexure 2.1.

B. Twelve-pulse rectifier:

Twelve-pulse rectifiers are used in VFDs, UPS systems, etc. and in higher-power applications because of economics. For the same rating, a 12-pulse system rectifier costs 40%–60% more than a six-pulse rectifier. It generates 11th and 13th order harmonics in a higher magnitude. A typical 12-pulse rectifier-generated current harmonic spectrum in % is shown in Fig. 2.10.

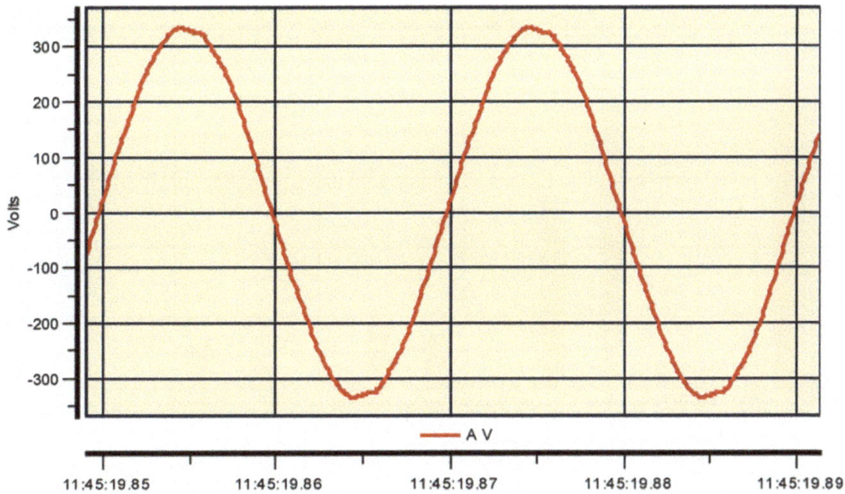

FIGURE 2.6 The instantaneous voltage of *R* phase to neutral.

From Fig. 2.10, the major harmonic order produced by a 12-pulse converter is 11th and 13th in a higher magnitude as compared to fifth and seventh order.

Example 2.6: A three-phase, 415 V, 160 kVA UPS system is monitored for 10 min between 18:02:22 and 18:12:47 h.

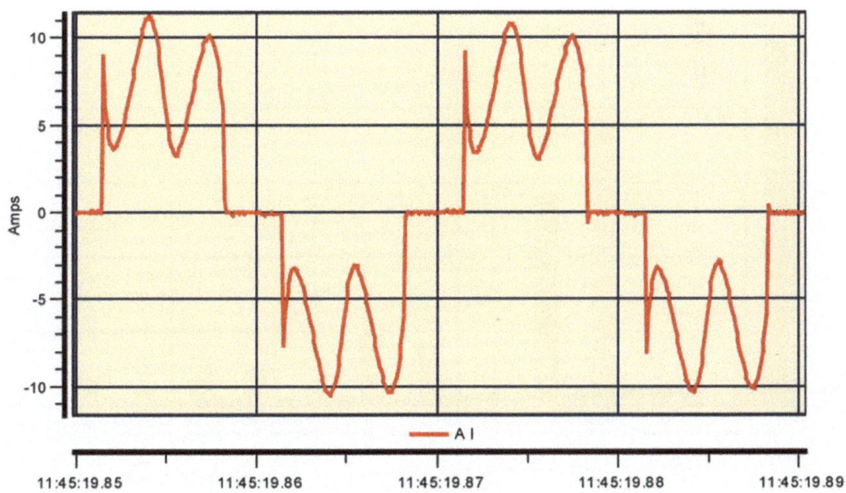

FIGURE 2.7 The instantaneous current of the *R* phase.

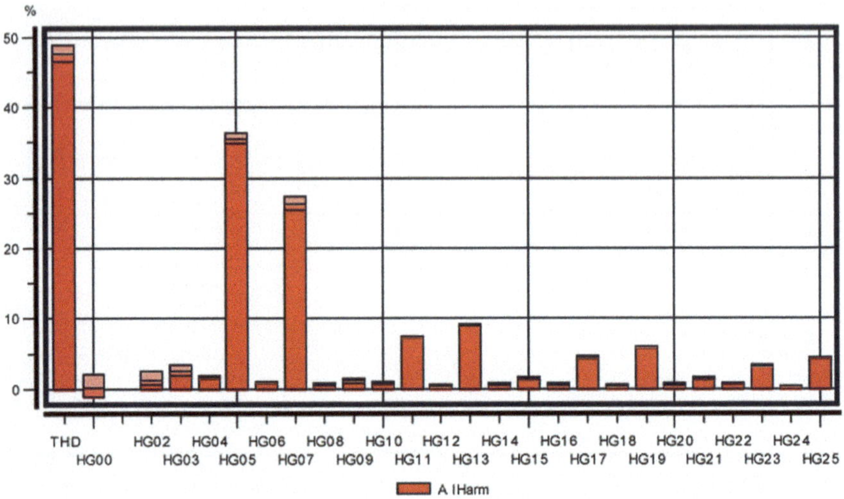

FIGURE 2.8 The current harmonic spectrum in % for the *R* phase. *THD*, Total harmonic distortion.

The current drawn by the rectifier circuit of the UPS system is 56 A per phase at 0.75 lag power factor. The instantaneous voltage of phase to neutral is shown in Fig. 2.11 and the instantaneous current is shown in Fig. 2.12.

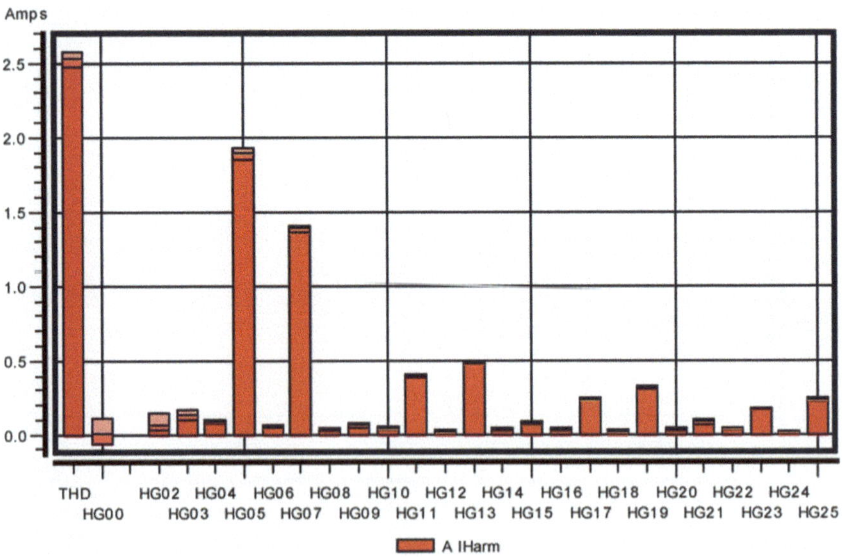

FIGURE 2.9 The current harmonic spectrum in A for the *R* phase. *THD*, Total harmonic distortion.

TABLE 2.2 Current harmonics generated by a six-pulse variable frequency drive.

Harmonic order	Harmonics in %	Harmonics in A
Current total harmonic distortion	47	2.5
3rd	3	0.14
5th	36	1.9
7th	26	1.39
9th	1	0.05
11th	7	0.39
13th	9	0.5
15th	2	0.07
17th	4	0.24
19th	6	0.32
23rd	3	0.18
25th	4	0.25

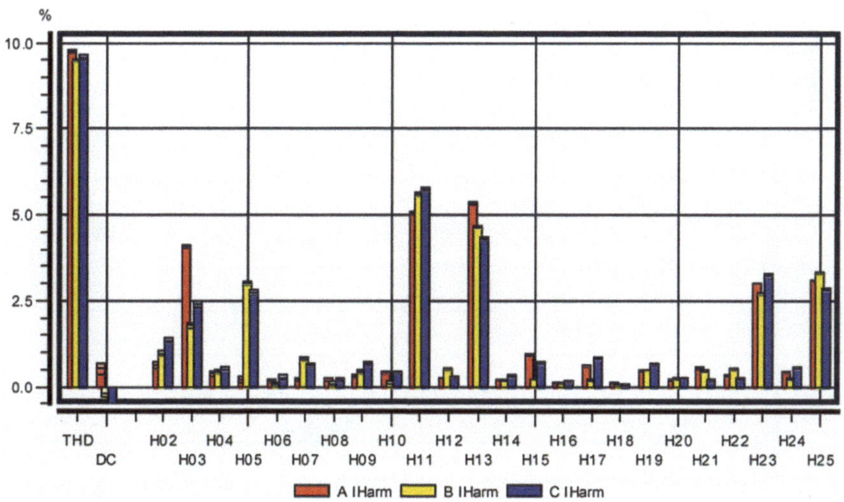

FIGURE 2.10 Current harmonic spectrum in % for a 12-pulse rectifier. *DC*, Direct current; *THD*, total harmonic distortion.

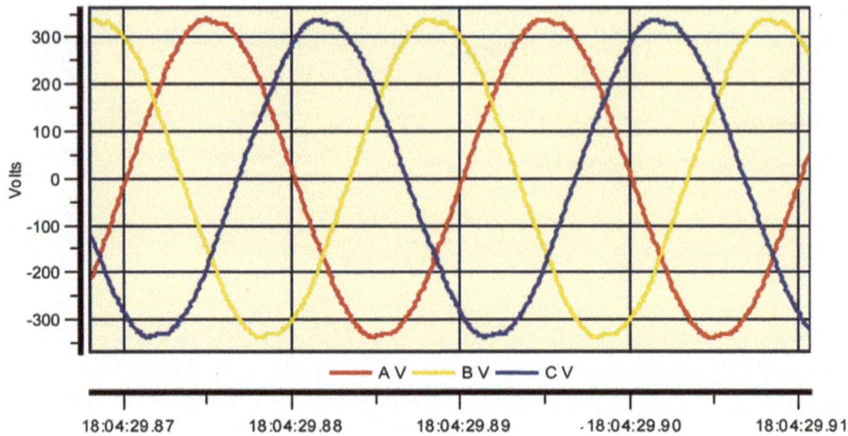

FIGURE 2.11 The instantaneous voltage of phase to neutral.

The corresponding current harmonic spectrum in % for Fig. 2.12 is shown in Fig. 2.13 and the current harmonic spectrum in absolute ampere is shown in Fig. 2.14.

From Figs. 2.13 and 2.14, THD is 10% and absolute harmonic current is 5.6 A. The individual major harmonic order and its magnitudes are listed in Table 2.3.

The current THD trend in % for 10 min monitoring duration is shown in Fig. 2.15.

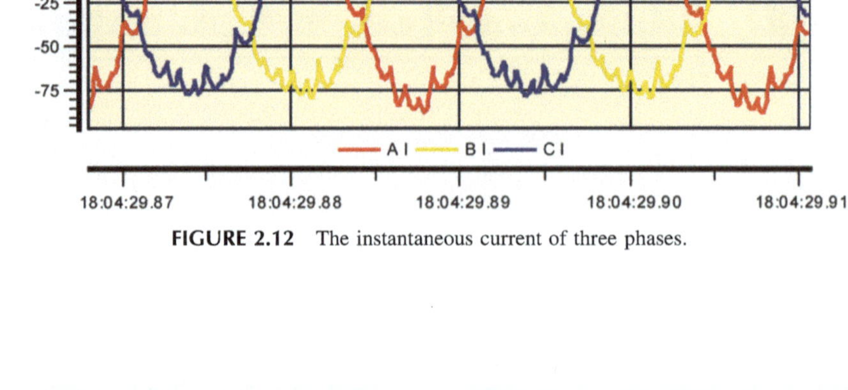

FIGURE 2.12 The instantaneous current of three phases.

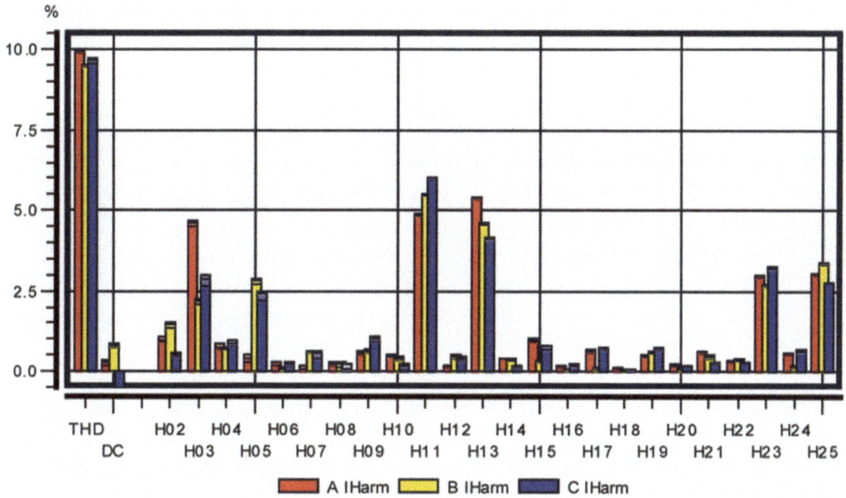

FIGURE 2.13 The current harmonic spectrum in %. *DC*, Direct current; *THD*, total harmonic distortion.

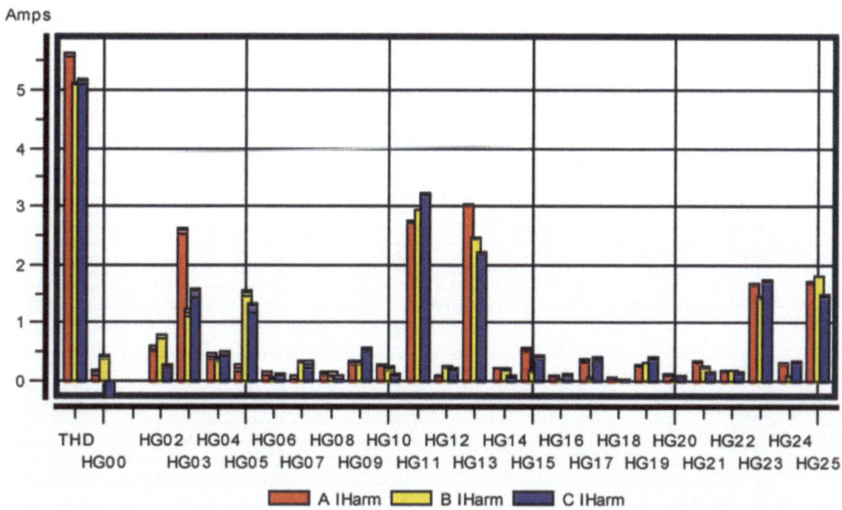

FIGURE 2.14 The current harmonic spectrum in A. *THD*, Total harmonic distortion.

2.4 Point of common coupling

The point of common coupling (PCC) is a common point or location where multiple customers and their equipment are connected to a utility power grid. IEEE standard 519-2014 [5] defines PCC as the point on a public power supply system electrically nearest to a particular load at which other loads could be connected. The PCC is a point located upstream of the considered plant electrical installation [6].

TABLE 2.3 Current harmonics generated by a 12-pulse uninterruptible power supply system.

Harmonic order	Harmonics in %	Harmonics in A
Current total harmonic distortion	10	5.6
3rd	4.5	2.7
5th	3	1.6
7th	0.8	0.4
9th	1	0.5
11th	6	3.2
13th	5.5	3
15th	1	0.5
17th	0.9	0.4
19th	0.9	0.4
23rd	3.3	1.8
25th	3.3	1.8

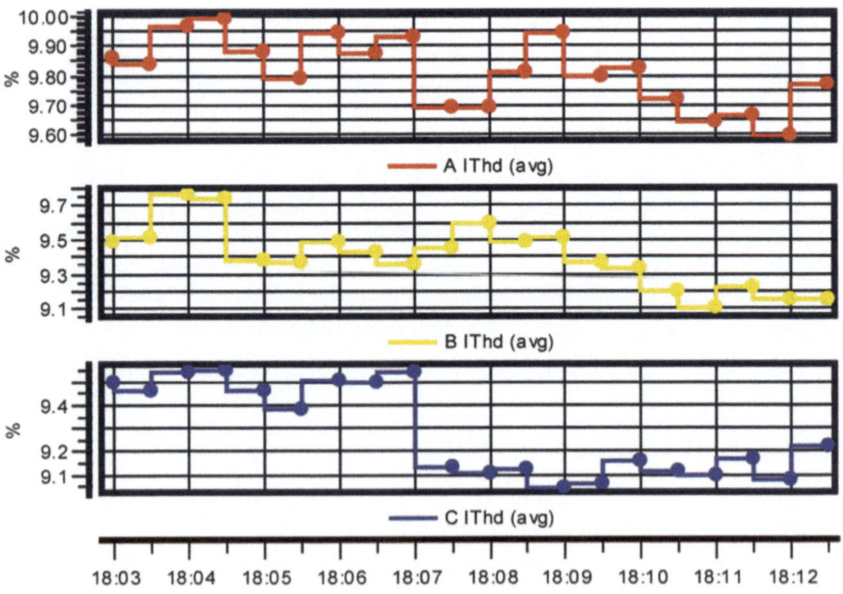

FIGURE 2.15 The current total harmonic distortion trend in %.

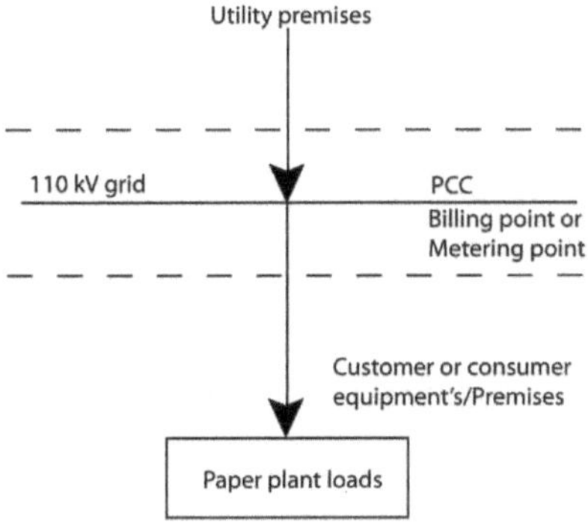

FIGURE 2.16 Point of common coupling (PCC) for single customer/consumer.

Example 2.7: A paper plant has a connected load of 10 MVA and receives power from a 110 kV grid supply. The electrical distribution of the industrial system is shown in Fig. 2.16.

The PCC for this paper plant is 110 kV. In most cases, a metering point or billing point is considered as the PCC.

Example 2.8: Two industrial plants have connected loads of 10 and 15 MVA, respectively, and receive power from a 110 kV grid supply. The electrical distribution of the industrial system is shown in Fig. 2.17.

The IEEE standard 519-2014 recommended harmonic limits both voltage harmonic distortion and current harmonic distortion at the PCC only. These limits should not be applied anywhere within the customer's locations, individual feeders, or equipment.

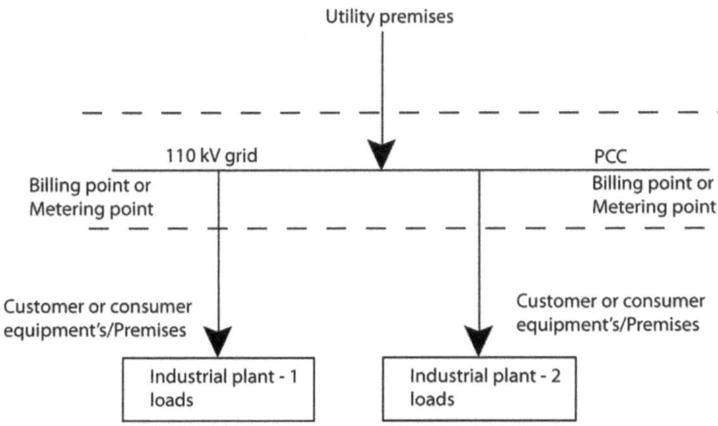

FIGURE 2.17 Point of common coupling (PCC) for two customers/consumers.

2.4.1 Source-side harmonics

Ideally, power sources should be sinusoidal in nature and free from harmonics. However, in a practical system, power sources no longer have sinusoidal characteristics and the minimal amount of harmonic content is the presence in the power source. Harmonics from the utility power supply can affect customer equipment. This means that customer loads powered by a distorted supply voltage affect the quality of the power supply.

2.4.2 Load-side harmonics

PCC is a common interconnection point for different customers connected to the same utility power supply. If current harmonic injections by customer loads are not limited, the power quality at the PCC will be affected. These current harmonic distortions at the PCC also affect the power quality of different customers connected to the same PCC, i.e., current harmonics injected by one customer affect the voltage quality at the PCC. Harmonic current is injected from customer 1 to the PCC as shown in Fig. 2.18 and this affects the voltage at the PCC.

The distorted voltage at the PCC affects other customers connected to the same PCC by supplying a distorted voltage supply, i.e., distorted voltage is supplied to customer 2 connected to the same PCC. Even if customer 2 loads are linear, they will draw the nonsinusoidal current from the voltage because of supply voltage distortion. Fig. 2.19 shows the distorted voltage at customer 2 connected loads.

Example 2.9: Two customers (customer 1 and customer 2) are connected to the same PCC at a 22 kV supply as shown in Fig. 2.18.

Customer 1 has higher nonlinear loads and customer 2 has linear loads. The instantaneous current wave shape drawn by customer 1 is shown in Fig. 2.20.

From Fig. 2.20 it is clearly visible that current drawn by customer 1 is nonsinusoidal and the corresponding current harmonic spectrum for Fig. 2.20 is shown in Fig. 2.21.

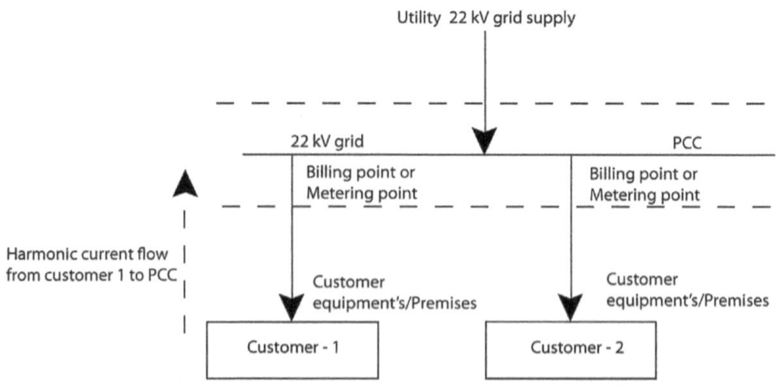

FIGURE 2.18 Harmonic current from customer 1 to point of common coupling (PCC).

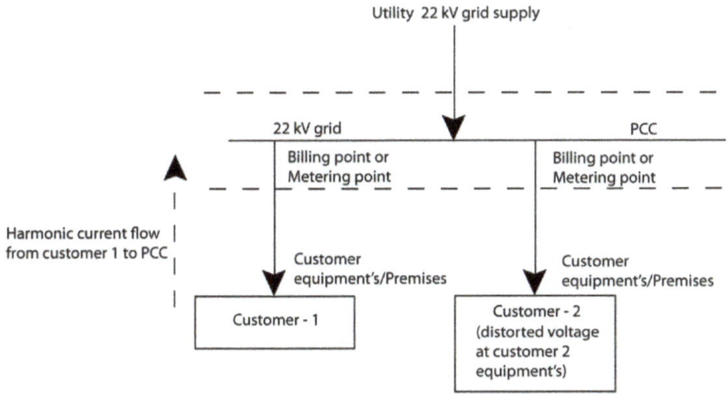

FIGURE 2.19 Distorted voltage at customer 2 equipment. *PCC*, Point of common coupling.

Due to distorted current drawn from the PCC, voltage quality at the PCC is no longer sinusoidal. The instantaneous phase-to-phase voltage wave shape at the 22 kV PCC is shown in Fig. 2.22.

From Fig. 2.22 it is clearly visible that voltage is distorted from the ideal sinusoidal wave shape. The corresponding voltage distortion at the PCC is 2.9% THD for the voltage wave shape shown in Fig. 2.22 and the individual voltage harmonic spectrum is shown in Fig. 2.23.

From Fig. 2.23, the voltage at the PCC is affected by current harmonic injection from customer 1. This distorted voltage is supplied to other customers in the PCC (customer 2). In this case, customer 2 has linear loads in their plant that not able to draw the linear current from the supply voltage due to distortion.

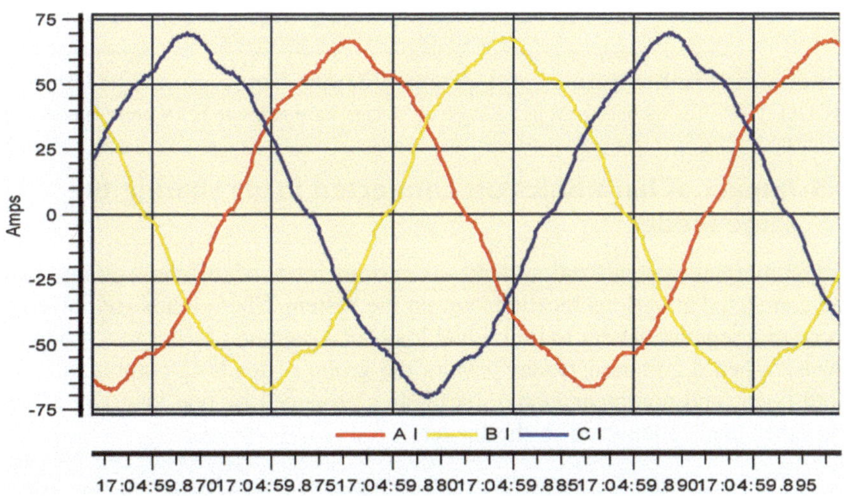

FIGURE 2.20 Instantaneous current wave shape of customer 1.

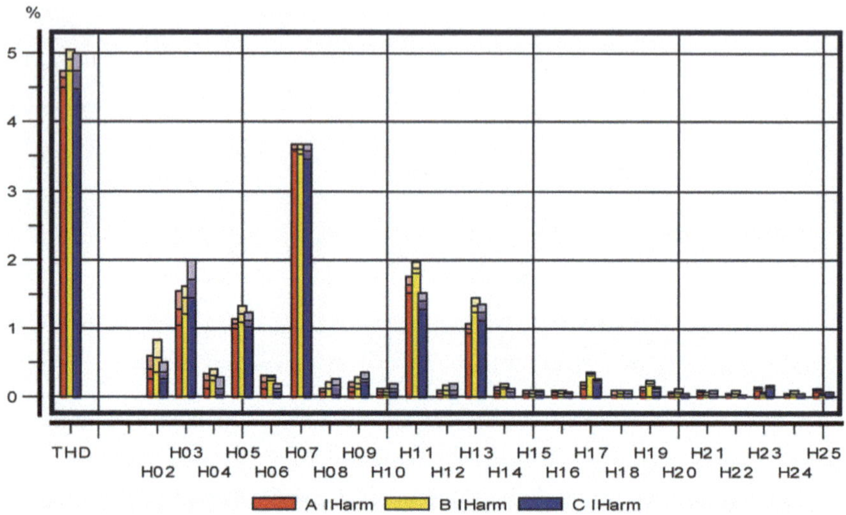

FIGURE 2.21 The current harmonic spectrum of customer 1. *THD*, Total harmonic distortion.

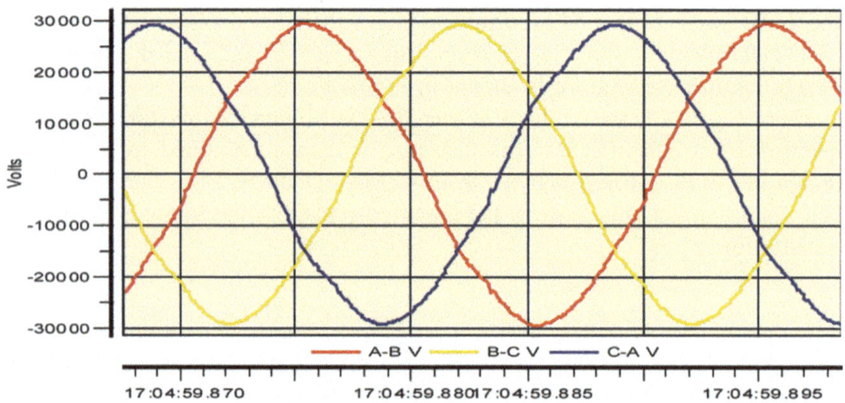

FIGURE 2.22 The instantaneous phase-to-phase voltage wave shape at 22 kV point of common coupling.

2.5 Impact of harmonics on connected loads sharing the same feeder

A commercial or industrial system has various loads of different ratings that are connected at various locations within the system. These loads are powered through a separate feeder to individual loads or a common feeder to a group of loads. When a common feeder powers the group of loads (a combination of both linear and nonlinear loads), harmonics produced by one load can affect the other loads connected to the feeder.

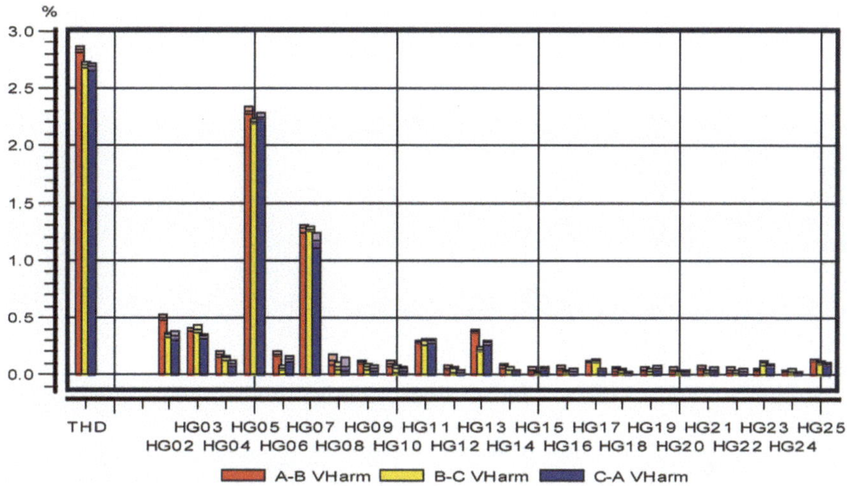

FIGURE 2.23 The voltage harmonic spectrum in % at 22 kV point of common coupling. *THD*, Total harmonic distortion.

Example 2.10: A motor control center (mcc) powers a combination of both linear and nonlinear loads as shown in Fig. 2.24.

From Fig. 2.24, nonlinear loads are 200 kW VFD and linear loads are 10 kW motor, 15 kW motor, 50 HP motor.

Compared to all the connected loads, 200 kW VFD is higher in power rating. The nonlinear current drawn by this VFD have an impact on supply voltage distortion. This distorted voltage has an impact on other linear (motor) loads connected to this feeder.

2.6 Impact of harmonics in multitenanted premises

In commercial power systems, due to various reasons such as reduced load demand, high cost for purchasing own transformer and power supply, space constraints, etc., multiple customers may be sharing power from the common transformer within a complex or park. In this case, a common transformer is the power source for all customers who are sharing the power infrastructure. In shared premises, harmonics produced by one customer affect the other customers.

Example 2.11: Three customers share the utility power supply through a 2500 kVA distribution transformer power infrastructure. These customers are customer A, customer B, and customer C and have power demands of 1250, 750, and 500 kVA, respectively. These three customers are located on different floors in the same building.

A single line diagram of electrical distribution is shown in Fig. 2.25.

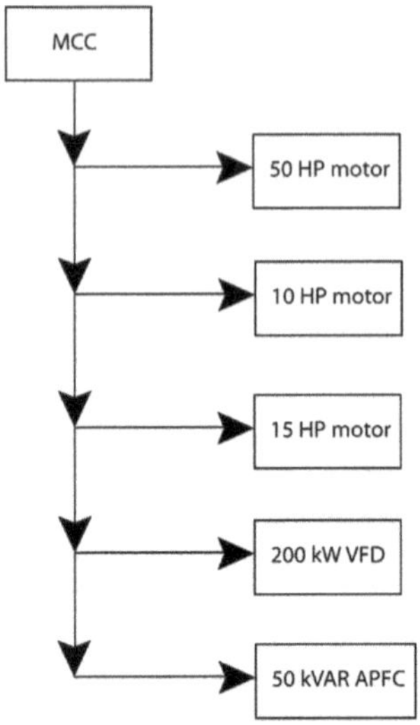

FIGURE 2.24 Single line diagram of electrical distribution. *APFC*, automatic power factor correction; *MCC*, motor control center; *VFD*, variable frequency drive.

The main low tension panel is the PCC for customer A, customer B, and customer C. These three customers have different types of loads with a combination of both linear and nonlinear loads in different power ratings. The loading percentages of these three customers are also different from each other from time to time. The current harmonics generated by one customer affect other customers in the shared premises. In this condition, harmonics generated by customer A affect the performance of customer B and customer C equipment and vice versa.

2.7 Evaluation of harmonics in the system

Generally, harmonics in the system are evaluated using THD for both voltage harmonic distortion and current harmonic distortion. IEEE Std 519-2014, using the total demand distortion (TDD) for evaluation of current harmonic distortion at the PCC, and IEEE Std 1547-2018, introduced new terminology for the evaluation of current harmonics called total rated harmonic distortion (TRD) [7].

2.7.1 Total harmonic distortion

THD is the ratio of the square root of the sum of all harmonic components except fundamental to the fundamental component [8]. The term THD is used to find the percentage of distortion from its fundamental wave shape. IEEE standard 519-2014 defines THD as the ratio of the root mean square of the

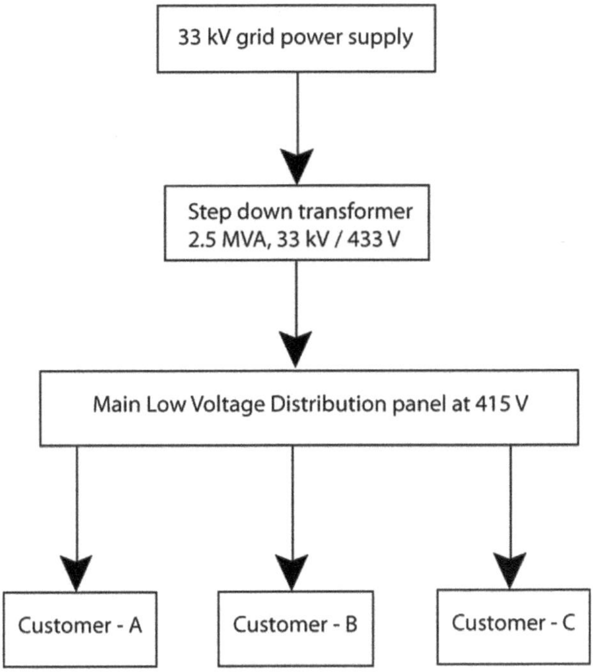

FIGURE 2.25 Single line diagram of electrical distribution.

harmonic content, considering harmonic components up to the 50th order and specifically excluding interharmonics, expressed as a percentage of the fundamental. Harmonic components of order greater than 50 may be included when necessary [5,6].

THD is used to evaluate harmonics in both voltage and current.

Voltage THD is the ratio of the square root of the sum of all voltage harmonic components except the fundamental voltage component to the fundamental voltage component. The expression for voltage harmonic distortion is shown in Eq. (2.3):

$$V_{THD} = \frac{\sqrt{\sum_{h=2}^{\infty} V_h^2}}{V_1} \tag{2.3}$$

where:

V_{THD} is total voltage harmonic distortion
V_h is voltage harmonic component in root mean square (RMS)
V_1 is a fundamental voltage in RMS.

The expression for voltage harmonic distortion in % is shown in Eq. (2.4):

$$V_{THD}(\%) = \frac{\sqrt{\sum_{h=2}^{\infty} V_h^2}}{V_1} x100(\%) \tag{2.4}$$

Example 2.12: Calculate the voltage THD from the individual voltage harmonics listed in Table 2.4 where the fundamental voltage is 240.54 V.

Solution:

From Eq. (2.4),

$$V_{THD}(\%) = \frac{\sqrt{\sum_{h=2}^{50} V_h^2}}{V_1} x100(\%)$$

where $V_{THD} = 2.6\%$.

Current THD is the ratio of the square root of the sum of all current harmonic components except the fundamental current component to the fundamental current component [1]. The expression for current harmonic distortion is shown in Eq. (2.5):

$$I_{THD} = \frac{\sqrt{\sum_{h=2}^{\infty} I_h^2}}{I_1} \tag{2.5}$$

where:

I_{THD} is total current harmonic distortion

I_h is the current harmonic component in RMS

I_1 is a fundamental current in RMS.

The expression for current harmonic distortion in % is shown in Eq. (2.6):

$$I_{THD}(\%) = \frac{\sqrt{\sum_{h=2}^{\infty} I_h^2}}{I_1} x100(\%) \tag{2.6}$$

Example 2.13: Calculate the current THD from the individual current harmonics listed in Table 2.5 where the fundamental current is 903.2 A.

Solution:

From Eq. (2.6),

$$I_{THD}(\%) = \frac{\sqrt{\sum_{h=2}^{50} I_h^2}}{I_1} x100(\%)$$

$I_{THD} = 5.1\%$.

TABLE 2.4 Individual voltage harmonic distortion.

Harmonic order	Voltage (V)	Harmonic order	Voltage (V)	Harmonic order	Voltage (V)	Harmonic order	Voltage (V)
H02	1.24	H15	0.06	H28	0.07	H41	0.03
H03	0.69	H16	0.09	H29	0.12	H42	0.04
H04	0.48	H17	0.16	H30	0.08	H43	0.06
H05	4.80	H18	0.13	H31	0.09	H44	0.05
H06	0.47	H19	0.06	H32	0.07	H45	0.03
H07	3.72	H20	0.09	H33	0.05	H46	0.04
H08	0.27	H21	0.17	H34	0.06	H47	0.05
H09	0.18	H22	0.07	H35	0.02	H48	0.04
H10	0.18	H23	0.16	H36	0.05	H49	0.04
H11	0.27	H24	0.08	H37	0.06	H50	0.04
H12	0.13	H25	0.04	H38	0.06		
H13	0.60	H26	0.06	H39	0.01		
H14	0.17	H27	0.14	H40	0.05		

TABLE 2.5 Individual current harmonic distortion.

Harmonic order	Current (A)	Harmonic order	Current (A)	Harmonic order	Current (A)	Harmonic order	Current (A)
H02	6.04	H15	0.32	H28	0.74	H41	0.47
H03	9.34	H16	0.65	H29	2.20	H42	0.20
H04	2.54	H17	1.21	H30	0.28	H43	0.51
H05	37.96	H18	0.43	H31	1.61	H44	0.22
H06	1.92	H19	0.68	H32	0.34	H45	0.32
H07	22.62	H20	0.52	H33	0.52	H46	0.30
H08	0.82	H21	0.45	H34	0.30	H47	0.21
H09	1.54	H22	0.50	H35	0.74	H48	0.20
H10	0.76	H23	3.69	H36	0.26	H49	0.23
H11	2.12	H24	0.57	H37	0.39	H50	0.12
H12	0.64	H25	1.37	H38	0.23		
H13	2.76	H26	0.23	H39	0.27		
H14	0.50	H27	0.26	H40	0.18		

Example 2.14: A single line diagram (SLD) of electrical distribution is shown in Fig. 2.26 [9]. The power quality analyzer is connected at the low voltage side of a 2.5 MVA transformer.

In this plant, UPS and the air-handling unit are nonlinear loads that inject the current harmonics into the system.

The voltage THD in % and individual voltage harmonic spectrum in % up to the 25th order for the *R* phase are shown in Fig. 2.27.

From Fig. 2.27, voltage THD is around 2.6% at the location.

The current THD in % and individual current harmonic spectrum in % up to the 25th order for the *R* phase are shown in Fig. 2.28.

From Fig. 2.28, current THD is around 5.1% on the feeder.

2.7.2 Total demand distortion

TDD is the ratio of the square root of the sum of all current harmonic components except fundamental current to the fundamental current of maximum demand [6]. IEEE standard 519-2014 [5] defines TDD as the ratio of the root mean square of the harmonic content, considering harmonic components up to the 50th order and specifically excluding interharmonics, expressed as a percentage of the maximum demand current. Harmonic components of order greater than 50 may be included when necessary.

$$\mathrm{TDD} = \frac{\sqrt{\sum_{h=2}^{50} I_h^2}}{I_L} \tag{2.7}$$

where:

TDD is total demand distortion

I_h is a current harmonic component in RMS

I_L is maximum demand load current in RMS at the PCC.

The maximum demand load current is calculated at the PCC as the sum of current corresponding to the maximum demand for the previous 12 months divided by 12:

$$\text{Max demand load current} = \frac{\text{Sum of max current of previous 12 months}}{12} \tag{2.8}$$

Example 2.15: An industrial customer receives a power supply of 33 kV at a PCC. The maximum demand current for the previous 12 months is listed in Table 2.6. Calculate the maximum demand load current.

Solution:

From Eq. (2.8),

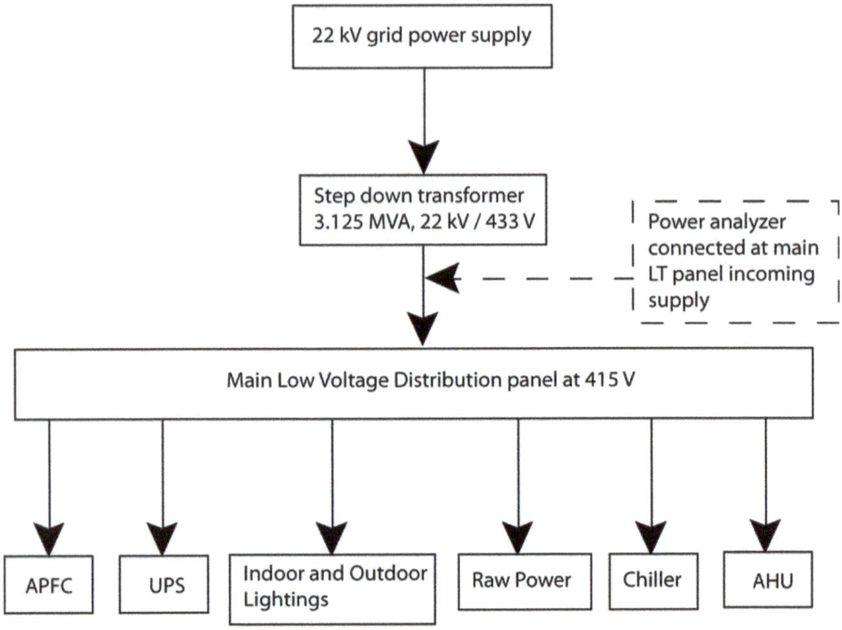

FIGURE 2.26 Single line diagram of electrical distribution. *AHU*, Air-handling unit; *APFC*, automatic power factor correction; *LT*, low tension; UPS, uninterruptible power supply.

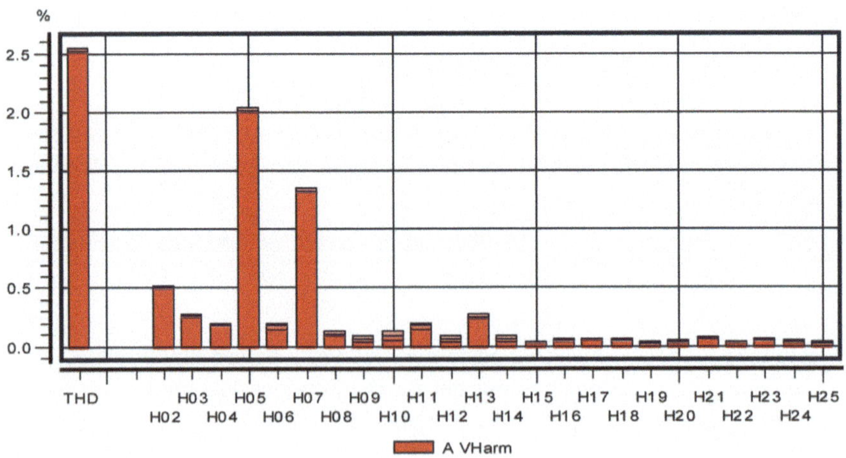

FIGURE 2.27 Voltage harmonic spectrum in %. *THD*, Total harmonic distortion.

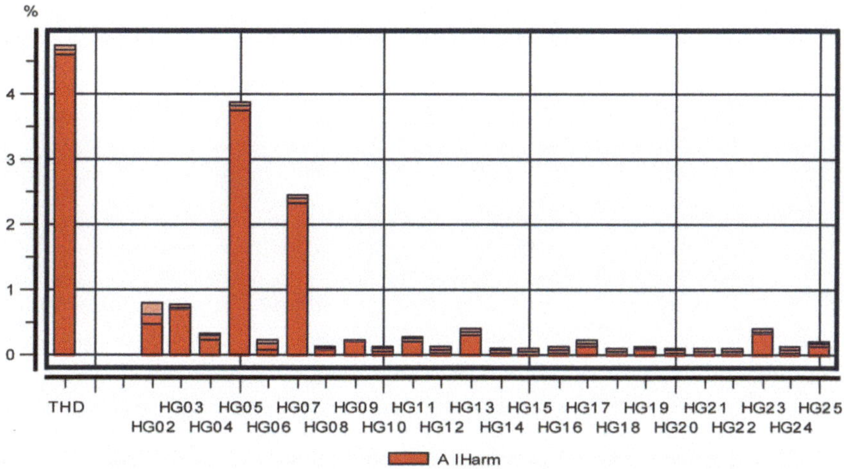

FIGURE 2.28 Current harmonic spectrum in %. *THD*, Total harmonic distortion.

Max demand load current

$$= \frac{82 + 87 + 89 + 95 + 100 + 96 + 94 + 92 + 91 + 90 + 88 + 87}{12}$$

where max demand load current = 90.92 A.

Example 2.16: An industrial customer receives a power supply of 110 kV at a PCC. Power quality measurement is carried out at 110 kV PCC and the individual current harmonic distortion is listed in Table 2.7; the maximum demand load current is 40 A. Find the TDD value at the PCC.

The TDD can be calculated as per Eq. (2.7):

$$\mathrm{TDD} = \frac{\sqrt{\sum_{h=2}^{50} I_H^2}}{I_L}$$

TDD = 0.1.

2.7.3 Total rated current distortion

TRD is introduced by IEEE standard 1547-2018. The evaluation of harmonics by TRD is different from THD and TDD in IEEE standard 519.

The definition of TRD as per IEEE standard 1547-2018 is the total root sum square of the current distortion components, including both harmonics and interharmonics created by the distributed energy resources (DER) unit to DER rated current capacity in percentage [10]. The expression for calculating TRD is given in Eq. (2.9):

TABLE 2.6 Maximum demand current for the previous 12 months.

Month	Maximum demand current (A)
January	82
February	87
March	89
April	95
May	100
June	96
July	94
August	92
September	91
October	90
November	88
December	87

TABLE 2.7 Summary of individual harmonic order.

Harmonic order	Current (A)	Harmonic order	Current (A)
H02	0.5	H27	0.1
H03	0.8	H28	0.1
H04	0.4	H29	0.1
H05	0.9	H30	0.0
H06	0.4	H31	0.2
H07	1.6	H32	0.1
H08	0.1	H33	0.0
H09	0.2	H34	0.0
H10	0.1	H35	0.1
H11	0.2	H36	0.0
H12	0.1	H37	0.2
H13	0.5	H38	0.1
H14	0.1	H39	0.0
H15	0.1	H40	0.0

TABLE 2.7 Summary of individual harmonic order.—cont'd

Harmonic order	Current (A)	Harmonic order	Current (A)
H16	0.0	H41	0.1
H17	0.1	H42	0.0
H18	0.1	H43	0.1
H19	0.3	H44	0.0
H20	0.1	H45	0.1
H21	0.1	H46	0.0
H22	0.1	H47	0.1
H23	0.3	H48	0.1
H24	0.1	H49	0.1
H25	0.3	H50	0.1
H26	0.1		

$$\mathrm{TRD}(\%) = \frac{\sqrt{I_{RMS}^2 - I_1^2}}{I_{RATED}} \mathrm{x}100\% \tag{2.9}$$

where:

I_{RMS} is the RMS of the DER current, including all frequency components at the reference point of applicability (RPA)

I_{RATED} is DER-rated current capacity

I_1 is fundamental current at the RPA.

The RPA is the point where the performance requirements of IEEE standard 1547 are evaluated; typically, the RPA will be the PCC.

Example 2.17: An SLD of a 10 MVA rated solar photovoltaic system as a DER unit connected to a 10 MVA transformer is shown in Fig. 2.29.

The rated current of the DER unit of 10 MVA at the PCC (33 kV) is 174.95 A.

Case 1: DER delivering a rated power of 10 MVA

The power analyzer connected at the PCC measures the fundamental current (I_1) and RMS current (I_{RMS}), including harmonic and interharmonic components, as follows:

I_1 is 174.8 A

I_{RMS} is 175.4 A

The TRD can be calculated as per Eq. (2.9):

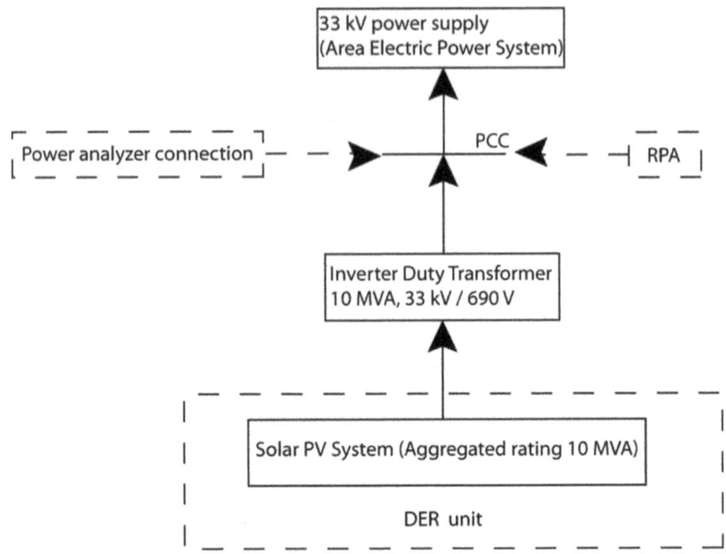

FIGURE 2.29 Single line diagram of DER power evacuation. *DER*, distributed energy resources; *PCC*, point of common coupling; *PV*, photovoltaic; *RPA*, reference point of applicability.

$$\mathrm{TRD}(\%) = \frac{\sqrt{175.4^2 - 174.8^2}}{174.95} \mathrm{x}100 = 8.28\%$$

TRD value at the RPA is 8.28%.

Case 2: DER delivering 75% of rated power—7.5 MVA

The power analyzer connected at the PCC measures the fundamental current (I_1) and RMS current (I_{RMS}), including harmonic and interharmonic components, as follows:

I_1 is 130.5 A

I_{RMS} is 131.5 A

The TRD can be calculated as per Eq. (2.9):

$$\mathrm{TRD}(\%) = \frac{\sqrt{131.5^2 - 130.5^2}}{174.95} \mathrm{x}100 = 9.25\%$$

TRD value at the RPA is 9.25%.

Case 3: DER delivering 25% of rated power—2.5 MVA

The power analyzer connected at the PCC measures the fundamental current (I_1) and RMS current (I_{RMS}), including harmonic and interharmonic components, as follows:

I_1 is 42 A

I_{RMS} is 44 A

The TRD can be calculated as per Eq. (2.9):

$$TRD(\%) = \frac{\sqrt{44^2 - 42^2}}{174.95} x100 = 7.49\%$$

TRD value at the RPA is 7.49%.

2.8 Resonance

Resonance is a condition in the power system. In this condition, system inductive reactance is equal to capacitive reactance at some harmonic frequency created by nonlinear loads. During this condition, cyclic energy transfers between the inductive and capacitive elements in the power system. This phenomenon is called resonance and the frequency at which the resonance phenomenon occurs is called resonance frequency [11]. The condition for resonance that occurs in the system is:

$X_L = X_C$

where:

X_L is inductive reactance

X_C is capacitive reactance.

The resonance frequency F_{res} is obtained from Eq. (2.10):

$$F_{res} = \frac{1}{2\pi\sqrt{LC}} \tag{2.10}$$

where

F_{res} is resonance frequency.

The major electrical equipment in power systems is inductive by their characteristics. The use of power factor correction capacitors and harmonic filters introduces capacitive characteristics into the system. The resonance condition occurs only because of multiple frequencies generated by nonlinear loads. At any one of the multiple frequencies, system inductive reactance is equal to capacitive reactance. Based on the system configuration, resonance conditions are classified into two types [12]:

- Series resonance
- Parallel resonance

2.8.1 Series resonance

Series resonance is formed by both inductive and capacitive reactance in series. Equivalent system impedance becomes series with utility power supply, transformers, and capacitors/filters. Series resonance provides a low impedance path to harmonic currents and results in unexpected harmonic current flow through the system equipment [12]. It is called current amplification in the system. The theoretical impedance for series resonance is shown in Eq. (2.11):

$$Z = j(X_L - X_C) = 0 \tag{2.11}$$

During series resonance conditions, excessive current flow may lead to the burning of cables and fuses, unnecessary relay operation, etc.

2.8.2 Parallel resonance

Parallel resonance is formed by both inductive and capacitive reactance in parallel. Equivalent system impedance becomes parallel with utility power supply, transformers, and capacitors/filters. Parallel resonance provides a high impedance path to harmonic voltage and results in excessive voltage harmonics across the equipment [12]. It is called voltage amplification in the system and results in capacitor and other equipment failure.

The theoretical impedance for parallel resonance is shown in Eq. (2.12):

$$Z = \frac{(X_L \cdot X_C)}{j(X_L - X_C)} = \infty \tag{2.12}$$

2.8.3 How to prevent the system resonance condition

There are several ways to overcome the system resonance condition. One of the easiest ways is by adding a reactor. Adding a reactor in series with each capacitor will change the system to inductive reactance at resonant frequencies and capacitive reactance at fundamental frequencies [11]. To achieve this, the capacitor and reactor should be tuned to a frequency below the lowest critical order of harmonics depending on the magnitude and order of harmonics present in the system.

Resonance can occur at any frequency, mostly fifth, seventh, 11th, and 13th order of harmonics or close to these frequencies because nonlinear loads inject the harmonic currents at the fifth, seventh, 11th, and 13th order of harmonics.

Resonant frequency introduced by the capacitor bank can be identified from Eq. (2.13) [5]:

$$F_{res} = F_{fund} X \sqrt{\frac{SC_{kVA}}{kVAR}} \tag{2.13}$$

where:

F_{res} is resonance frequency

F_{fund} is fundamental frequency

SC_{kVA} is short circuit capacity of the system in kVA

kVAR is reactive power from the capacitor in kilovolt ampere reactive.

The harmonic order that introduces resonant frequency from the capacitor bank can be identified from Eq. (2.14):

$$h = \sqrt{\frac{SC_{kVA}}{kVAR}} \tag{2.14}$$

where:

h is harmonic order

SC_{kVA} is short circuit capacity of the system in kVA

kVAR is reactive power from the capacitor in kilovolt ampere reactive.

Example 2.18: A plant has an 11 kV/433 V, 50 Hz, 2500 kVA transformer with an impedance of 6.8%. The plant is consuming 1806 kW active power at 0.85 power factor. To improve the power factor from 0.85 to 0.99 a capacitor bank of 300 kVAR is required. Find the likelihood of the system resonance condition occurring.

Solution:

SC_{kVA} is short circuit capacity at 36,764.

From Eq. (2.13) the resonant frequency when adding the capacitor is:

$F_{res} = 553$ Hz

From Eq. (2.14) the harmonic order is:

$h = 11$

When adding the 300 kVAR capacitor bank to the plant, the resonance condition will arise at a frequency of 553 Hz.

2.9 Annexure 2.1

Example A.2.1: A 3ϕ, 415 V, 500 kW VFD is monitored for 30 min between 11:31:19 and 12:01:20 hours.

The drive is consuming a maximum of 475 kW; the current is 700 A per phase and is operating at a 0.94 lag power factor. The instantaneous voltage of phase to neutral is shown in Fig. A.2.1 and the instantaneous current is shown in Fig. A.2.2.

The corresponding current harmonic spectrum in % is shown in Fig. A.2.3 and the current harmonic spectrum in absolute ampere is shown in Fig. A.2.4

From Fig. A.2.3, THD is 27% and individual major harmonic order and its magnitudes are listed in Table A.2.1.

The current THD trend in % for the 30 min monitoring duration is shown in Fig. A.2.5.

Example A.2.2: A 3ϕ, 415 V, 160 kVA UPS is monitored for 58 min between 13:25:53 and 14:23:05 h.

The rectifier circuit draws a current of 19 A per phase and operates at 0.86 lag power factor. The instantaneous voltage of phase to neutral is shown in Fig. A.2.6 and the instantaneous current is shown in Fig. A.2.7.

The corresponding current harmonic spectrum in % is shown in Fig. A.2.8 and the current harmonic spectrum in absolute ampere is shown in Fig. A.2.9.

From Fig. A.2.8, THD is 26% and individual major harmonic order and its magnitudes are listed in Table A.2.2.

The current THD trend in % for the monitoring duration of 58 min is shown in Fig. A.2.10.

Example A.2.3: A 3ϕ, 415 V elevator (capacity of maximum eight person or 560 kg) is monitored for 10 min between 12:15:35 and 12:25:19 h.

The maximum current drawn by the elevator per phase is 15 A at 0.8 lag power factor. The instantaneous voltage of the *R* phase to neutral is shown in Fig. A.2.11 and the instantaneous current of the *R* phase is shown in Fig. A.2.12.

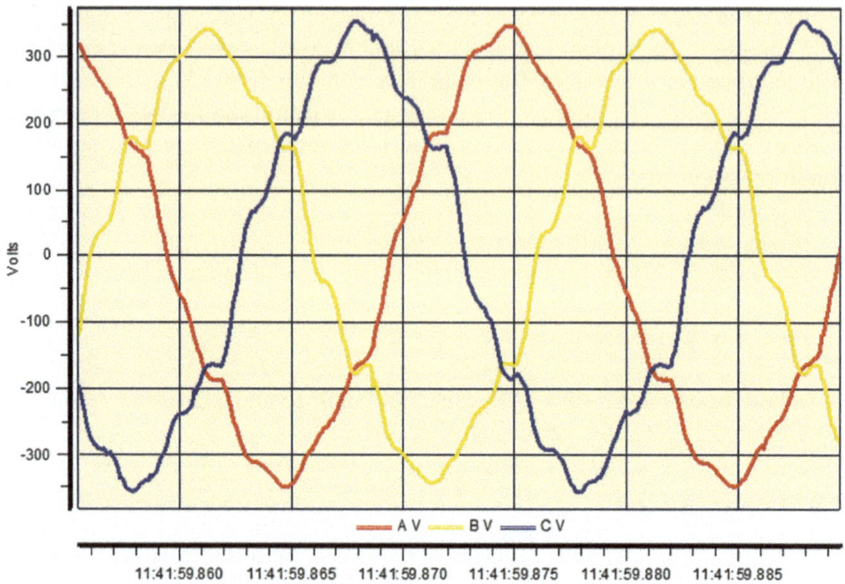

FIGURE A.2.1 The instantaneous voltage of phase to neutral.

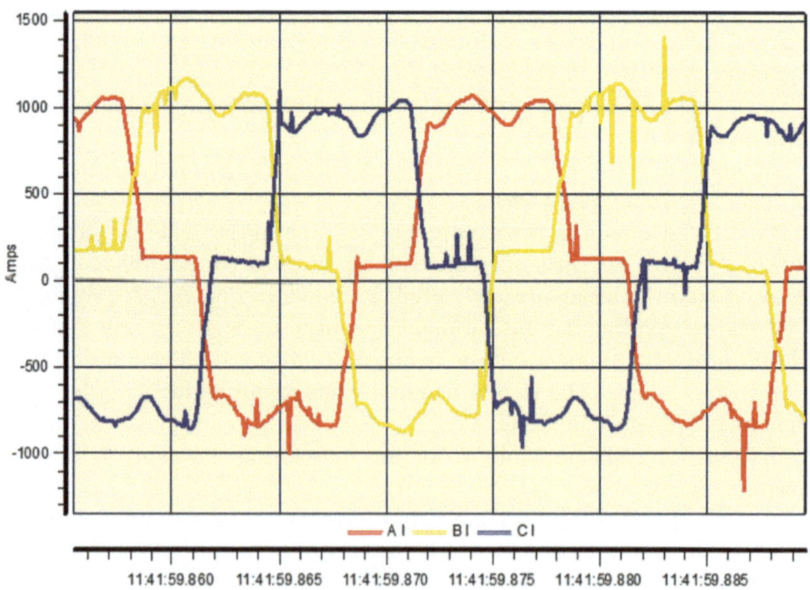

FIGURE A.2.2 The instantaneous current of three phases.

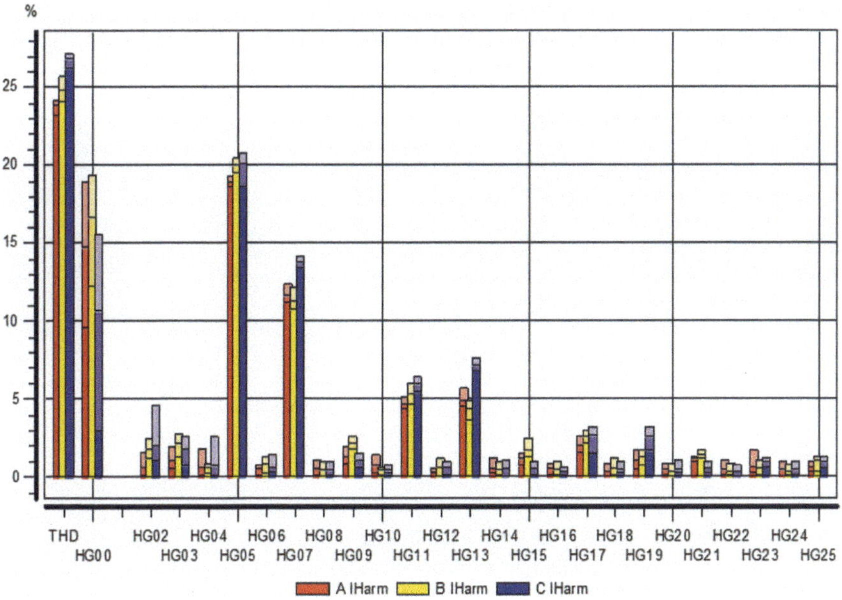

FIGURE A.2.3 The current harmonic spectrum in %.

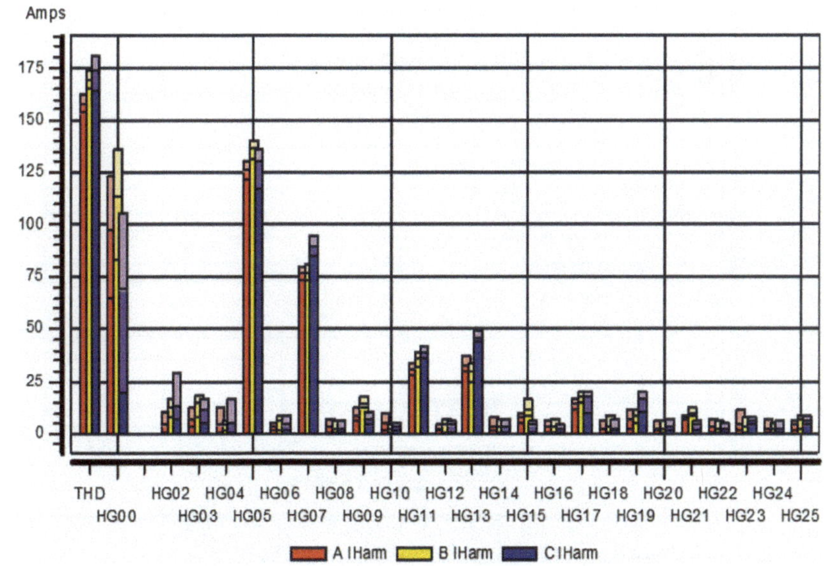

FIGURE A.2.4 The current harmonic spectrum in A.

TABLE A.2.1 Current harmonics generated by 3ϕ variable frequency drive.

Harmonic order	Harmonics in %	Harmonics in A
Current total harmonic distortion	27	176
3rd	3	20
5th	21	140
7th	14	95
9th	3	20
11th	6.5	42.4
13th	7.5	50
15th	2.5	17
17th	3	20.4
19th	3	20.4
23rd	2	11.2
25th	1.5	8.7

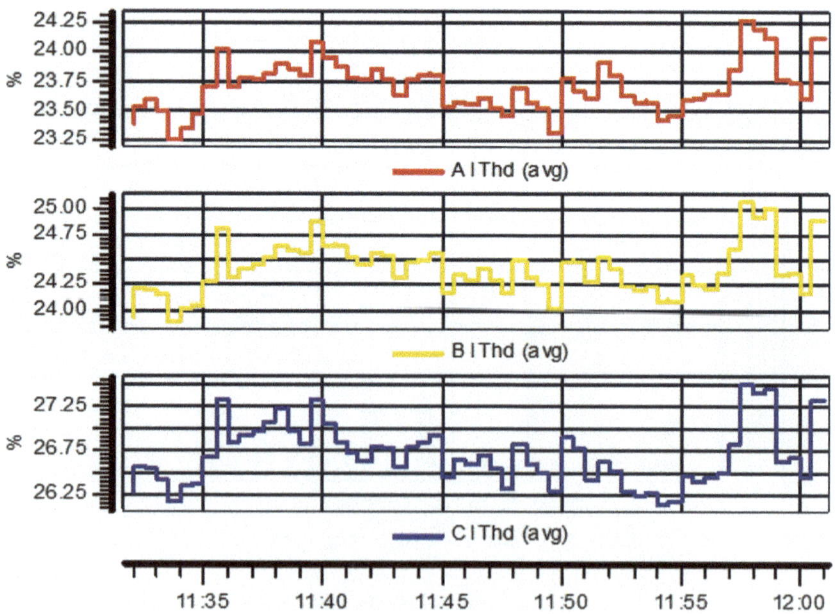

FIGURE A.2.5 The current total harmonic distortion trend in %.

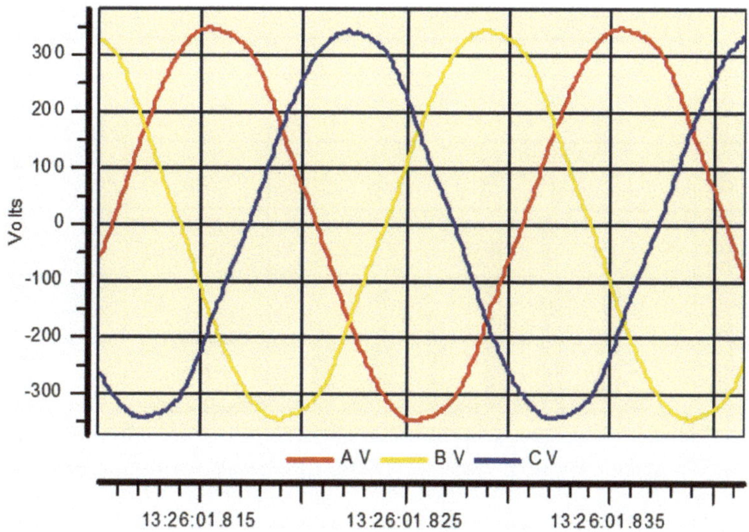

FIGURE A.2.6 The instantaneous voltage of phase to neutral.

The corresponding current harmonic spectrum in % is shown in Fig. A.2.13 and the current harmonic spectrum in absolute ampere is shown in Fig. A.2.14.

From Fig. A.2.13, THD is 55.28% and individual major harmonic order and its magnitudes are listed in Table A.2.3.

The current THD trend in % for a monitoring duration of 10 min is shown in Fig. A.2.15.

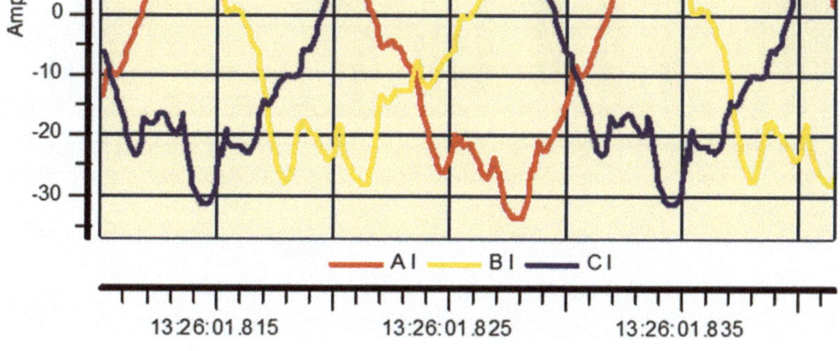

FIGURE A.2.7 The instantaneous current of three phases.

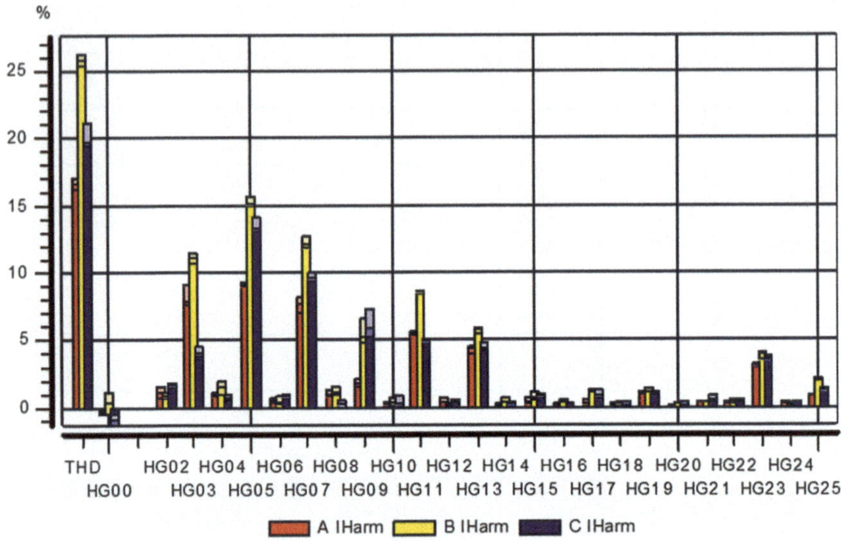

FIGURE A.2.8 The current harmonic spectrum in %.

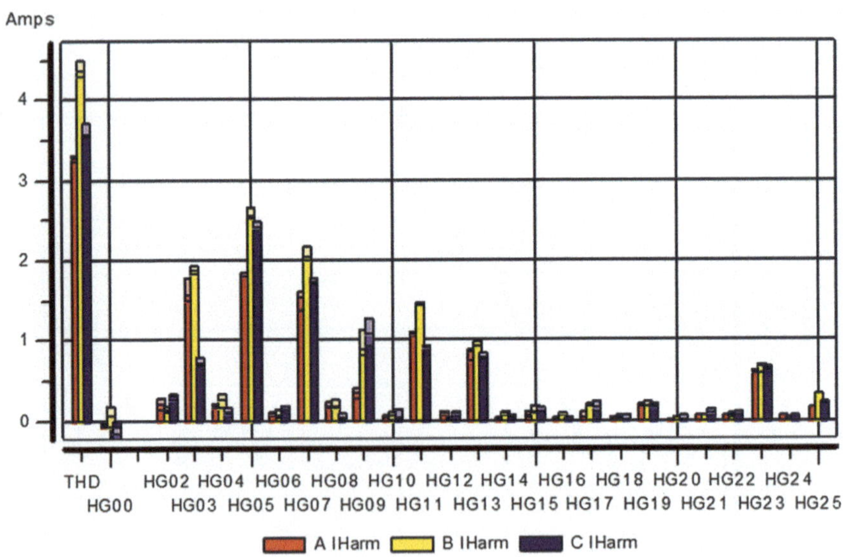

FIGURE A.2.9 The current harmonic spectrum in A.

TABLE A.2.2 Current harmonics generated by 3ϕ uninterruptible power supply.

Harmonic order	Harmonics in %	Harmonics in A
Current total harmonic distortion	26	5
3rd	11.5	1.9
5th	15.5	2.7
7th	13	2.2
9th	7.5	1.3
11th	9	1.5
13th	6	1
15th	1	0.2
17th	1	0.2
19th	1.5	0.25
23rd	4	0.7
25th	2	0.4

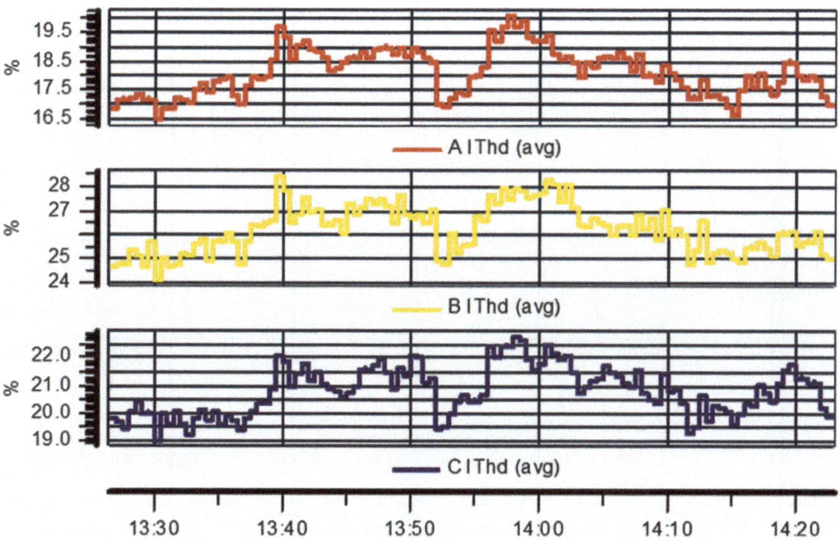

FIGURE A.2.10 The current total harmonic distortion trend in %.

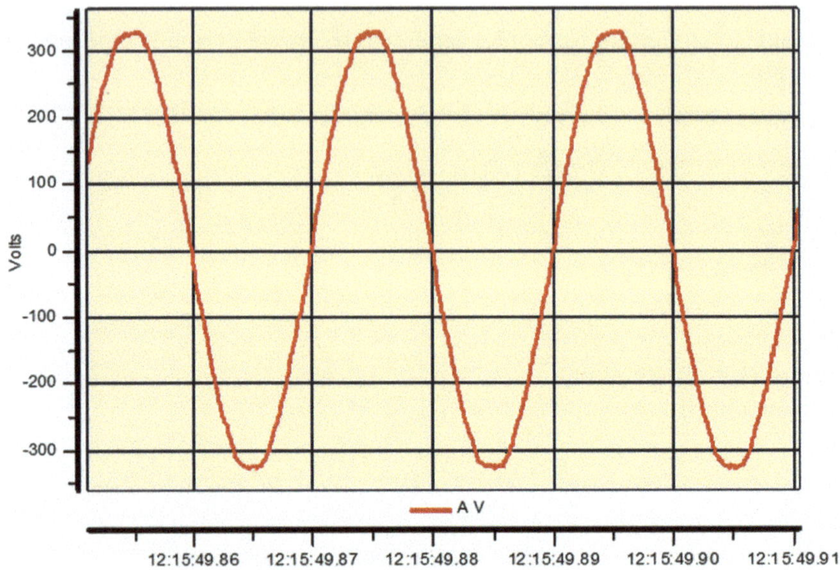

FIGURE A.2.11 The instantaneous voltage of phase to neutral.

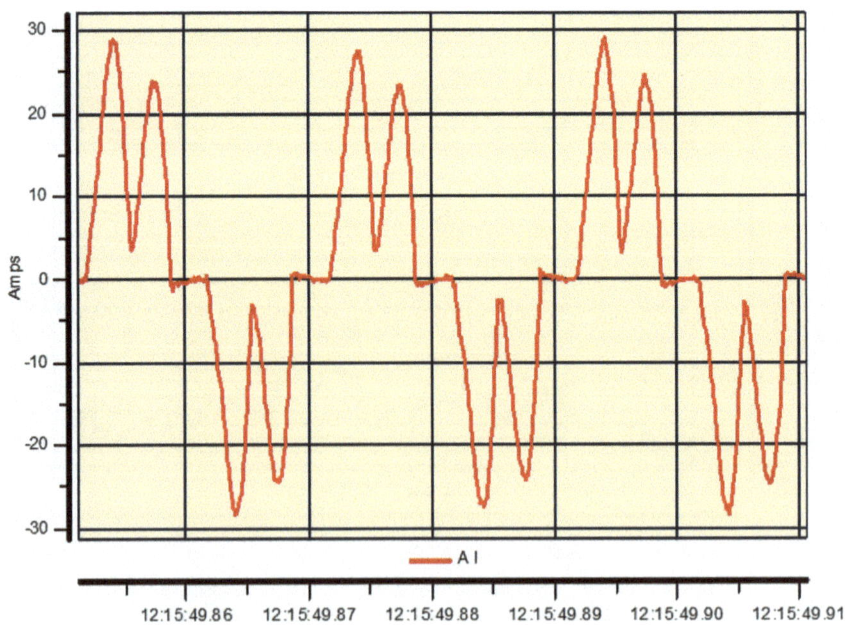

FIGURE A.2.12 The instantaneous current of the *R* phase.

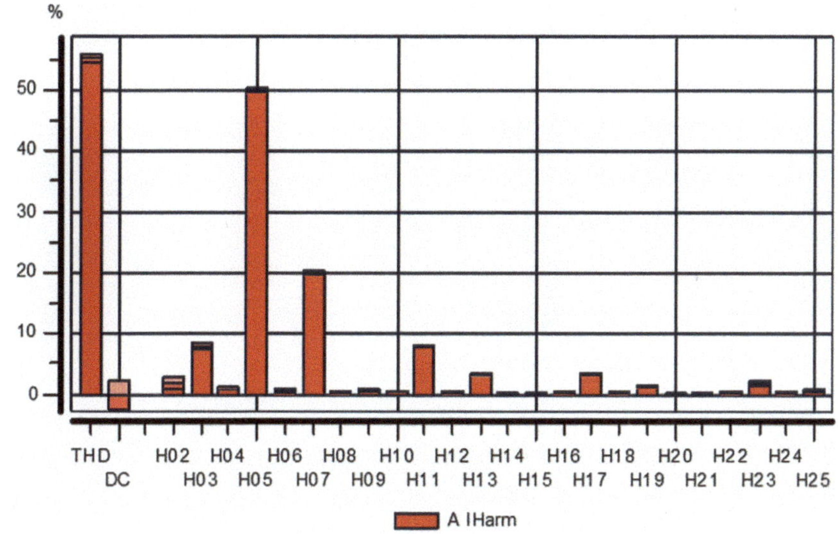

FIGURE A.2.13 The current harmonic spectrum in %.

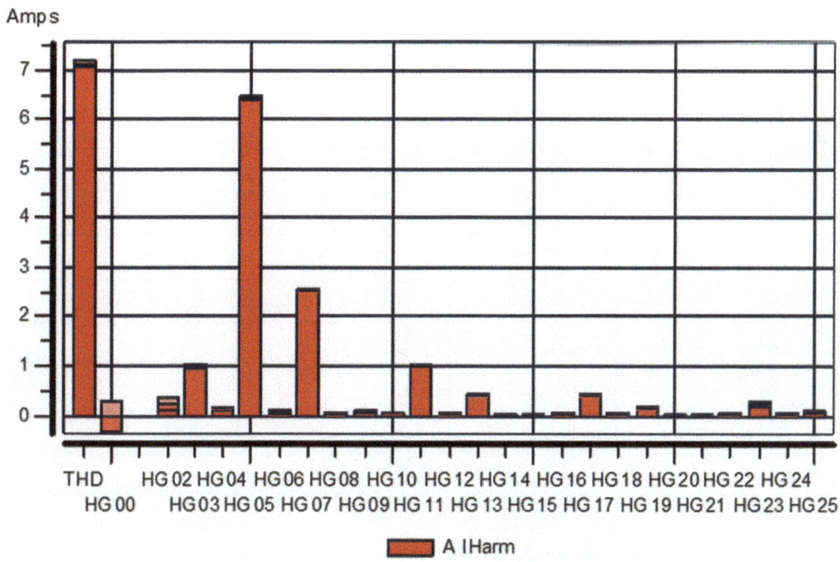

FIGURE A.2.14 The current harmonic spectrum in A.

TABLE A.2.3 Current harmonics generated by an elevator.

Harmonic order	Harmonics in %	Harmonics in A
Current total harmonic distortion	55.28	7.12
DC	2.2	0.28
3rd	7.8	1
5th	49.98	6.44
7th	19.89	2.56
9th	0.75	0.09
11th	7.9	1.01
13th	3.24	0.41
15th	0.16	0.02
17th	3.3	0.43
19th	1.38	0.17
23rd	1.7	0.23
25th	0.8	0.1

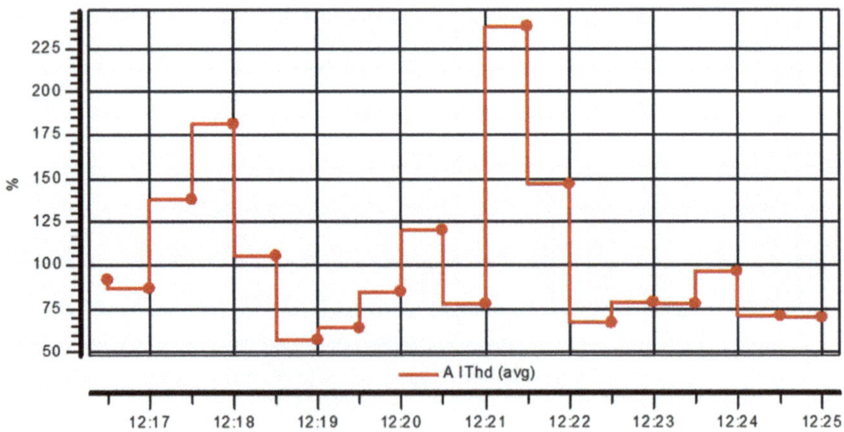

FIGURE A.2.15 The current total harmonic distortion trend in %.

References

[1] C. Sankaran, Power Quality, CRC Press, Boca Raton, 2002.
[2] C. Sharmeela, B. Surendhar, P. Sivaraman, R. Mahendran, Harmonic Reduction and Voltage Flicker Mitigation by Using Static VAR Compensator and Shunt Passive Filter for Steel Plant, in: 5th International Exhibition & Conference 2015, GRIDTECH 2015, April 8 – 10, 2015, pp. 524–528. New Delhi.
[3] B. Angelo, Handbook of Power Quality, John Wiley & Sons, United States of America, 2007.
[4] E. Ali, A. Nasiri, B. Stoyan, Bekiarov, Uninterruptible Power Supplies and Active Filters, CRC Press, United States of America, 2005.
[5] IEEE Std 519 - 2014, IEEE Recommended Practice for Harmonic Control in Electric Power Systems.
[6] W. Barry, Kennedy, Power Quality Primer, McGraw-Hill, New York, 2000.
[7] Power Factor Correction and Harmonic Filtering in Electrical Plants, Technical Application Papers, ABB, July 2018.
[8] IEEE Std 1159 – 2019, IEEE Recommended Practice for Monitoring Electric Power Quality.
[9] P. Sivaraman, C. Sharmeela, Introduction to the electric distribution system, in: B. Kahn, H.H. Alhelou, Ghassan (Eds.), Handbook of Research on New Solutions and Technologies in Electrical Distribution Networks, IGI Global, Hershey, PA, 2020, pp. 1–31.
[10] IEEE Std 1547 – 2018, IEEE Standard for Interconnection and Interoperability of Distributed Energy Resources with Associated Electric Power Systems Interfaces.
[11] P. Sivaraman, C. Sharmeela, Existing issues associated with electric distribution system, in: B. Kahn, H.H. Alhelou, Ghassan (Eds.), Handbook of Research on New Solutions and Technologies in Electrical Distribution Networks, IGI Global, Hershey, PA, 2020, pp. 1–31.
[12] IEEE Std 399 – 1997, IEEE Recommended Practice for Industrial and Commercial Power System Analysis.

Further reading

[1] P. Sivaraman, C. Sharmeela and D. P. Kothari, "Enhancing the voltage profile in distribution system with 40 GW of solar PV rooftop in Indian grid by 2022: a review" 1st International Conference on Large Scale Grid Integration Renewable Energy in India, September, 2017, New Delhi, India.

Chapter 3

Power quality problems with renewable energy integration

Vasundhara Mahajan[1], Pramod Agarwal[2], Hari Om Gupta[3]
[1]*Department of Electrical Engineering, Sardar Vallabhbhai National Institute of Technology (SVNIT), Surat, Gujarat, India;* [2]*Department of Electrical Engineering, Indian Institute of Technology, Roorkee, Uttarakhand, India;* [3]*Jaypee Institute of Technology, Greater Noida, Uttar Pradesh, India*

Chapter outline

3.1 Motivation **105**
3.2 Renewable energy sources and their integration **107**
3.3 Power quality analysis with RES **107**
3.3.1 Voltage fluctuations 108
3.3.2 Voltage transients 108
3.3.3 Voltage unbalance/sag/swell 108
3.3.4 Origins of harmonics 109
3.3.5 Consequences/impacts of harmonics 109
3.4 Mathematical modeling of power systems with RES **110**
3.5 Artificial intelligence-based controllers **112**
3.5.1 Artificial neural network-based controller 112
3.5.2 ANN controller for separation of events related to PQ of RES 115
3.5.3 ANN controller for harmonic separation 116
3.5.4 Fuzzy logic controller 118
3.5.5 Fuzzy logic controller for DC link capacitor voltage matching 119
3.5.6 Fuzzy logic controller for current adaptation and modulation 121
3.6 Simulation results and analysis **123**
3.7 Conclusion **129**
References **129**

3.1 Motivation

Recently, sophisticated and advanced technology has been increasingly implemented due to its immense benefits in terms of lifestyle, business, infrastructure, and health. Accepting these aids, however, intensifies reliance on electrical power and often that power has to be completely free from interruption and disruption of the system to function efficiently. Modern

Power Quality in Modern Power Systems. https://doi.org/10.1016/B978-0-12-823346-7.00011-6

civilization demands the utilization of clean and efficient electrical systems. The continuity and quality of supply is desirable with all advancements. This can be ensured by addressing power quality (PQ)-related difficulties. This imposes the elimination or minimization of disturbances causing impurity.

The power system network is developing in terms of advanced technology and renewable energy sources (RES). The popularity of RES is increasing exponentially due to their free availability and fewer tariffs. The integration of RES has brought a great revolution. Now, RES technology can be easily incorporated in any network ranging from small households to power grids. Expansion in technology has provided easy interfacing options with greater reliability and competence. These RES have their drawbacks, however, such as uncertainty of availability and intermittent power. This limits their performance to some extent. This randomness and variation also cause some PQ problems such as voltage flicker, sag, and swell. The converter interface for DC/AC conversion causes harmonic problems. These problems need attention and should be mitigated appropriately. The filters are designed to mitigate PQ issues in the power system network.

The use of two-level inverters becomes challenging at medium- to high-voltage levels as it requires an interfacing transformer. This increases the rating of the transformer with an increase in voltage level. This leads to multilevel inverters (MLI). The MLI extends system rating with enhanced performance. Higher levels in MLI give almost sinusoidal voltages at lower switching frequency with reduced switching stress. The rating of MLI can be augmented by adding levels. Researchers have unvaryingly modified, tested, and implemented various MLI configurations for a large number of applications for PQ enhancement. Harmonic mitigation, flexible AC transmission system, and drives are the major applications. The number of levels in MLI is selected from an application perspective. For MLI, the modulation scheme plays an important role in switching and hence in harmonic distortion at the output. Pulse width modulation schemes are modified from time to time depending upon application and use [3].

Control schemes have proportional integral (PI) or proportional integral differential (PID) controllers. The adaptability of these controllers is less and hence more efficient controllers are required. Therefore artificial intelligence (AI)-based controllers have been developed. The AI-based control scheme is more flexible and accommodates most of the arbitrariness. Recently, these have been utilized in many practical industrial applications, as these controllers can work with inaccurate and random inputs; they can handle disturbances; they are more robust; and so on. Therefore it is a viable candidate controller for power electronics fields. Cutting-edge computer tools are fast in solving intricate problems. AI controllers are based on soft computing and have gained popularity due to flexibility and effectiveness. AI is smart processing for complex control. AI controllers estimate an imprecise dataset by iterative processing, therefore they do not necessitate the exact mathematical model for design. Fuzzy and neural networks are the most explored avenues of AI [1,2].

This chapter highlights RES integration to the grid and its PQ issues. Details of major disturbances are presented, mathematical modeling is described, and AI-based controller design and its implementation is demonstrated.

3.2 Renewable energy sources and their integration

The need for cleaner and safer environments has strongly motivated the use of RES. Wind turbine generators and solar energy-based power producers are very popular these days, and advanced technology permits wider use. The integration of RES necessitates flexibility and adaptability for successful operation. There is a need for new considerations for grid planning, reserves, transmission, voltage control, and inertia response [3,4].

3.3 Power quality analysis with RES

The propagation/proliferation of microelectronic and sophisticated automatic processors in a wide range of modern equipment in every aspect of electrical utilization in industrial, business, medical, as well as household applications has amplified the vulnerability of such equipment to PQ problems. PQ issues take account of all probable situations in which the supply waveform and load waveform diverge from the sinusoidal waveform along with variation in fundamental frequency and magnitude beyond permissible limits. These problems include a variety of electrical disturbances, which may be of short and/or long duration with small and/or high amplitude. These disturbances may originate in several forms and affect various sensitive loads, but might not affect the conventional apparatus; however, in present-day systems they may cause major damage to sophisticated and high precision equipment and affect productivity. This necessitates the application of PQ improvement devices to maintain the standards of power supply parameters for uninterrupted and steady flow of power.

The widely used phrase "power quality" includes all characteristics of events that deviate from normal operation in a system. PQ is an umbrella concept for power system disturbances such as instantaneous interruptions, transients (oscillatory/impulsive), voltage variations, voltage flicker, voltage unbalance, notching, and current/voltage harmonic distortions. Among all power line instabilities, harmonics are the most deteriorating condition to PQ because they are in a steady-state condition and have severe effects on the system. There are many PQ disturbances caused by RES due to their uncertainty and unpredictability [3,4]. The variable nature of input and output causes voltage flickering, fluctuation, sag, and swell. This leads to instability of the entire system. Voltage fluctuations are due to the intermittency of wind speed and solar irradiation. The integration of RES can be limited by the flicker-level standardized limits. The direct coupling of RES, without electrical energy storage, injects intermittent power. Therefore voltage variations can occur at the point of common coupling (PCC). The connection of RES to the low-voltage grid can experience voltage variations and harmonics. The details of these PQ issues with RES are as follows.

3.3.1 Voltage fluctuations

Voltage fluctuation occurs due to the output power variations depending upon the source type, generator characteristics, and impedance of the network. Flicker coefficient is the indicator of severity of fluctuation. This is influenced by averaging annual wind speed, the point of turbine installation, and the phase of short circuit impedance. The flicker coefficient and its specifications are either given by manufacturers or can be determined. Depending on the voltage level the angle varies from 30 to 85 degrees for medium- and high-voltage networks, respectively. The wind turbine on/off sequence produces the voltage fluctuations. These can be evaluated as per the emission limits guidelines for fluctuating loads in medium-voltage and high-voltage networks. The allowed variation is $\pm 10\%$ for most countries. Some countries have tighter limits for voltage variations. There can be overlap of load modification variations that can result in wider limit bands.

3.3.2 Voltage transients

There are mechanical switches in the capacitor banks used for reactive power for induction generators. There can be transients due to these mechanical switches. The sensitive equipment of the system can be damaged due to these transients and frequent switching.

3.3.3 Voltage unbalance/sag/swell

There can be lengthy deviation from the actual set point of voltage amplitude, including overvoltage and undervoltage. This may occur due to variable wind speed or solar irradiation. The power network has small resistance and large inductance, and the reactive power affects the voltage deviation. In addition, in an RES-integrated power grid, the deviation in the distribution network can be caused by load and system voltage level, as well as the short circuit capacity of the PCC.

3.3.4 Origins of harmonics

The power electronics interface of RES causes harmonics. This can further propagate and cause additional problems. All these devices produce harmonic distortion, which may be very small individually, but after passing through a system it may grow significantly in magnitude and cause trouble in the whole system.

The harmonic problem in power systems was noticed in the early 1920 and 1930s. At that time, transformers were the main sources of harmonics and inductive interference with open wire telephone systems was a main problem. Harmonic filtering by some means in distribution feeders was achieved at that time. The problem of harmonics identification and extermination has gained wide interest due to their prominent presence and deleterious effects on PQ. Power electronic devices and most other equipment create harmonics. A three-phase converter is shown in Fig. 3.1. It has significant odd order harmonics and higher harmonic distortion. A typical current waveform and its harmonic spectrum with smoothing capacitance is shown in Fig. 3.2. This converter has 85.66% total harmonic distortion with fifth and seventh order harmonics as the most dominating. The other higher order odd harmonics are also found to be present in the harmonic spectrum.

3.3.5 Consequences/impacts of harmonics

In modern situations, harmonic distortion can have major unfavorable effects on electrical distribution systems. It causes consumption of more energy and hence lowers system capacity. This can harm electrical distribution systems and devices operating on these systems. The harmonic causes additional losses and consequential overheating. The losses are iron loss, copper loss, and dielectric loss. This can cause a rise in temperature and hence reduces the life of equipment.

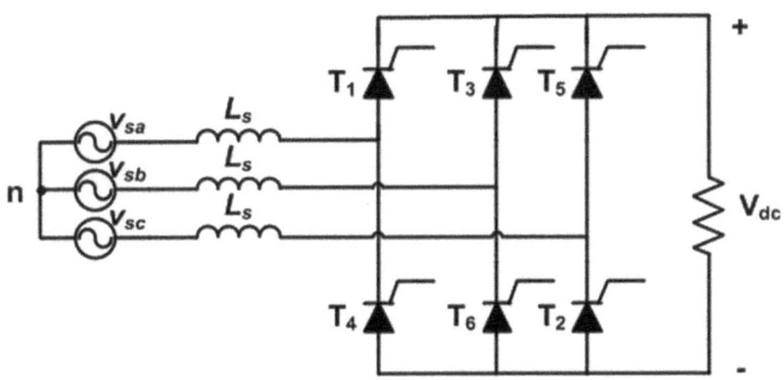

FIGURE 3.1 Three-phase six-pulse controlled converter.

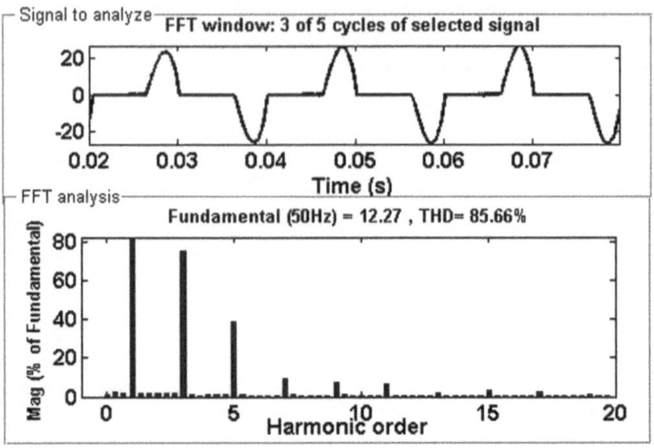

FIGURE 3.2 Three-phase power converter current waveform and its harmonic spectrum. *FFT*, Fast Fourier transform; *THD*, total harmonic distortion.

Harmonics can cause electromagnetic interference with advanced communication systems through inductive coupling. The measuring devices show errors in measuring energy in harmonic conditions because of their structure and various algorithms. The latest digital meters have power electronics-based devices for signal measurement, which may cause error due to nonlinearity and disturbances caused by harmonics. These meters tend to have negative values at harmonic frequencies and errors increase with increases in frequency. Similarly, digital relays and their control mechanisms depend on the sampling data and zero crossing, which may be adversely affected due to harmonics in both voltage and current. Power systems are susceptible to defective behavior due to harmonics. Harmonic distortion is observed to be different for different types of converters/inverters.

3.4 Mathematical modeling of power systems with RES

To avoid voltage collapse of the RES system, reactive current corresponding to sag is required. Wind generators should be capable of handling a three-phase fault with approximately 15% sag limits. There has to be a provision for feeding the current into the grid during faults. The DC link voltage has to be maintained.

The voltage disturbances and harmonics are modeled as Eq. 3.1:

$$\mathrm{V_{disturbance\_RES}} = \{\mathrm{V_{sag}(t), V_{swell}(t), V_{flicker}(t), V_{hamonics}(t), I_{harmonics}(t)}\} \quad (3.1)$$

Harmonics modeling and segregation:

The load current with its harmonic composition in Fourier series form is Eq. 3.2:

$$
\begin{aligned}
i_{L\_phase\_a}(t) &= A_{0\_phase\_a} + \\
&\sum_{h_r=1}^{\infty}\left[A_{h_{r\_phase\_a}} * \cos\left(h_{r\_phase\_a}\omega_0 t\right) + B_{h_{r\_phase\_a}} * \sin\left(h_{r\_phase\_a}\omega_0 t\right)\right] \\
i_{L\_phase\_b}(t) &= A_{0\_phase\_b} + \\
&\sum_{h_r=1}^{\infty}\left[A_{h_{r\_phase\_b}} * \cos\left(h_{r\_phase\_b}\omega_0 t - \frac{2\pi}{3}\right) + B_{h\_phase\_b} * \sin\left(h_{r\_phase\_b}\omega_0 t - \frac{2\pi}{3}\right)\right] \\
i_{L\_phase\_c}(t) &= A_{0\_phase\_c} + \\
&\sum_{h_r=1}^{\infty}\left[A_{h_{r\_phase\_c}} * \cos\left(h_{r\_phase\_c}\omega_0 t + \frac{2\pi}{3}\right) + B_{h\_phase\_c} * \sin\left(h_{r\_phase\_c}\omega_0 t + \frac{2\pi}{3}\right)\right]
\end{aligned}
\tag{3.2}
$$

Instantaneous power theory (IPT) gives instantaneous power and its constituents as expressed by Eq. 3.3:

$$
\begin{bmatrix} p \\ q \end{bmatrix} = \begin{bmatrix} p_{con\_c} + p_{osc\_c} \\ q_{con\_c} + q_{osc\_c} \end{bmatrix} = \begin{bmatrix} v_\alpha & v_\beta \\ v_\beta & -v_\alpha \end{bmatrix} \begin{bmatrix} i_\alpha \\ i_\beta \end{bmatrix} \tag{3.3}
$$

The expression for modified active power is given as Eq. 3.4:

$$
\begin{aligned}
p = p_{con\_c} &+ \sum_{h_{r\_abc}=2}^{N} p_{(h_{r\_abc}-1)} * \cos\left(\left(h_{r\_abc} - 1\right)\omega_0 t - \phi_{d1} + \varphi_{dh_{r\_abc}}\right) \\
&+ \sum_{h_{r\_abc}=1}^{N} p_{(h_{r\_abc}+1)} * \cos\left(\left(h_{r\_abc} + 1\right)\omega_0 t + \phi_{d1} + \varphi_{dh_{r\_abc}}\right)
\end{aligned}
\tag{3.4}
$$

where: h_r is harmonic order, N is the highest order of harmonic to be considered, d is positive-sequence indicative, d' is negative-sequence indicative.

The coefficients are defined as Eq. 3.5:

$$
\begin{aligned}
p_{con\_c} &= 3 * V_{d\_1} * I_{d\_1} * \cos\left(\phi_{d\_1} - \varphi_{d\_1}\right) \\
p_{(h_r-1)} &= 3 * V_{d\_1} * I_{d\_h_{r\_abc}} \\
p_{(h_r+1)} &= 3 * V_{d\_1} * I_{d\_h_{r\_abc}}
\end{aligned}
\tag{3.5}
$$

3.5 Artificial intelligence-based controllers

The response of a controller is determined by its adaptability and response time. PI and PID controllers require system-specific tuning and designing, which makes them challenging to sync with system dynamics. Therefore more advanced controllers are required. This is achieved by AI and its variants. This accommodates the variations more efficiently and quickly. AI gained popularity due to flexibility and effectiveness. The use of cutting-edge methods permitted the design of these robust controllers. Recently, these controllers have become viable candidate controllers for power electronics and RES. There are many classifications for AI controllers such as expert systems, neural networks, and fuzzy and genetic algorithms [5].

A neural network learns by training; fuzzy logic is useful with uncertainty and qualitative information [6,7]. An expert system has a set of predefined rules derived from experience. A genetics-based algorithm is in line with Darwin's concept of "survival of the fittest." Furthermore, there can be hybrid AI, which is a combination of two or more such methods such as neurofuzzy, neurogenetic, or genetic fuzzy [8–11]. These techniques are capable of handling most of the complex nonlinear problems that are challenging to design with conventional methods [12,13].

Over the last two decades many AI-based controllers have been explored for systems with RES. AI-based controllers achieve their behavior by following continuous learning and training, and can process large numbers of inputs and outputs making them suitable for multivariable control [14,15].

The artificial neural network (ANN) is utilized for harmonic extraction and separation. The fuzzy logic-based controller (FLC) is implemented for current adjustment and DC voltage regulation [1]. The neural network controller separates the constant or oscillating components from nonsinusoidal voltage/current signals and adjusts by knowledge and teaching [16,17]. This chapter focuses on AI for harmonic estimation and compensation. Generally, a distorted wave is processed by Fourier transformation for separating components into their constituents. It is then passed through high- or low-pass filters for extracting fundamental components. This complicates the control structure and a time delay is also introduced due to mathematical calculation. This prohibits real-time implementation of the scheme. The complete process can be simplified and instigated in real time by using ANN-based controllers [2,18–22].

3.5.1 Artificial neural network-based controller

The ANN is the extension of computation, which is biologically inspired. It embodies computational paradigms, based on biological metaphor, to mimic the processing of the human brain. The output of a neuron is expressed as Eq. 3.6:

$$y = w_1x_1 + w_2x_2 + \ldots + w_nx_n + b_t \tag{3.6}$$

where input vector x is assigned with synaptic weight (w) to obtain output y. Here, b_t is the bias or offset. A positive synaptic weight implies activation, whereas a negative weight implies deactivation of the input. The absolute value of the synaptic weight defines the strength of the connection.

The classification of a neural network as per the type of neuron connection of input and output signals is categorized as shown in Fig. 3.3. The simple and generally used neural network is a feedforward multilayer. The multilayered and other extended varieties are applied for various practical applications. The neurons of a network are structured in layers to form a multilayered ANN as shown in Fig. 3.4.

Assuming ANN training with NP input patterns, the squared error of output for all the output layer neurons is given by Eq. 3.7:

$$E_{np} = (d^{np} - y^{np})^2 = \sum_{m=1}^{Q} \left(d_m^{np} - y_m^{np}\right)^2 \tag{3.7}$$

where: d is expected output of the mth neuron of the output layer$y$ is corresponding actual output Q dimension of the output vector and NP is number of patterns.

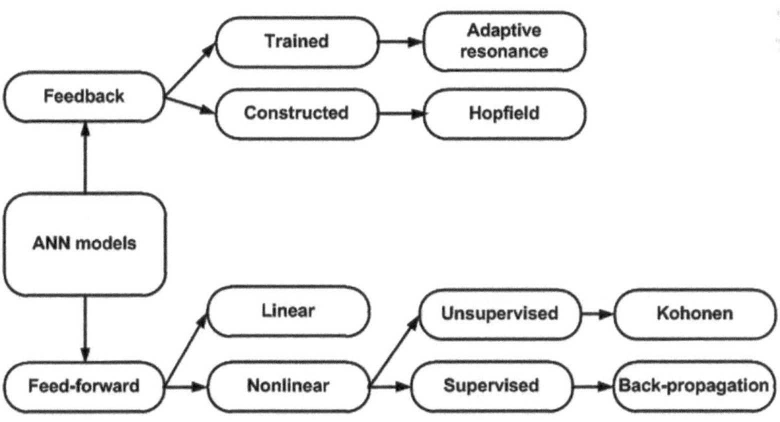

FIGURE 3.3 Types of artificial neural network (ANN).

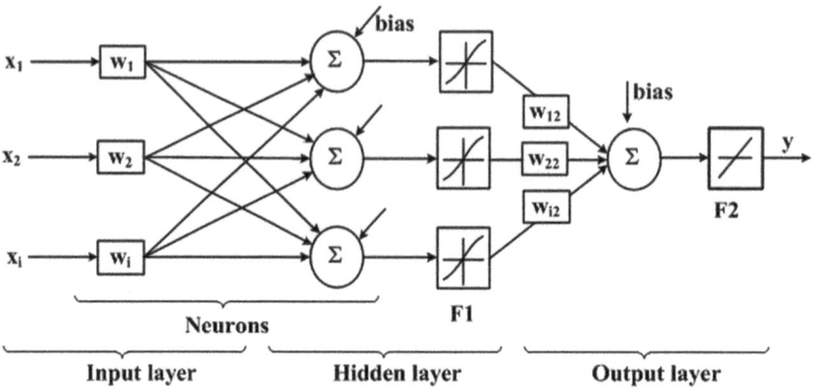

FIGURE 3.4 Structure of basic artificial neural network.

The sum of squared error (SSE) for NP patterns is then given by Eq. 3.8:

$$SSE = E = \sum_{np=1}^{NP} E_{np} = \sum_{np=1}^{NP} \sum_{m=1}^{Q} \left(d_m^{np} - y_m^{np}\right)^2 \tag{3.8}$$

The weights are updated to minimize SSE by the gradient descent method. The weights are updated as Eq. 3.9:

$$\begin{aligned} W_{ij}(t+1) &= W_{ij}(t) - \eta\left(\frac{\delta E_{np}}{\delta W_{ij}(t)}\right) \\ momentum &= \mu[W_{ij}(k) - W_{ij}(k-1)] \end{aligned} \tag{3.9}$$

where: η is learning rate, $W_{ij}(t+1)$ is new weight between neuron pair $i - jW_{ij}(t)$ is weight at the tth instance.

For all training patterns (NP), the weights are continuously updated. The momentum term ensures that error should converge to a global minimum. Backpropagation is made adaptive by Eq. 3.10:

$$\eta(k+1) = u \ . \ \eta(k) \tag{3.10}$$

where u is less than 1.

A number of backpropagation training algorithms are available in the literature. Here, the Levenberg–Marquardt backpropagation algorithm is used because of its fast convergence. This uses the resulting derivatives for the weight modifications. In this the model output errors are calculated in forward pass and weights are updated in regressive pass to equalize the model response with actual response. A flowchart of ANN algorithm implementation is shown in Fig. 3.5.

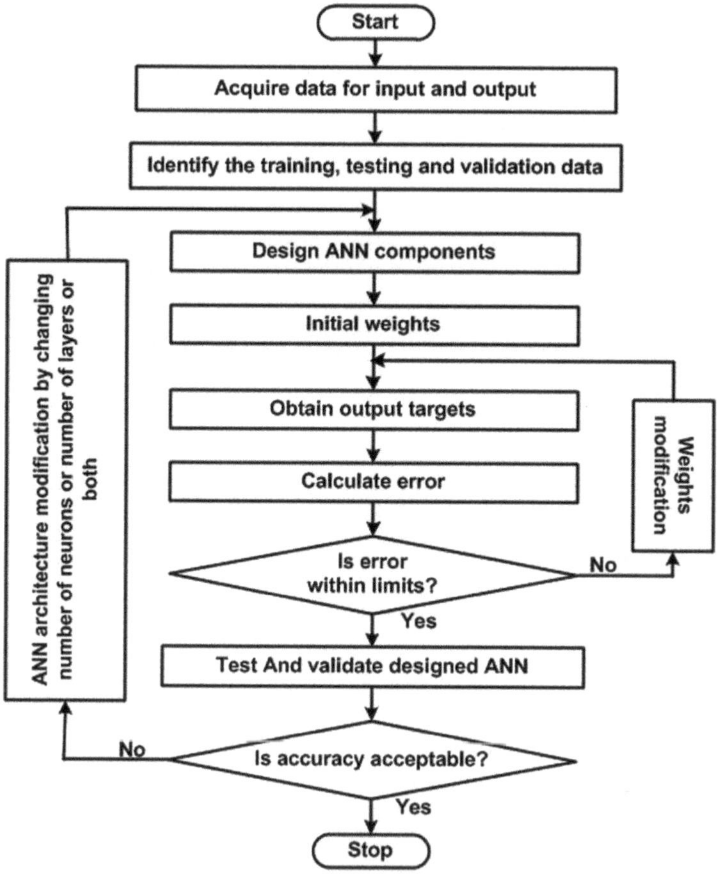

FIGURE 3.5 Basic flow chart of artificial neural network (ANN)-based controller.

3.5.2 ANN controller for separation of events related to PQ of RES

This ANN has historical data of PQ events as inputs and outputs for training and validation. The parameters for each event are set and separated for identification with its typical characteristics and detrimental effects [23]. The ANN has patterns for matching and evaluation of criticality of each event. During normal operation the ANN will provide regular control commands for execution. However, for any disturbance, the ANN will take some decisions and control the severity of event, if any. The details of this ANN are shown in Fig. 3.6.

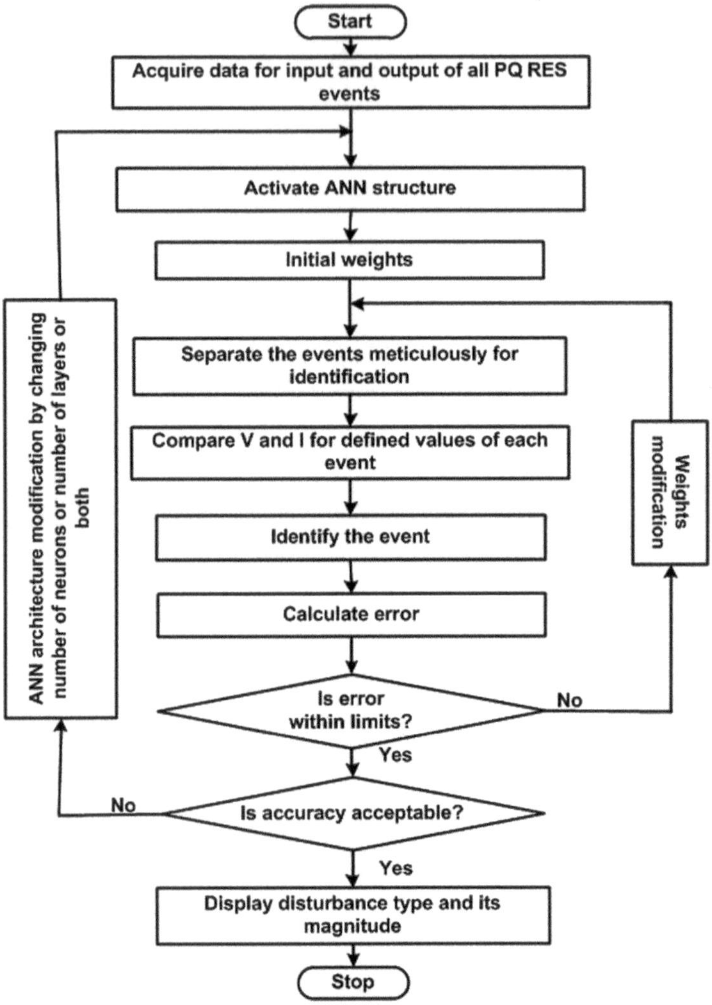

FIGURE 3.6 Flow chart for artificial neural network (ANN) controller for separation of events related to power quality (PQ) of renewable energy sources (RES).

3.5.3 ANN controller for harmonic separation

This controller has an ANN structure based on IPT [24–28]. The IPT is modified by using the ANN and the symmetrical components are used for harmonic separation. The ANN controller replaces the traditional low-pass filter, which separates the constant and oscillating components from

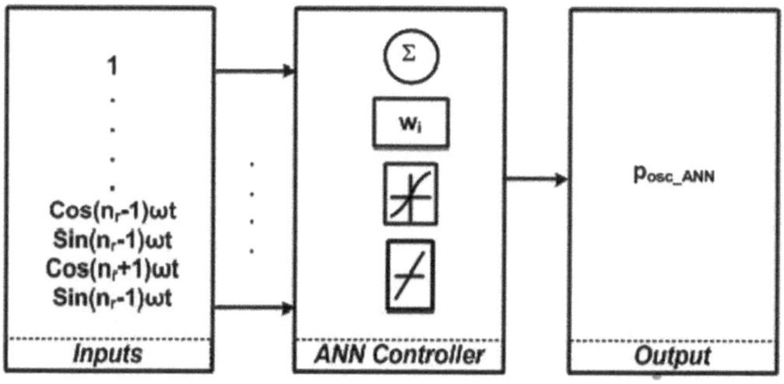

FIGURE 3.7 Artificial neural network (ANN) structure blocks for harmonic separation.

distorted signals. The active power is disintegrated into stable constant and oscillating or variable components by the ANN. The linear mapping of input and output of the ANN gives oscillating components of instantaneous active power extracted as shown in Fig. 3.7 by using the ANN. The expression for mapping is given as Eq. 3.11:

$$y(k) = x(k) * w(k) = p_{osc\_ANN} \tag{3.11}$$

The input vector is expressed as Eq. 3.12:

$$x(j) = \begin{bmatrix} 1 \\ . \\ . \\ \cos(n_r - 1)\omega t \\ \sin(n_r - 1)\omega t \\ \cos(n_r + 1)\omega t \\ \sin(n_r + 1)\omega t \end{bmatrix} \tag{3.12}$$

The ANN with these inputs is accomplished, confirmed, and validated. The updates for weights and neuron count are decided based on desired accuracy. This is then extended for harmonics mining using the method just described. The hidden layer uses sigmoid activation function (F_1) to trace the nonlinearity of a system and a linear activation function (F_2) is used in the output layer to convert the processed data into a linear dataset corresponding to disturbance. The functions are defined as Eq. 3.13:

$$F_1 = \frac{1}{1 + e^{-x}}, \quad F_2 = x \tag{3.13}$$

The use of bias in a neural network increases the capacity of the network to solve problems. The bias also configures the changes appropriately so that the neutralization of simultaneous manipulation weight can be handled. This controls the behavior of the layer. The frequently used value of bias is 1.

3.5.4 Fuzzy logic controller

FLCs were first introduced by Professor Lotfi Zadeh [29] in 1965 and have gained popularity due to their adaptability and high performance. The FLC uses linguistic or qualitative information for its control algorithm [30]. Hence, it overcomes the difficulty of crisp information in terms of numerical values. The FLC works in certain ranges of information and infers from the rule base. The rule base is a representation of historical data, and its consequences and actions taken or required.

The fuzzy logic control is a range-to-point or range-to-range control. FLCs are multivalued rationality based and include approximation instead of defined and accurate values. These controllers are highly suitable for multivariable and inconsistent processes, where the exact measurement of input and its effect is difficult. FLCs may have single or multiple inputs and outputs depending upon the application. Fuzzy sets and fuzzy reasoning are used in an FLC for its design and execution. The design of an FLC includes "if … then …" rules, which are based on available data history and/or expert's knowledge. These rules are a linguistic description of systems' behavior and characteristics with their imprecise numeric value. Fig. 3.8 has three main stages or blocks:

1. Fuzzification
2. Inference or interpretation
3. Defuzzification

The functioning of each block is described as follows:

1. Fuzzification.

In the first step of FLC, the imprecise values of input data are converted into fuzzy, a variable that is called fuzzification. This uses a universe of discourse defined by membership functions. The relationship function may be

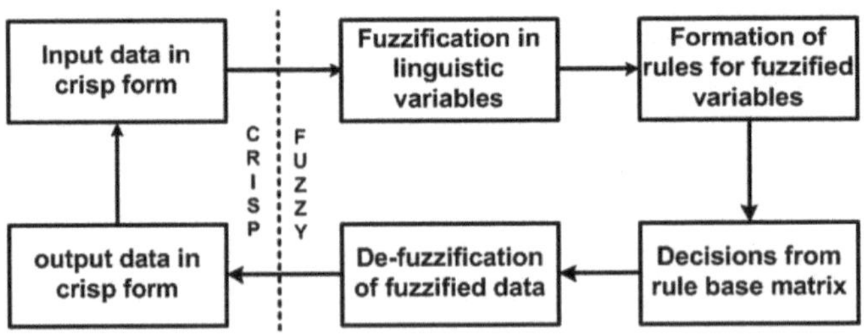

FIGURE 3.8 Basic structure of a fuzzy logic-based controller.

a triangular, exponential, or Gaussian type. Additionally, for more complicated systems more than one membership function in combination may be used. The performance of membership functions depends upon the system in which it is to be implemented. However, the triangular membership function is the most common.

2. Inference or interpretation

In the second step of FLC the rules are formulated in terms linguistic variables based on system performance and previous experience. The interpreter is used for decision making for the given input and rule base.

3. Defuzzification

In the final step of FLC the decision taken by the inference block is converted into its corresponding numerical values and is called defuzzification. This block interfaces the controller to the actual system by providing precise values for further processing.

This is the elementary scheme of FLC, which is modified and designed according to input and output. This chapter describes two FLCs:

1. FLC for DC side capacitor voltage matching/regulation
2. FLC for modulation by current error mapping

The FLC controller parameter details are given in Table 3.1. The set of rules and separate design procedure for each FLC are described in a later section.

3.5.5 Fuzzy logic controller for DC link capacitor voltage matching

There can be two-level or multilevel inverters at the RES interface. These inverters require matching of DC capacitor voltage. The three-phase five-level cascaded H-bridge multilevel inverter has three capacitances on the DC side of each phase. Fig. 3.9 shows the FLC of this scheme, in which the voltage (V_{DC}) difference component corresponding to the PQ issue is calculated and compensated.

TABLE 3.1 Design parameters for fuzzy logic-based controller (FLC) controllers.

FLC	Relationship function type	Linguistic variables	Dimension of rule matrix
DC side voltage matching	Triangular	7	7×7
Current error mapping	Triangular	5	5×5

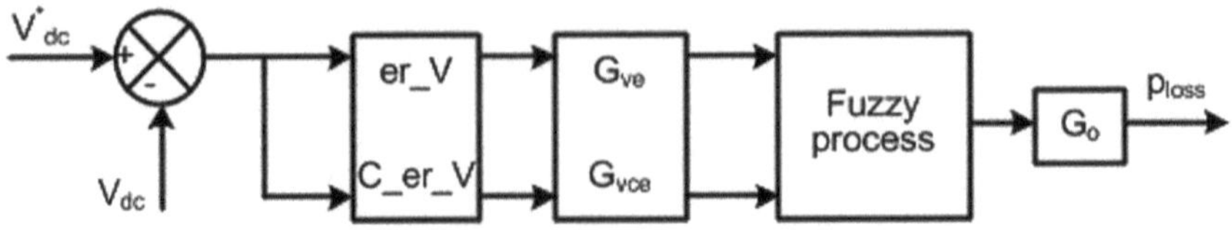

FIGURE 3.9 Fuzzy logic-based controller for DC voltage matching.

In this the total error (*er*) is tracked by calculating the error (*er_V*) and change in error (*c_er_V*) for voltage at the *j*th instant given by Eq. 3.14-3.16:

$$er\_v_j = \Delta V_{dc,j} = V^*_{dc,j} - V_{dc,j} \quad (3.14)$$

$$c\_er\_v_j = \Delta er_j = \Delta V_{dc,j} - \Delta V_{dc,(j-1)} \quad (3.15)$$

$$\Delta p_{PQ\_loss,j} = \Delta p_{PQ\_loss,(j-1)} + \lambda * \Delta p_{PQ-loss,j} \quad (3.16)$$

where $\Delta p_{PQ\_loss,j}$ is the active power loss component inferred by the FLC at the *j*th sampling time and λ is gain factor.

The input and output have a triangular relationship function with seven linguistic variables as shown in Fig. 3.10 and defined as:

1. More negative (MN)
2. Less negative (LN)
3. Negative (N)
4. Zero (Z)
5. Positive (P)
6. Less positive (LP)
7. More positive (MP)

The inference engine executes the fuzzy variables *er_V* and *c_er_V* and p_{PQ_loss} based on control rules of rule matrix of the order of (7 × 7) as given

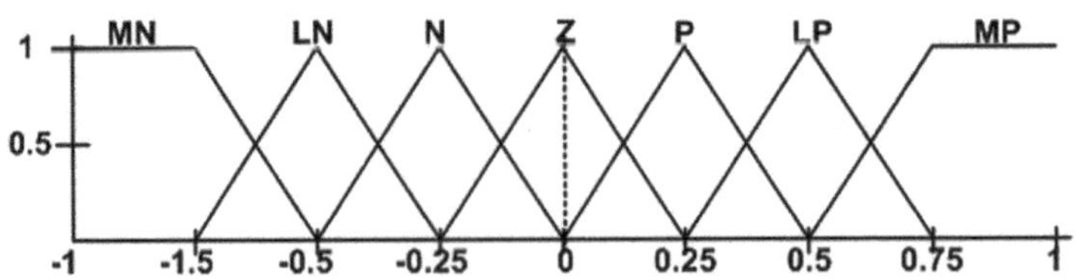

FIGURE 3.10 Triangular relationship functions for DC side voltage regulation.

TABLE 3.2 Rule base for DC voltage control.

Change in voltage error (c_er_V)	Error in voltage (er_V)						
	MN	**LN**	**N**	**Z**	**P**	**LP**	**MP**
MN	*MN*	*MN*	*MN*	*MN*	*LN*	*N*	*Z*
LN	*MN*	*MN*	*MN*	*LN*	*N*	*Z*	*P*
N	*MN*	*LN*	*LN*	*N*	*Z*	*P*	*LP*
Z	*MN*	*LN*	*N*	*Z*	*P*	*LP*	*MP*
P	*LN*	*N*	*Z*	*P*	*LP*	*MP*	*MP*
LP	*N*	*Z*	*P*	*MP*	*MP*	*MP*	*MP*
MP	*Z*	*P*	*LP*	*MP*	*MP*	*MP*	*MP*

in Table 3.2. The actions are inferred from already available data logic based on history. The definition is as follows:

- When the error is more on the positive side, then more reference is required.
- When the error is less on the positive side and error change is higher, then constant reference is required.
- When the error is zero, then there is no change in reference.
- When the error is found to be on the negative side (less or more), the reference also has to be negative.

The controller parameters are normalized through matching factors B_{er_V}, $B_{c\_er\_V}$, and B_o depending upon the system constraints.

3.5.6 Fuzzy logic controller for current adaptation and modulation

Fig. 3.11 shows the FLC for current error to augment the mitigation performance during PQ disturbances. This has a triangular relationship function

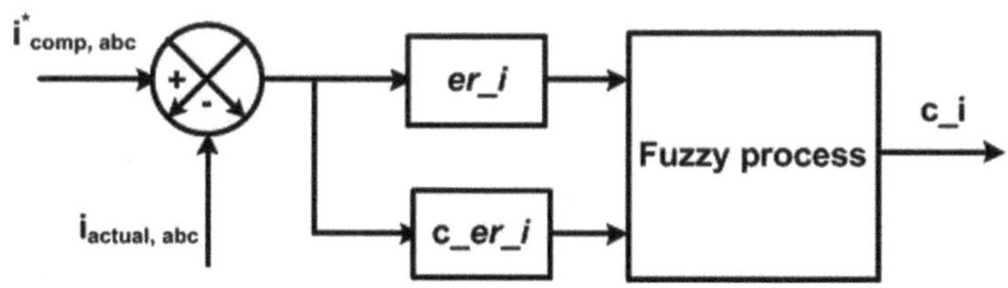

FIGURE 3.11 Fuzzy logic-based controller for current error adaptation and modulation.

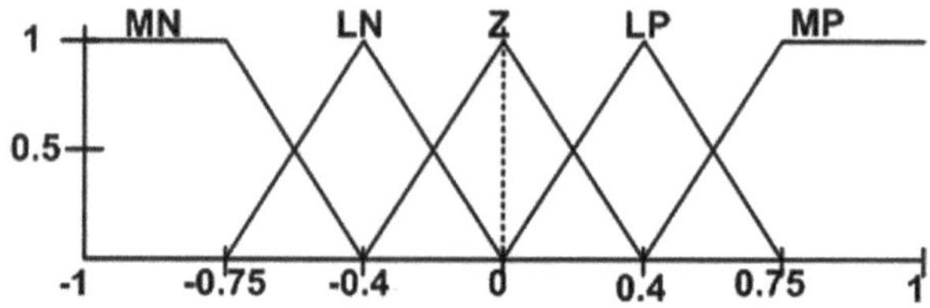

FIGURE 3.12 Triangular relationship functions for current error adaptation and modulation.

(Fig. 3.12) for input and output to incorporate the disturbance ambiguity. It has current error (*er_i*) and rate of change of error in current (*c_er_i*) as inputs and the output is the amended current signal (*c_i*). The rule matrix (5×5) given in Table 3.3 has *five* linguistic variables defined as:

1. More negative (MN)
2. Less negative (LN)
3. Zero (Z)
4. Less positive (LP)
5. More positive (MP)

The tracing rules are as follows:

- For small negative error and its change, the compensation is negative.
- For positive error and positive/zero change, the compensation is positive.
- For no error and no change, no compensation.
- For negative error and positive change, the compensation is zero.

TABLE 3.3 Rule base for current error adaptation and modulation.

Rate of change of error in current (*c_er_i*)	Error in current (*er_i*)					
		MN	**LN**	**Z**	**LP**	**MP**
	MN	*MN*	*MN*	*MN*	*LN*	*Z*
	LN	*MN*	*LN*	*LN*	*Z*	*LP*
	Z	*MN*	*LN*	*Z*	*LP*	*MP*
	LP	*LN*	*Z*	*LP*	*LP*	*MP*
	MP	*Z*	*LP*	*MP*	*MP*	*MP*

There can be other versions of error rules for system-specific changes and adaptation. The relationship function can also be changed for simpler implementation of FLC.

3.6 Simulation results and analysis

The described FLC and neural network control structure system with RES is simulated in MATLAB 2009a. This is a distributed generation scheme, which has both RES and nonlinear load. The *nftool* is utilized for harmonic extraction and ANN design. The *fuzzy* tool is utilized for fuzzy logic controller design. The generalized block diagram of the system is shown in Fig. 3.13. The system data is given in Table 3.4. The neural network pseudocode is shown in Fig. 3.14. The ANN training states are shown in Fig. 3.15, which indicate that this has taken 58 epochs to train and validate harmonic separation. Fig. 3.16 shows that best validation performance is achieved at epoch 52. This might vary with number of neurons. Here, 20 neurons are considered. Figs. 3.17 and 3.18 show the system performance waveforms when RL or RC load is connected as nonlinear load. The fast Fourier transform analysis for without and with AI-based controller is shown in Figs. 3.19–3.22. The RL load harmonic distortion is reduced to 2.27% (Fig. 3.20) from 20.67% (Fig. 3.19), whereas RC load harmonic distortion is found to be reduced to 2.24% (Fig. 3.22) from 26.33% (Fig. 3.21). This validates and establishes the described AI-based controller's efficiency.

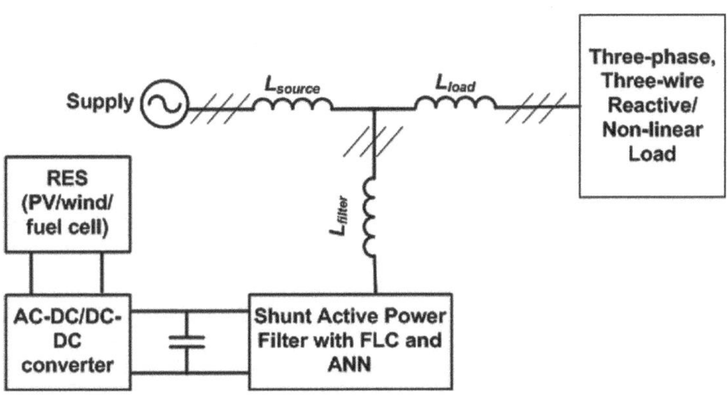

FIGURE 3.13 System with renewable energy sources (RES) and harmonic filter for simulation. *ANN*, Artificial neural network; *FLC*, fuzzy logic-based controller; *PV*, photovoltaic.

TABLE 3.4 System data for simulation.

Parameters	Value	Parameters	Value
Grid voltage and frequency	230 V (rms) at 50 Hz	Grid side inductance	2 mH
Boost converter input capacitor	550 μF	Boost converter inductor	4.5 mH
DC bus capacitor	5000 μF	RES voltage	300 V
Inverter side inductance	3 mH	Load side of uncontrolled converter	15 Ω, 120 μF and 20 Ω, 50 mH
Switching frequency	1 kHz	Sample time	10 μs

RES, Renewable energy source.

```
function net = create_fit_net(inputs,targets)
%CREATE_FIT_NET Creates and trains a fitting neural network.
%
% NET = CREATE_FIT_NET(INPUTS,TARGETS) takes these arguments:
%   INPUTS - RxQ matrix of Q R-element input samples
%   TARGETS - SxQ matrix of Q S-element associated target samples
% arranged as columns, and returns these results:
%   NET - The trained neural network
%
% For example, to solve the Simple Fit dataset problem with this
function:
%
%   load simplefit_dataset
%   net = create_fit_net(simplefitInputs,simplefitTargets);
%   simplefitOutputs = sim(net,simplefitInputs);
%
% To reproduce the results you obtained in NFTOOL:
%
%   net = create_fit_net(harmin,harmout);

% Create Network
numHiddenNeurons = 20;  % Adjust as desired
net = newfit(inputs,targets,numHiddenNeurons);
net.divideParam.trainRatio = 70/100;  % Adjust as desired
net.divideParam.valRatio = 15/100;  % Adjust as desired
net.divideParam.testRatio = 15/100;  % Adjust as desired

% Train and Apply Network
[net,tr] = train(net,inputs,targets);
outputs = sim(net,inputs);

% Plot
plotperf(tr)
plotfit(net,inputs,targets)
plotregression(targets,outputs)
```

FIGURE 3.14 Pseudocode for neural network fitting.

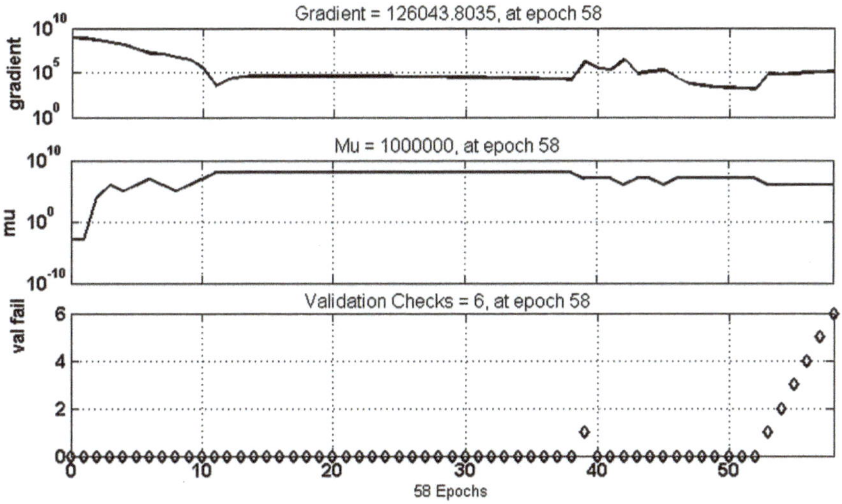

FIGURE 3.15 Artificial neural network training states.

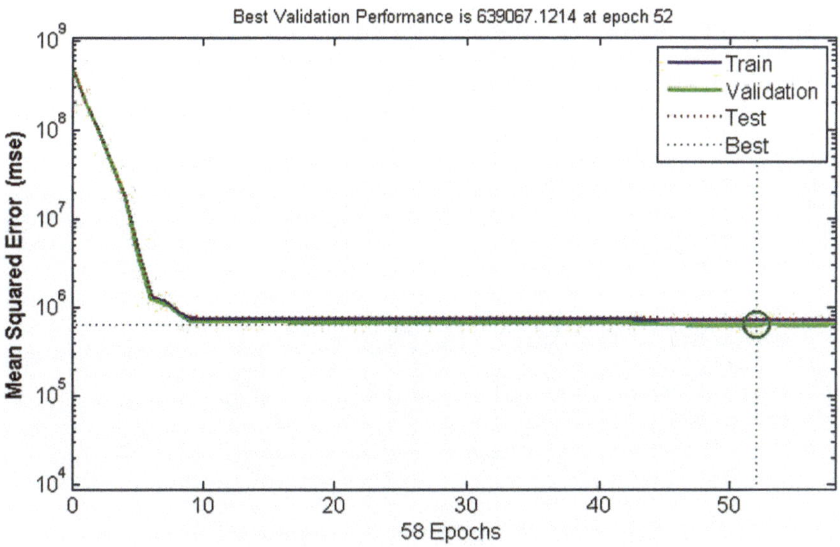

FIGURE 3.16 Artificial neural network validation stage.

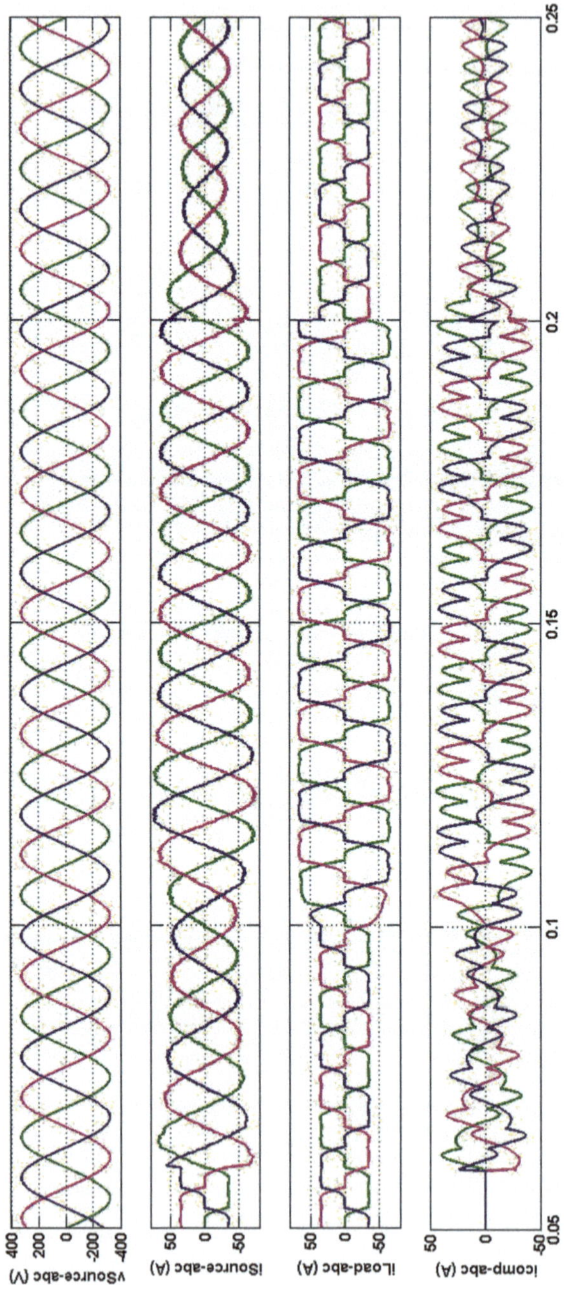

FIGURE 3.17 System performance waveforms with artificial intelligence controllers for RL load.

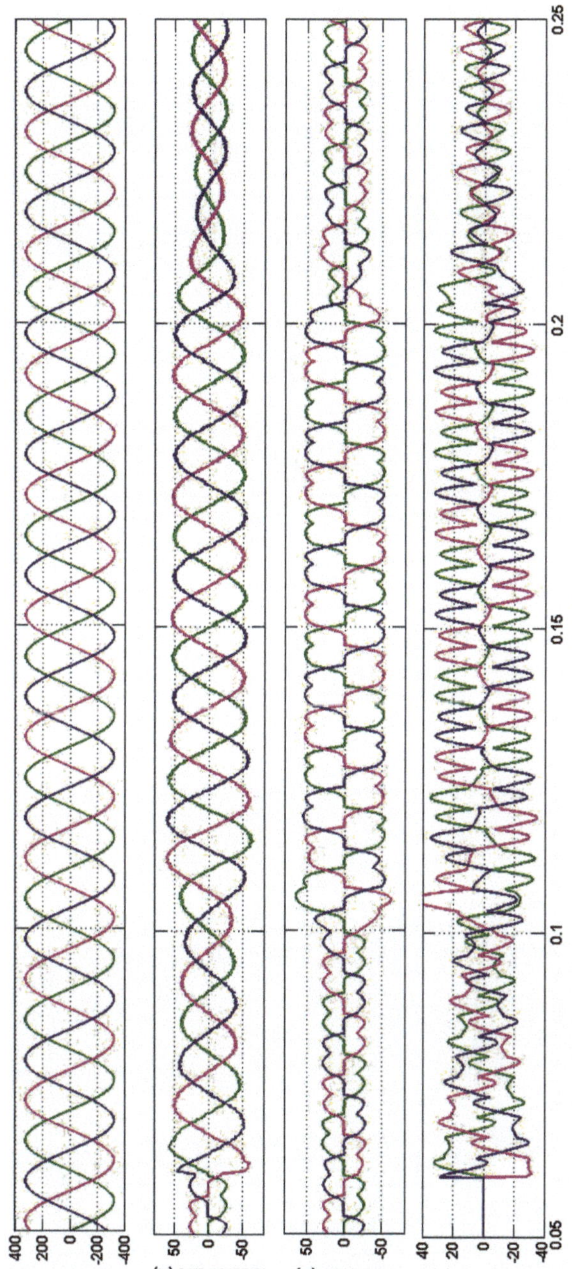

FIGURE 3.18 System performance waveforms with artificial intelligence controllers for RC load.

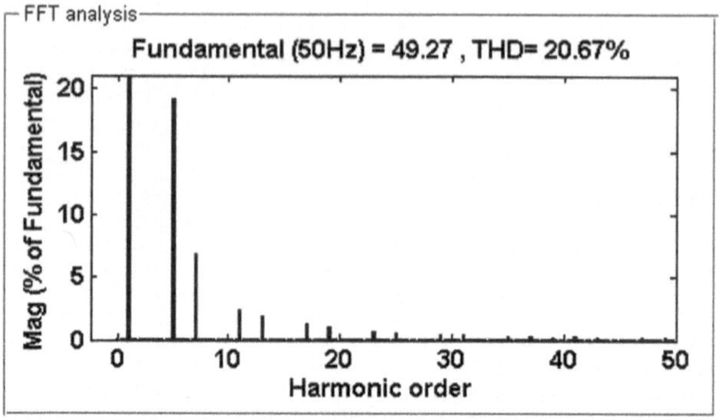

FIGURE 3.19 System current fast Fourier transform without artificial intelligence controllers for RL load.

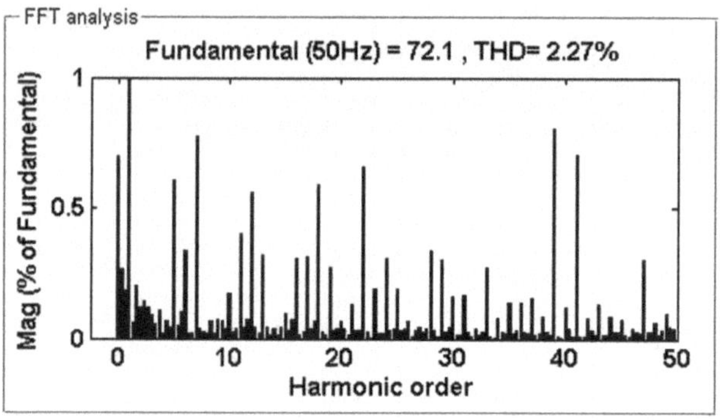

FIGURE 3.20 System current fast Fourier transform with artificial intelligence controllers for RL load.

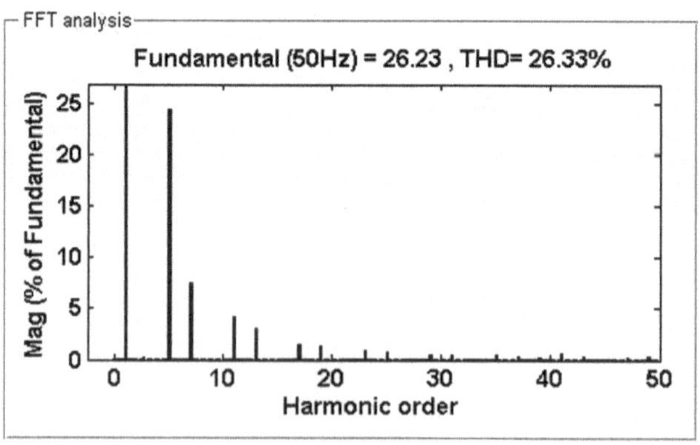

FIGURE 3.21 System current fast Fourier transform without artificial intelligence controllers for RC load.

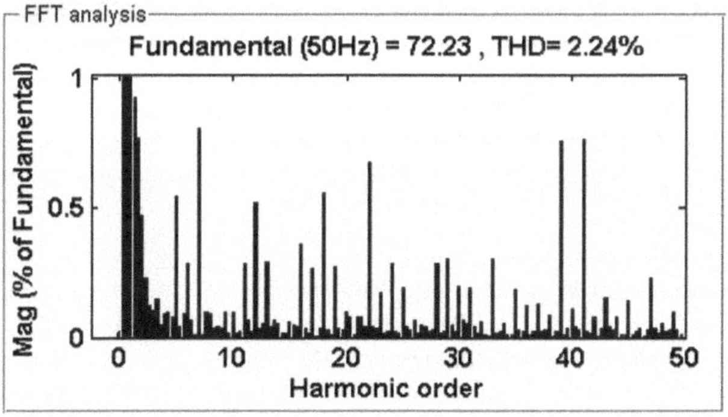

FIGURE 3.22 System current fast Fourier transform with artificial intelligence controllers for RC load.

3.7 Conclusion

This chapter presented a detailed modeling and AI implementation of PQ issues with RES and its grid integration. RES are a better solution for today's scenario. Therefore the chapter provided an advanced method of controlling and operation.

References

[1] R.H. Alabbasi, S.M. Salih, Control of cascade multilevel inverter using fuzzy logic technique, in: 2nd IEEE International Symposium on Power Electronics for Distributed Generation Systems (PEDG), 2010, pp. 96–101.

[2] N. Altin, A. Sefa, dSPACE based adaptive neuro-fuzzy controller of grid interactive inverter, Energy Convers. Manag. 56 (2012) 130–139.

[3] F. Blaabjerg, R. Teodorescu, M. Liserre, A.V. Timbus, Overview of control and grid synchronization for distributed power generation systems, IEEE Trans. Ind. Electron. 53 (2006) 1398–1409.

[4] X. Liang, "Emerging power quality challenges due to integration of renewable energy sources," IEEE Trans. Ind. Appl., vol. 53, pp. 855-866, March-April 2017.

[5] B. Ozpineci, L.M. Tolbert, J.N. Chiasson, Harmonic optimization of multilevel converters using genetic algorithms, IEEE Power Electron. Lett. 3 (2005) 92–95.

[6] H. Djeghloud, H. Benalla, Harmonics cancellation using SVPWM shunt active power filter, Compel 26 (2007) 114–126.

[7] N.K. Nguyen, D.O. Abdeslam, P. Wira, D. Flieller, J. Merckle, Artificial neural networks for harmonic currents identification in active power filtering schemes, in: Industrial Electronics, 2008. IECON 2008. 34th Annual Conference of IEEE, 2008, pp. 2696–2701.

[8] C. Cecati, A. Dell'Aquila, M. Liserre, A. Ometto, A fuzzy-logic-based controller for active rectifier, IEEE Trans. Ind. Appl. 39 (2003) 105–112.

[9] S. Saad, L. Zellouma, Fuzzy logic controller for three-level shunt active filter compensating harmonics and reactive power, Elec. Power Syst. Res. 79 (2009) 1337–1341.

[10] A. Hamadi, K. Al-Haddad, S. Rahmani, H. Kanaan, Comparison of fuzzy logic and proportional integral controller of voltage source active filter compensating current harmonics and power factor, in: IEEE International Conference on Industrial Technology, 2004, pp. 645–650.

[11] M. Farrokhi, S. Jamali, S.A. Mousavi, Fuzzy logic based indirect current control of the shunt active power filter, in: 39th International Universities Power Engineering Conference, 2004, pp. 489–493.

[12] D. Flieller, D. Ould Abdeslam, P. Wira, J. MercklÃ, Distortions identification and compensation based on artificial neural networks using symmetrical components of the voltages and the currents, Elec. Power Syst. Res. 79 (2009) 1145–1154.

[13] C.F. Nascimento, A. Goedtel, A.B. Dietrich, Harmonic distortion monitoring for nonlinear loads using neural-network-method, Appl. Soft Comput. 13 (2013) 475–482.

[14] P.K. Dash, S.K. Panda, A.C. Liew, B. Mishra, R.K. Jena, A new approach to monitoring electric power quality, Elec. Power Syst. Res. 46 (1998) 11–20.

[15] F. Temurtas, R. Gunturkun, N. Yumusak, H. Temurtas, Harmonic detection using feed forward and recurrent neural networks for active filters, Elec. Power Syst. Res. 72 (2004) 33–40.

[16] K. Vardar, E. Akpınar, T. Sürgevil, Evaluation of reference current extraction methods for DSP implementation in active power filters, Elec. Power Syst. Res. 79 (2009) 1342–1352.

[17] D.O. Abdeslam, P. Wira, J. MercklÃ, D. Flieller, Y.-A. Chapuis, A unified artificial neural network architecture for active power filters, IEEE Trans. Ind. Electron. 54 (2007) 61–76.

[18] H. Akagi, H. Fujita, S. Yonetani, Y. Kondo, A 6.6-kV transformerless STATCOM based on a five-level diode-clamped PWM converter: system design and experimentation of a 200-V 10-kVA laboratory model, IEEE Trans. Ind. Appl. 44 (2008) 672–680.

[19] V. Mahajan, P. Agarwal, H.O. Gupta, An artificial intelligence based controller for multilevel harmonic filter, Int. J. Electr. Power Energy Syst. 58 (2013) 170–180.

[20] V. Mahajan, P. Agarwal, H.O. Gupta, dSPACE implementation of cascaded H-bridge inverter for harmonics minimization using artificial-intelligence, Compel 33 (2014) 2053–2081.

[21] V. Mahajan, P. Agarwal, H.O. Gupta, Implementation of high-voltage multilevel harmonic filter based on rotated carrier modulation and artificial intelligence-based controllers, Arabian J. Sci. Eng. 39 (2014) 7127–7143.

[22] V. Mahajan, P. Agarwal, H.O. Gupta, Neural network and fuzzy logic controllers for three-phase three-level shunt active power filter, in: IEEE Workshop on Computational Intelligence: Theories, Applications and Future Directions (WCI), 2015, pp. 1–6.

[23] A. Bhattacharya, C. Chakraborty, A shunt active power filter with enhanced performance using ANN-based predictive and adaptive controllers, IEEE Trans. Ind. Electron. 58 (Feb. 2011) 421–428.

[24] G. Tsengenes, T. Nathenas, G. Adamidis, A three-level space vector modulated grid connected inverter with control scheme based on instantaneous power theory, Simulat. Model. Pract. Theor. 25 (2012) 134–147.

[25] H. Akagi, Trends in active power line conditioners, IEEE Trans. Power Electron. 9 (1994) 263–268.

[26] H. Akagi, Active harmonic filters, Proc. IEEE 93 (2005) 2128–2141.

[27] H. Akagi, The state-of-the-art of active filters for power conditioning, in: European Conference on Power Electronics and Applications, 2005, pp. 1–15.

[28] H. Akagi, E.H. Watanabe, M. Aredes, Instantaneous Power Theory and Applications to Power Conditioning, IEEE Press, Wiley-Interscience, John Wiley and Sons, Inc., 2007.

[29] L.A. Zadeh, G.J. Klir, B. Yuan, Fuzzy Sets, Fuzzy Logic, and Fuzzy Systems: Selected Papers, vol. 6, World Scientific Publishing Company Incorporated, 1996.

[30] F. Jurado, M. Valverde, Genetic fuzzy control applied to the inverter of solid oxide fuel cell for power quality improvement, Elec. Power Syst. Res. 76 (2005) 93–105.

Chapter 4

Fault ride-through (FRT) capability and current FRT methods in photovoltaic-based distributed generators

Gökay Bayrak[1], Davood Ghaderi[1], P. Sanjeevikumar[2]
[1]*Bursa Technical University, Engineering and Natural Sciences, Department of Electrical and Electronics Engineering, Bursa, Turkey;* [2]*Esbjerg Energy Section, Department of Energy Technology, Aalborg University, Esbjerg, Denmark*

Chapter outline

4.1 Introduction **133**
4.2 Fault ride-through strategies for PV power plants **136**
4.3 Fault ride-through methods in PV power plants **137**
4.3.1 External FRT methods 139
4.3.1.1 Energy storage-based methods 139
4.3.1.2 Power electronics-based method 139
4.3.1.3 FACTS-based methods 140
4.3.2 External FRT methods 143
4.3.2.1 Inverter-resident FRT methods 143
4.4 Assessment of current FRT methods for PV power plants **146**
4.5 Conclusion **147**
References **147**

4.1 Introduction

According to the grid regulations used before distributed generation (DG) and microgrid system applications, system breakers became active during failures occurring on the grid side, and the load–grid connection had to be disconnected to prevent damaging the devices in the system. However, due to the fact that DGs, which use renewables, allow bidirectional power flow, updates have been made in the current grid regulation. These existing regulations, known as grid code, which are called new grid regulations, aim to detect the malfunction as soon as possible [1,2]. There are various standards for identifying island

Power Quality in Modern Power Systems. https://doi.org/10.1016/B978-0-12-823346-7.00008-6

mode operation, voltage sags, voltage swells, and other power quality disturbance events. Besides, during breakdown, the system must continue operating without losing the connection between the renewables-based DGs and the grid. The aim is to minimize the impact of failure problems in a power system. For this reason, the subject of fault ride through (FRT) has emerged as an essential study topic in recent years in the literature.

IEEE Std. 1547 defines the abnormal operating conditions for a DG by limiting the fault duration to disconnect the grid in time. This standard indicates voltage and frequency threshold values for a DG to provide a stable operation. DGs must be separated from the grid as soon as possible (maximum 2 s), and a DG must be operated stably in case of any failure in the grid, according to IEEE Std. 1547. The allowed maximum fault disconnection time consists of the relay detection time and the breaker disconnection time. Table 4.1 shows these limitations, considering the operating conditions of DGs. Thus the tripping time of DGs must be provided before the fault clearing time is reached. The monitoring of DGs is a significant issue because of this condition. DGs must be operated according to specified regulations in IEEE 1547 and disconnected from the grid in time. The safe operation of DGs depends on following the regulations of defined standards.

After potential failure of the system to increase the grid stability of the DG system and for reliable operation, various electrical criteria have been determined to reconnect the DG system to the grid. Grid codes determined by each country define these criteria, and the DG system connected to the grid must comply with these criteria [3,4]. These features, also called FRT, are defined as the capacity of DG to reconnect after any failures.

Rapid disconnection of photovoltaic (PV) power plants in grid faults affects grid stability, especially in PV power plants. Therefore the requirements of FRT implemented by grid codes have some restrictions for PV systems related to change of grid voltages and grid frequencies. Fig. 4.1 shows the FRT requirement desired by grid codes. This approach is necessary to ensure that

TABLE 4.1 Voltage and frequency disconnection requirements in IEEE Std. 1547.

Voltage range (% nominal)	Max. clearing time (s)	Frequency range (Hz)	Max. clearing time (s)
$V < 50\%$	0.16	$f > 60.5$	0.16
$50\% = V < 88\%$	2.0	$f < 57.0$	0.16
$110\% < V < 120\%$	1.0	$57 < f < 59.8$	Adjustable between 0.16 and 300
$V \geq 120\%$	0.16		

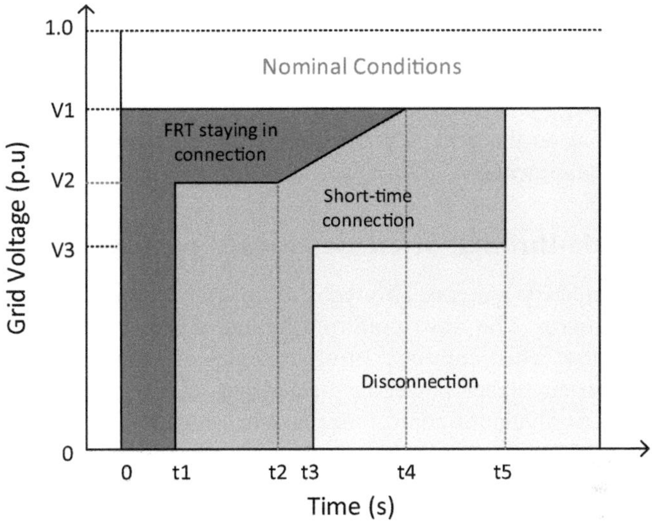

FIGURE 4.1 Fault ride-through (FRT) requirement desired by the grid codes.

there is no power mismatch due to voltage sags or voltage swells. Grid codes request the provision of reactive power to fix the stability of the grid and support voltage recovery by operating the PV system in a proper FRT operation [5]. The FRT requirement desired by the grid codes is generally summarized in Fig. 4.1.

Voltage functions versus time are generally used in voltage ride-through capabilities of DGs in many countries' grid codes. This approach is a general FRT method for abnormal conditions, but it is not suitable or efficient for defining occurring fault types. The unbalanced or symmetrical faults on all phases are completely applied for the UK's and Denmark's grid codes for providing ride-through capability. However, the desired requirements are more flexible for some faults in Spanish grid codes. A three-phase symmetrical fault with a 20% voltage drop from the nominal value is required to be tolerated for 500 ms in the Spanish grid code. However, when the voltage drop is 60% in two-phase-to-ground faults, only a generator is allowed for operation. Frequency ride through is also an essential criterion for grid codes. For example, frequency limits change between 49.8 and 50.2 Hz in the Turkish grid code. European Network of Transmission System Operators for Electricity (ENTSO-E) requirements define two voltage/time profiles for German grid codes, and most European countries use ENTSO-E requirements to provide FRT capability of DGs.

Depending on the FRT requirements specified by the grid code in case of any voltage collapse, the fast detection of the DG system when it is running due to malfunction during continuous operation, protection of the inverter against overcurrents due to fault currents, and protection of the inverter DC link and capacitors from overvoltages are based on voltage drop. As a result, it is necessary to supply the desired amount of reactive current into the grid to restore the voltage and to ensure that the PV system is in a stable condition when connecting to the grid. Various FRT methods have been developed to meet these requirements.

4.2 Fault ride-through strategies for PV power plants

In grid-connected PV systems, two types of inverters are used: voltage control and current control. The most preferred among these are current-controlled inverters, because when current control strategies are applied with a current-controlled inverter, higher power factor and less harmonic distortion can be obtained. We can divide the control mechanism structure of inverters into three classes: an inner loop, an outer loop, and other control [6,7].

In the inner loop model, a feedforward strategy has been adopted to separate the active and reactive currents. The inner loop compensates for the active current generated by the outer loop control [8]. DC link voltage regulation generally is the outer loop of the inverters, by supplying a reference value of active current to the internal loop control. DC link voltage (V_{dc}) is fixed to its nominal value with this structure. A conventional proportional integral (PI) controller is used to correct the DC link voltage. Other controls and protection models are classified as maximum power point (MPP) monitoring control, islanding mode protection, flexible, reactive power support integration requirements, and managing power quality. These structures can be added to the system in both single and two-stage topologies to increase the capacity of PV power plants. Especially, PV power plants with FRT features should be created using these topologies. In application, 80%–85% of current-controlled inverters are used, while only 10%–15% of voltage-controlled inverters are used [8,9].

According to today's modern grid regulations, system harmonics are required to work in a corrected and determined manner during the transition of faults in large-scale PV plants. In the grid-connected PV system shown in Fig. 4.2, control strategies that are either embedded in the inverter or independent from the inverter are generally recommended. The control system should be able to take the following measures to meet FRT requirements when a voltage drop occurs [10,11]:

- When the PV system is out of nominal operation, the proposed control method must detect the fault as soon as possible and accurately.
- When an overcurrent has occurred on the AC side of the PV inverter, the proposed control method must protect the DG and all of the electrical equipment.

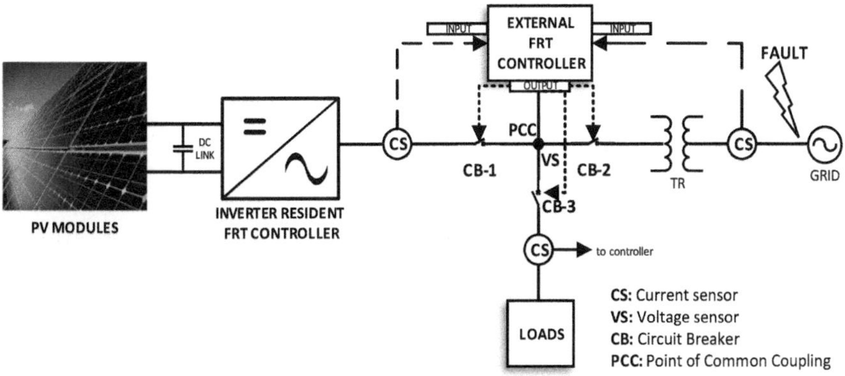

FIGURE 4.2 General fault ride-through (FRT) strategies for photovoltaic (PV) power plants.

- The proposed controller method must also protect the inverter and DC link capacitor from an overvoltage that occurred on the DC side of the inverter.
- The controller must adjust DG operation by providing voltage recovery and the desired reactive power to the grid by taking into account the grid code requirements.
- The controller must provide stable operation of the PV system by connecting the DG to the grid in nominal operating conditions.

The PV system must have two operating modes to meet FRT requirements. These two modes are normal operating in steady-state conditions and temporary operation with FRT capacity control when a voltage drop occurs due to grid faults [12]. Besides, the sag detection method is decisive in defining the overall performance of PV-based DGs in any fault conditions of the grid [13].

In PV systems, MPP monitoring is used to check whether maximum power is available from the system, thereby predicting both system efficiency and the possibility of failure. While in FRT mode created using external devices, the MPP should work regularly. In contrast, in modified inverter-based FRT mode, it is desired to switch to an operating state without MPP monitoring. However, in the case of overvoltage, MPP should stay in follow mode, and active power should be produced according to the depth of failure for system stability.

4.3 Fault ride-through methods in PV power plants

In this chapter, FRT methods used in grid-connected PV systems are classified, and the methods used in practice are examined. The classification of current FRT methods used in PV power plants is shown in Fig. 4.3. The methods used for FRT support are generally classified as internal and external methods.

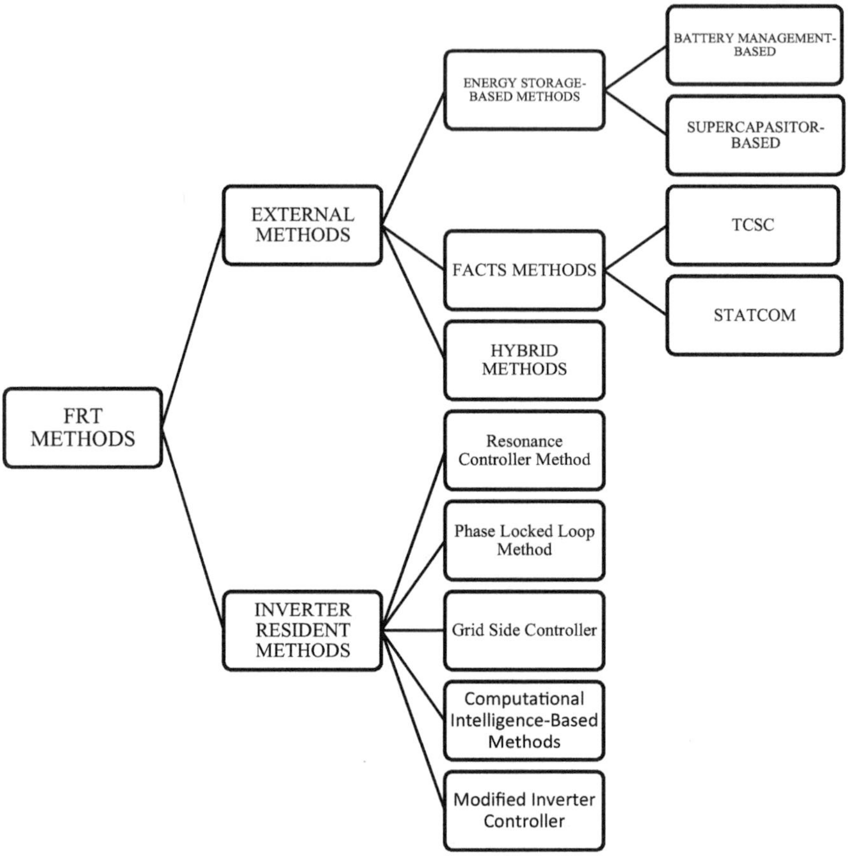

FIGURE 4.3 A classification of fault ride-through (FRT) methods used in photovoltaic power plants. *FACTS*, Flexible alternating current transmission system; *STATCOM*, static synchronous compensator; *TCSC*, thyristor-controlled static compensator.

Methods that use external devices are usually energy storage system (ESS), brake chopper protection circuits, flexible alternating current transmission system (FACTS) methods using thyristor-controlled static compensator (TCSC), and static synchronous compensator (STATCOM) [14–16].

In the ESS method, DC voltage storage is carried out on the DC side to protect the system from overvoltage during the fault on the grid side and until the fault is eliminated. In the brake chopper circuit, PV power plants generate excess energy, and this energy is damped on this resistance. The inverter will be protected against overvoltage because of the increase in DC link voltage. FACTS devices are also an alternative solution for protecting PV systems that provide reactive power when desired from the grid by connecting to the power system in any faults.

Besides, some methods make improvements in the system's FRT capacity through modifications made directly on the inverter or the line, without integrating an external device into the system. In the modified inverter controller (MIC) method, the primary purpose in the realization of internal strategies embedded in the inverter is to minimize the cost and complexity of the system while developing FRT capacity [17–19]. The MIC method provides the safety of DG and electrical equipment and stops the inverter from being connected to the grid. This method also ensures all electrical parameters return to nominal operating conditions in almost 0 s compared to the fault, and provides reactive and active power control and grid support in different types of fault. The MIC method also has a high response time and can produce results without low/overshooting. Overall, the MIC method is the most efficient FRT method, as it efficiently provides all grid regulatory requirements at a low cost. Computational methods are different methods used to increase the FRT capacity of the PV system, such as fuzzy logic control (FLC) and particle swarm optimization (PSO) [20].

4.3.1 External FRT methods

4.3.1.1 Energy storage-based methods

One of the strategies applied for the stability of the system and to avoid damage to devices during the transition of malfunction in PV systems is ESS. In the ESS method, DC voltage storage is carried out on the DC side to protect the system from overvoltage during the fault on the grid side and until the fault is eliminated. In this strategy, a connection is provided to the lowering-amplifier converter. It is then applied to the ESS device from the line on the DC side of the inverter. Thus DC voltage storage is carried out on the DC side to protect the system from overvoltage during the fault on the grid side and until the fault is eliminated [21,22]. Fig. 4.4 shows the general operation of energy storage-based FRT methods in PV power plants. We provide this storage process with two different energy depot system (EDS) methods:

- **Battery ESS:** After the converter, energy storage is carried out through the connected batteries.
- **Supercapacitor ESS:** After the converter, energy storage is carried out through the connected supercapacitor.

4.3.1.2 Power electronics-based method

This aims to dampen the overvoltage occurring on the DC side of the inverter due to a fault on the grid side. Therefore the brake chopper circuit is installed parallel with the DC link capacitor. This circuit consists of one isolated double gate transistor and a chopper resistor connected in series [23,24]. As soon as the malfunction occurs, the G (gate) of the transistor in the switching task will

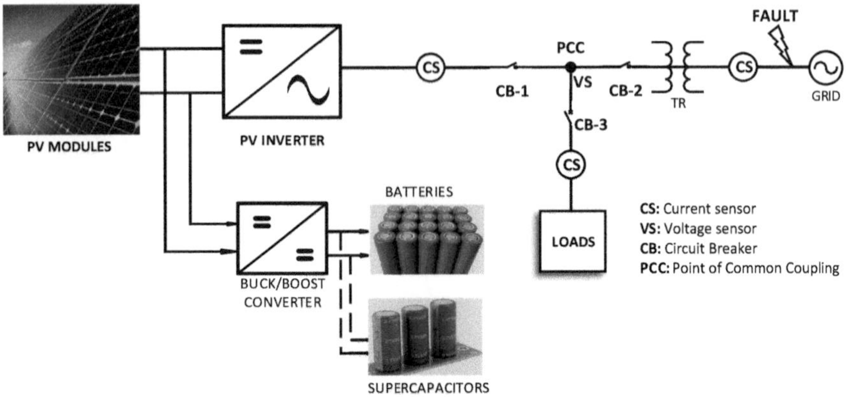

FIGURE 4.4 General operation of energy storage-based fault ride-through methods in photovoltaic (PV) power plants.

be triggered and will start passing a current through it. Excess energy generated by the PV generators will be damped on this resistance. In this way, the inverter will be protected against overvoltage by increasing the DC link voltage. Fig. 4.5 shows the general operation of power electronics-based FRT methods in PV power plants.

4.3.1.3 FACTS-based methods

FACTS devices are an alternative solution for protecting PV systems that provide reactive power when desired from the grid by connecting to the power system in any faults.

FACTS is a power electronics-based system and other static hardware that provides control of one or more alternating current transmission system parameters to increase controllability and power transfer capacity. A FACTS controller is faster than traditional mechanical controllers (shunt capacitor, series capacitor, phase shifter, etc.) since it is a power electronics-based

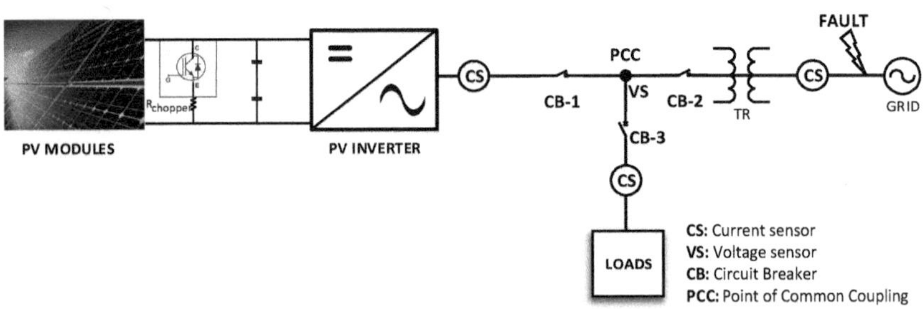

FIGURE 4.5 General operation of power electronics-based fault ride-through methods in photovoltaic (PV) power plants.

application. It increases the stable operating limits of transmission systems when properly adjusted. These devices provide control of the power system through appropriate compensation of grid parameters such as serial impedance, current, voltage, and reactive and active power of the transmission line [25,26]. FACTS devices have two main purposes:

- To increase the power transfer capacity of transmission systems.
- To provide control of power flow over defined transmission routes.

The first aim is to increase the power flow to a particular line up to the thermal limits of the transmission line. At the same time, system stability must be maintained through proper real-time control of the power flow during and after system errors. The second aim controls the current in a line and directs the power flow to the selected transmission corridors. Also, for this purpose, the primary power flow may change rapidly to a secondary route that can be used to provide all the desired power transmission in the system under unexpected conditions. It can be said that the performance of these two main objectives will significantly increase the use of transmission systems and will play a significant role in facilitating customization with minimal requirements for new transmission lines [27].

TCSC: TCSC is a typical serial-connected FACTS device. TCSC configurations use a fixed capacitor, a thyristor-controlled reactor (TCR), and a capacitor (C) connected in parallel with them, as shown in Fig. 4.6. The TCR-C combination allows fundamental frequency capacitive reactance to be properly controlled over a wide range [28].

A metal oxide varistor is placed parallel to the capacitor C to protect it from overvoltage levels. TCSC operations are classified into three modes:

1. Thyristors in full transmission (bypass mode): Most of the line current flows over the TCR in this mode, and it has a small inductive impedance.
2. Thyristors blocked: TCSC impedance is only equal to capacitor reactance.

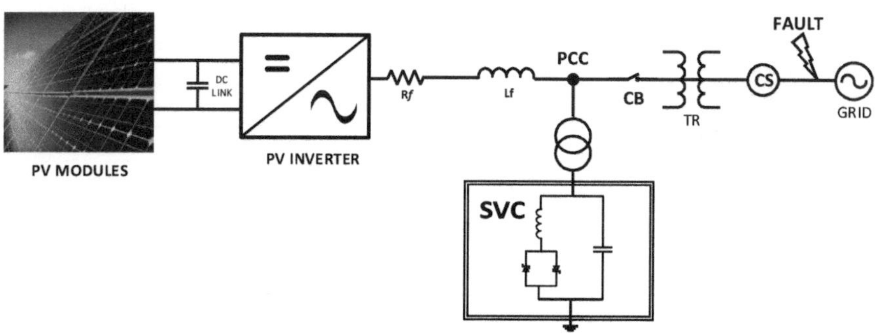

FIGURE 4.6 General operation of static VAR compensator (SVC)-based fault ride-through methods in photovoltaic (PV) power plants.

3. Vernier mode (where thyristors are partially in transmission): In this mode, TCSC is operated in a capacitive or inductive zone. This mode is a working mode where the TCSC's dynamic operation is ensured by the constant change of thyristor trigger angles. This mode is used to prevent subsynchronous resonance (SSR) [11,29].

TCSC can change the reactance of thyristors quickly and adequately by controlling their trigger angles. TCSC can directly regulate power flow along desired paths and allow the system to work closer to the line limits. Due to its fast and flexible editing capability, it can improve the dynamic performance and transient stability of the power system. It is also used to extinguish SSR and power oscillations.

STATCOM: STATCOM is a FACTS device that is shunted to the transmission line. It is also called an advanced static VAR compensator in the literature. Fig. 4.7 shows the STATCOM circuit structure. Simply, a STATCOM connection transformer consists of a converter and a DC capacitor. Here, V represents the AC system voltage and V_o means the STATCOM output voltage. It can be used for reactive shunt compensation, such as thyristor controlled static VAR compensator (TCSVC) [30,31]. It supplies reactive power to the gridline or consumes reactive power from the transmission line to regulate the bus voltage to which it is connected.

Balanced three-phase voltages at the STATCOM output are obtained at controllable amplitude at the fundamental frequency. The device is generally reactive when it operates with a power exchange between the AC system. The reactive power exchange between the STATCOM and the AC system is regulated by controlling the magnitude of the converter output voltage.

When STATCOM output voltage exceeds the AC system voltage ($V_o > V$), the current (I) flows from device to power system via reactance of the transformer. This current generates a reactive power for the device transmission line. When the circuit losses are neglected because the STATCOM only exchanges reactive power, an output voltage of the converter is in phase with the grid voltage. Therefore the converter output current is 90 degrees ahead of

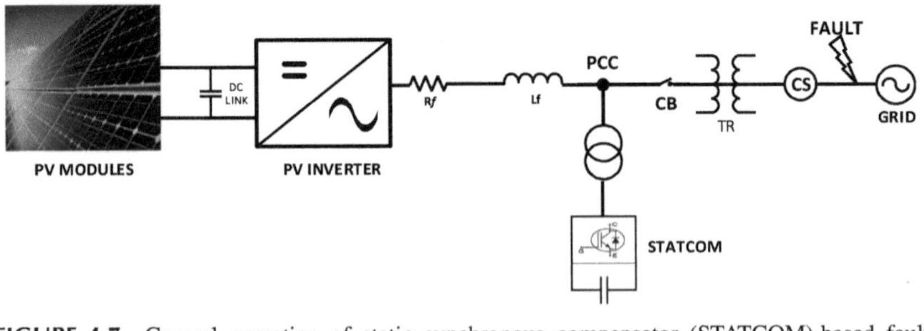

FIGURE 4.7 General operation of static synchronous compensator (STATCOM)-based fault ride-through methods in photovoltaic (PV) power plants.

the transmission line voltage. In this case, the device has worked in capacitive mode. If the output voltage of the STATCOM is less than grid voltage ($V_o < V$), the current flows from the AC system to the STATCOM via the device and the transformer reactance consumes reactive power from the transmission line. The converter current is 90 degrees behind the transmission line voltage.

The STATCOM is capable of supporting the voltage of hybrid power systems, providing reactive power, and thus increasing the reactive power control capacity [32]. In general, these devices increase both the system's FRT capacity and the cost and complexity of the system since they are added as an external device. However, they are useful in injecting reactive current.

4.3.2 External FRT methods

The FRT models mentioned so far have been made using an external device. In this section, without an external device integrated into the system, improvements will be made in the system's FRT capacity through modifications to be made directly on the inverter or the line. The main purpose of realization strategies in this section is to present the FRT capacity and to achieve the lowest cost and complexity of the system.

4.3.2.1 Inverter-resident FRT methods

FRT strategy in inverter-resident methods for PV systems involves calculating power references to provide desired reactive currents under different fault conditions, as outlined in grid codes. Overvoltage and overcurrent are also handled in this strategy with decreasing active power and limiting peak current, respectively [20,33].

1. **Maximum power point tracking (MPPT) methods:** The high penetration of the PV system creates a negative impact on the grid. Issues such as stability, power quality, efficiency, and reliability become even more critical to provide full support to the grid-connected PV system. When the FRT-based control strategy based on solar energy is connected to the grid, the power generation technique is used to provide better stability to the system. They are also called low-voltage ride through (LVRT). LVRT does not apply to a small-scale PV system. Other controllers such as MPPT and solar monitoring are more focused on these applications for stable operation. Fig. 4.8 shows the MPPT controller, which is an inverter-resident FRT method for PV power plants.
2. **Phase-locked loop (PLL) method:** However, due to reactive power compensation, the FRT-based control technique is applied [34]. The grid code must be maintained for a renewable energy-based distributed system. The need for an FRT-based control strategy is necessary on a large scale to

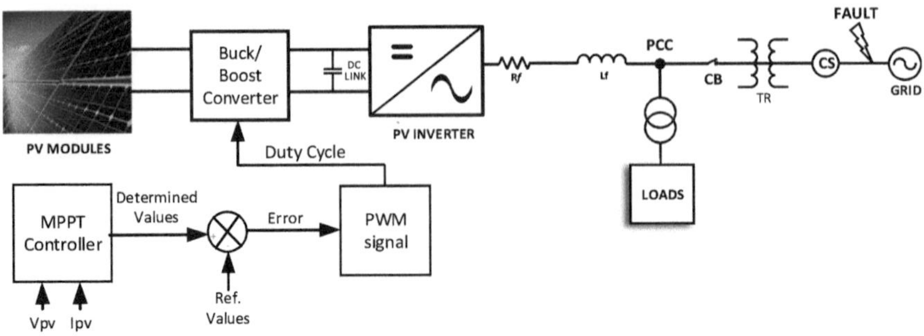

FIGURE 4.8 Inverter-resident fault ride-through method (maximum power point tracking controller) for photovoltaic (PV) power plants. *PWM*, Pulse-width modulation.

eliminate the resulting grid failure as soon as possible. PV application, the grid-connected single-phase PV system, contains two control loops:

- An internal current control loop to overcome the power quality problem and protect the system with high surge current.
- An external voltage control loop in which the desired current reference is generated to control the internal loop using the power and voltage obtained.

Stability of the entire control system depends on accuracy and speed of the voltage sag detection and the synchronization of the system with the grid in an incorrect condition. Sag detection techniques are essential to determine the efficiency and stability of the system. The sagging detection technique is an important method used to increase the effectiveness of the resonance controller, PLL controller, mean square root (RMS) method, and peak voltage method.

3. **Resonance controller method:** Resonance controller is the standard control technique like the PI controller. The resonance controller is based on the proportional resonance approach used to control the DC signal of the system. In the three-phase system, the direct quadrature (d-q) conversion method is used to convert the signal to DC [35]. The resonance control strategy can track the gain in infinite gain at the resonance frequency, thereby allowing low-pass conversion. The PLL behavior of the PV system fails during grid failure or asymmetric behavior of grid synchronization. In the PLL method, the stability of the system is affected by the sudden change in grid condition, since Park transform is based on the d-q conversion. The strategy is carried out in two stages. First, the PLL is stabilized under an asymmetrical grid condition, and then the current is guaranteed to be within the specified code limit.

4. **RMS method:** The RMS method is based on the determination of phase voltages. The voltage between the phases is calculated over a half period. Thus the voltage drop can be determined by determining the value of each phase. In the peak voltage method, the absolute voltage of each period is calculated, and the voltage drop over time is recorded. It is similar to the peak value method to detect an error where the parameters are calculated for the half cycle [5]. Among all the sagging detection methods, the RMS-based voltage sagging detection method acts fast to detect the voltage drop and complies with the standard grid code.
5. **Grid side controller (GSC) method:** The GSC-based control strategy also includes the reactive power balancing method. When a voltage drop is detected, the PV system enters the LVRT mode. The voltage drop must withstand the specified time to follow the grid code. During a voltage drop, the control system must deliver reactive power to the grid to support grid recovery. LVRT is mainly used for large-scale applications to support the grid and protect the system [6,25]. Reactive power injection into the grid when the LVRT is running can stabilize the grid voltage. Reactive power injection methods mainly used in three-phase PV applications are as follows:
   - Continuous active power control strategy
   - Power factor control strategy
   - Positive and negative sequence control strategy
   - Continuous reactive power control strategy

The presence of interaction between the voltage sequence and the current sequence under mains failure can cause an unstable mains state. Thus the oscillation will be present in both the active and reactive power control strategies. A zero sequence control pathway is added, which increases the system's freedom of control and eliminates oscillation, thereby increasing the stability of the system to overcome this. Injection of reactive power in a single-phase PV system is performed using two control strategies:

- Peak flow control strategy
- Active current control strategy

6. **Computational intelligence-based methods:** In these methods, various approaches are used to increase the FRT capacity of the PV system. FLC, PSO, and artificial neural networks are among the known methods [2,34,35].

A nonlinear control method for the full-bridge converter and its transient behavior is used in the PSO method to develop FRT, and a chopper circuit is used for the transient response. However, drooping is alleviated without the injection of reactive current and results in oscillation and overshoot. A vector control graphic is used for DC link voltage control and safe passage of the fault [20].

It is aimed to train the system with artificial neural networks and to respond quickly to incoming faults. However, the fact that the training process requires long processes to be experimentally verified and the difficulty of applying all the failures encountered in practice increase the duration of the control with artificial neural networks.

Although computational intelligent-based strategies partially meet FRT requirements, the desired efficiency in FRT control cannot be fully achieved since it does not address reactive current injection and increases the complexity of system control.

4.4 Assessment of current FRT methods for PV power plants

In this section, a general comparison of existing FRT methods for PV power plants is given in the literature. Which method or methods have higher performance in practice by comparing the existing FRT methods was investigated. Table 4.2 shows a comparison of current FRT methods used in PV power plants.

FACTS devices are an alternative solution for protecting PV systems that provide reactive power when desired from the grid by connecting to the power system in any faults. FACTS controllers are faster than traditional mechanical controllers (shunt capacitor, series capacitor, phase shifter, etc.) since they are power electronics-based applications. These controllers increase the stable operating limits of the transmission systems when properly adjusted. STATCOM is capable of supporting the voltage of the hybrid power systems, providing reactive power, and thus increasing reactive power control capacity. In general, these devices increase the system's FRT capacity and increase the

TABLE 4.2 A comparison of current fault ride-through (FRT) methods used in photovoltaic power plants.

FRT method	Reliability	Response time	Complexity	Cost
PI/PID	Low	Slow	Low	Low
Sliding mode control	Medium	Medium	High	High
Feedback	Low	Medium	Low	Low
MIC method	High	High	Medium	Low
Fuzzy logic	High	Fast	High	Medium
Neural networks	High	Fast	High	Medium
Hybrid computational	High	Very fast	Medium	High

MIC, Modified inverter controller; *PI/PID,* proportional integral/proportional integral derivative.

cost and complexity of the system since they are added as an external device. However, they are useful in injecting reactive current.

Another high-cost FRT strategy is energy storage-based FRT systems, because these methods are equipped with expensive supercapacitors and batteries that require inspection and regular maintenance, respectively. In the ESS method, DC voltage storage is carried out on the DC side to protect the system from overvoltage during the fault on the grid side and until the fault is eliminated. In the brake chopper circuit, excess energy generated by PV generators is damped on this resistance, and the inverter will be protected against overvoltage due to the increase in DC link voltage. Compared to FACTS methods, FRT methods based on energy storage are less expensive among FRT control methods based on external devices, as they contain fewer switches and use higher-power resistors.

Computational methods are different methods used to increase the FRT capacity of the PV system, such as FLC and PSO. Although computational intelligent-based strategies partially meet FRT requirements, the desired efficiency in FRT control cannot be fully achieved because it does not address reactive current injection and increases the complexity of system control.

These methods make improvements in the system's FRT capacity through modifications made directly on the inverter or the line, without integrating an external device into the system.

4.5 Conclusion

In this chapter, the best methods that can be applied for PV systems by analyzing the FRT methods currently used were determined. Energy storage-based methods are expensive, and the life cycle of used units is short. Besides, energy storage-based methods are easy to implement. The total cost and complexity of FACTS-based methods were found to be the highest, among others. Regarding grid regulation compatibility, inverter-resident methods are highly effective, and a modified inverter controller method has been found to be the best solution among the existing methods.

Inverter-resident control strategies have the advantage of low cost and fast calculation time. The FRT strategy is also required for the regular operation of the PV system during a mains failure condition. For this reason, it is the most preferred research area among researchers and industries.

References

[1] M. Islam, M. Huda, J. Hasan, M.A.H. Sadi, A. AbuHussein, T.K. Roy, M. Mahmud, Fault ride through capability improvement of DFIG based wind farm using nonlinear controller based bridge-type flux coupling non-superconducting fault current limiter, Energies 13 (7) (2020) 1696.

[2] M.H. Qais, H.M. Hasanien, S. Alghuwainem, Whale optimization algorithm-based Sugeno fuzzy logic controller for fault ride-through improvement of grid-connected variable speed wind generators, Eng. Appl. Artif. Intell. 87 (2020) 103328.
[3] Y. Gu, Y. Huang, Q. Wu, C. Li, H. Zhao, Y. Zhan, Isolation and protection of the motor-generator pair system for fault ride-through of renewable energy generation systems, IEEE Access 8 (2020) 13251−13258.
[4] D. Yoosefian, R.M. Chabanloo, Protection of distribution network considering fault ride through requirements of wind parks, Elec. Power Syst. Res. 178 (2020) 106019.
[5] I. Hamdan, A.M. Ibrahim, O. Noureldeen, Modified STATCOM control strategy for fault ride-through capability enhancement of grid-connected PV/wind hybrid power system during voltage sag, SN Appli. Sci. 2 (3) (2020) 364.
[6] A.A. Hossam-Eldin, E.N. Abdallah, M.S. Elgamal, K.M. AboRas, Fault ride-through of grid-connected THIPWM fired DCMLI-based DFIG using parallel switched feedback-controlled DVR, IET Gener. Transm. Distrib. 14 (6) (2020) 945−954.
[7] R. Georgious, J. Garcia, M. Sumner, S. Saeed, P. Garcia, Fault ride-through power electronic topologies for hybrid energy storage systems, Energies 13 (1) (2020) 257.
[8] B. Han, Y. Li, Simulation test of a DC fault current limiter for fault ride-through problem of low-voltage DC distribution, Energies 13 (7) (2020) 1753.
[9] H. Li, T. Zheng, S. Huang, Z. Tang, H. Cao, A fault ride through strategy of unified power flow controller and its coordination with protection, Elec. Power Syst. Res. 184 (2020) 106323.
[10] J.H. Lee, S.G. Jeon, D.K. Kim, J.S. Oh, J.E. Kim, Temporary fault ride-through method in power distribution systems with distributed generations based on PCS, Energies 13 (5) (2020) 1123.
[11] A.E. Amorim, D. Carletti, J.F. Fardin, L.F. Encarnação, D.S. Simonetti, A new hybrid multilevel converter for DFIG-based wind turbines fault ride-through and transient stability enhancement, Electr. Eng. (2020) 1−16.
[12] A.S. Yunus, A. Abu-Siada, M.A. Masoum, M.F. El-Naggar, J.X. Jin, Enhancement of DFIG LVRT capability during extreme short-wind gust events using SMES technology, IEEE Access. 8 (2020) 47264−47271.
[13] V.R. Babu, A. Ganapathi, Low voltage ride through (LVRT) capability enhancement of axial flux induction generator-based wind energy conversion system, in: Innovations in Electrical and Electronics Engineering, Springer, Singapore, 2020, pp. 405−416.
[14] M.M. Moghimian, M. Radmehr, M. Firouzi, Series resonance fault current limiter (SRFCL) with MOV for LVRT enhancement in DFIG-based wind farms, Elec. Power Compon. Syst. (2020) 1−12.
[15] A. Yılmaz, G. Bayrak, A real-time UWT-based intelligent fault detection method for PV-based microgrids, Elec. Power Syst. Res. 177 (2019) 105984.
[16] B. Qin, H. Li, X. Zhou, J. Li, W. Liu, Low-voltage ride-through techniques in DFIG-based wind turbines: a review, Appl. Sci. 10 (6) (2020) 2154.
[17] H. Dong, N. Chen, X. Li, H. Li, Improved adaptive robust control for low voltage ride-through of front-end speed regulation wind turbine, IEEE Access. 8 (2020) 55438−55446.
[18] M. Firouzi, Low-voltage ride-through (LVRT) capability enhancement of DFIG-based wind farm by using bridge-type superconducting fault current limiter (BTSFCL), J. Power Technol. 99 (4) (2020) 245−253.
[19] S. Shakeri, S. Esmaeili, M.H.R. Koochi, Determining accurate area of vulnerability for reliable voltage sag assessment considering wind turbine ride-through capability, Int. J. Electr. Power Energy Syst. 119 (2020) 105875.

[20] M.E.B. Aguilar, D.V. Coury, R. Reginatto, R.M. Monaro, Multi-objective PSO applied to PI control of DFIG wind turbine under electrical fault conditions, Elec. Power Syst. Res. 180 (2020) 106081.

[21] A. Alhejji, Capability enhancement of DC-link voltage in a single-stage grid-connected photovoltaic power system, Int. J. Renew. Energy Resour. 10 (1) (2020) 474–485.

[22] J. Luo, H. Zhao, S. Gao, M. Han, A low voltage ride through strategy of DFIG based on explicit model predictive control, Int. J. Electr. Power Energy Syst. 119 (2020) 105783.

[23] A. Sabir, S. Ibrir, A robust control scheme for grid-connected photovoltaic converters with low-voltage ride-through ability without phase-locked loop, ISA Trans. 96 (2020) 287–298.

[24] A. Bamigbade, V. Khadkikar, M. Al Hosani, H.H. Zeineldin, M.S. El Moursi, Gain compensation approach for low-voltage ride-through and dynamic performance improvement of three-phase type-3 PLL, IET Power Electron. 13 (8) (2020) 1613–1621, https://doi.org/10.1049/iet-pel.2019.1174.

[25] O.S. Elazab, M. Debouza, H.M. Hasanien, S.M. Muyeen, A. Al-Durra, Salp swarm algorithm-based optimal control scheme for LVRT capability improvement of grid-connected photovoltaic power plants: design and experimental validation, IET Renew. Power Gener. 14 (4) (2020) 591–599.

[26] L. Chen, G. Li, H. Chen, M. Ding, X. Zhang, Y. Li, Y. Tang, Investigation of a modified flux-coupling-type SFCL for low-voltage ride-through fulfillment of a virtual synchronous generator, IEEE Trans. Appl. Supercond. 30 (4) (2020) 1–6.

[27] R. Meng, Y. Du, X. Han, L. Wang, P. Wang, Coordinated control of series compensation link and bus interface converter in the AC–DC hybrid microgrid, J. Power Electron. 20 (2) (2020) 590–600.

[28] M. Rafiee, M. Aghamohammadi, An investigation of the influences of the voltage sag on the doubly fed induction generator using tuned PI controllers, Teh. Vjesn. 27 (1) (2020) 149–156.

[29] G. Bayrak, Wavelet transform-based fault detection method for hydrogen energy-based distributed generators, Int. J. Hydro. Energy 43 (44) (2018) 20293–20308.

[30] M.I. Mosaad, Direct power control of SRG-based WECSs using optimised fractional-order PI controller, IET Electr. Power Appl. 14 (3) (2020) 409–417.

[31] P. Verma, P. Gupta, Proactive stabilization of grid faults in DFIG based wind farm using bridge type fault current limiter based on NMPC, Energy Sources, Part A Recovery, Util. Environ. Eff. (2020) 1–20.

[32] M. Songkin, N.N. Barsoum, F. Wong, Impact assessment of large-scale solar photovoltaic integration on sabah grid system, in: Computational Science and Technology, Springer, Singapore, 2020, pp. 659–668.

[33] Alias, W. N. H. A., Sujod, M. Z., & Kamari, N. A. M. (2020). DC-link protection for grid-connected photovoltaic system: a review. In InECCE2019 (pp. 725-736). Springer, Singapore.

[34] G. Bayrak, D. Ghaderi, An improved step-up converter with a developed real-timefuzzy-based MPPT controller for PV-based residential applications, Int. Trans. Electr. Energ. Syst. 29 (2019) e12140.

[35] L. Benaaouinate, M. Khafallah, D. Voyer, A. Mesbahi, T. Bouragba, Nonlinear control based on fuzzy logic for a wind energy conversion system connected to the grid, Int. J. Renew. Energy Resour. 10 (1) (2020) 193–204.

Chapter 5

Power quality problems associated with electric vehicle charging infrastructure

P. Sivaraman[1], C. Sharmeela[2]

[1]*Leading Engineering Organisation, Chennai, Tamil Nadu, India;* [2]*Department of Electrical and Electronics Engineering, College of Engineering Guindy, Anna University, Chennai, Tamil Nadu, India*

Chapter outline

5.1 Introduction 151
5.2 Types of charging station 152
5.3 Power quality problems associated with EVSE and its impacts 154
5.3.1 Power quality problems from EV charging station into the grid 154
5.3.2 Power quality problems from grid into EV charging station 157
5.3.3 Case study: harmonics analysis for a DC fast charger of 150 kW rating 157
5.4 Mitigation of impact of higher penetration of EVs into distribution system 158
5.4.1 Energy storage system 159
5.4.2 Distributed FACTS devices 160
5.4.3 Demand response 160
5.5 Conclusion 160
References 161
Further reading 161

5.1 Introduction

An electric vehicle (EV) charging station, also referred as electric vehicle supply equipment (EVSE), supplies electric power to recharge the batteries in plug-in electric vehicles (PEVs) and other EVs. Charging stations are classified depending upon the nature of the charging current, speed of charging, and location of the charging stations. Based on the nature of the charging current, EVSE is classified into AC charging and DC charging depending on where this AC/DC conversion takes place. As known, batteries can be charged only with

Power Quality in Modern Power Systems. https://doi.org/10.1016/B978-0-12-823346-7.00005-0

DC electricity. If AC/DC conversion takes place on-board (inside the vehicle), then it is called AC charging and if it takes place off-board (inside EVSE), then it is called DC charging. Power quality (PQ) is the necessity for reliable operation of equipment in modern power systems. Power electronics-based converters used in EVSE introduce PQ problems such as harmonics, DC injection, and flicker. Higher penetration of EVs into the distribution system will increase the peak power demand requirements. This chapter discusses PQ problems associated with EV charging infrastructure, impact of PQ problems on both grid and EVSE, and presents a case study on harmonics analysis for a DC fast charger of 150 kW rating.

5.2 Types of charging station

A typical single line diagram (SLD) of an EV public charging station is shown in Fig. 5.1.

A typical EV public charging station consists of grid power source, dedicated transformer, power quality meter, switchgear panels, chargers, etc., as shown in Fig. 5.1. Power quality meters are required at the point of common coupling as per IEC 61000-4-30 and these meters will have features for measuring harmonics, including supraharmonics, voltage sag and swell, and voltage flicker and disruptions [2]. Depending on where this AC/DC conversion takes place, EVSE is classified as AC charging and DC charging. If AC/DC conversion takes place on-board (inside the vehicle), then it is called AC charging. A typical SLD of AC charging is shown in Fig. 5.2.

If AC/DC conversion takes place off-board (inside EVSE), then it is called DC charging. A typical SLD of DC charging is shown in Fig. 5.3.

Typical charger types for both fast and slow charging are given in Table 5.1 [3].

Based on the speed of charging or time taken to reach 100% state of charge (SoC) of the battery, the charging station can be classified as slow charging and fast charging. Slow charging implies overnight charging or charging that takes about 6–8 h or above to charge an EV battery pack. A charging station setup other than slow charging is referred to as fast charging, which comparatively takes less time. Fast charging systems are generally described as chargers providing high-power output capable of charging a vehicle to 80% in as little as 15 min. Such charge points are common at present and are in the range of 10–50 kW [13]. Charging solutions for 100 kW level may be available in the medium term. Speed of charging depends on the power rating of the EVSE and battery capacity. The higher the power rating, the higher the speed of charge. If charging time is less than 30 min, the charger is usually referred to as an ultrafast/superfast/flash charging/rapid charger.

Charging stations can also be classified as home/residential charging and public charging station based on where the charging process is carried out. Home charging is commonly practiced at home because it is economical. It

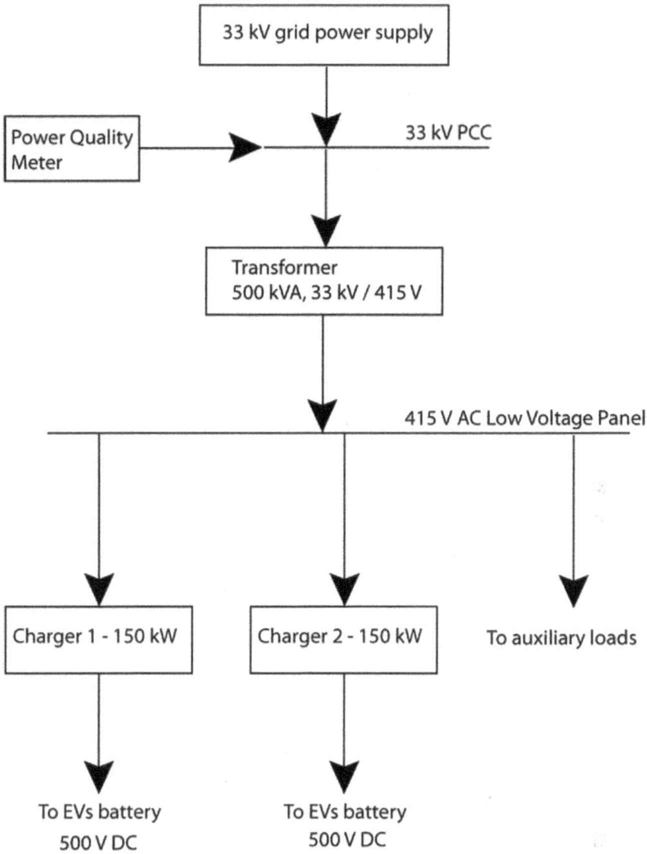

FIGURE 5.1 Typical single line diagram of an electric vehicle (EV) public charging station. *PCC*, Point of common coupling.

usually involves slow chargers because of the flexibility of available time for charging. EV owners prefer to charge their vehicles overnight in their home EVSE because electricity tariffs are lower compared to commercial or industrial electricity tariffs on which they are billed in the public charging infrastructure. EV owners roughly perform 80% of the charging in their home charging station. Public charging stations are the charging facility for EVs installed and operated by a utility and other the private companies licensed by the utility.

Vehicle to grid (V2G) supplies power back to the grid during peak hours as a prosumer. Prosumer means, at same point of connection (PoC), consuming the electric power from grid and supplying power back to the grid [2]. V2G will reduce peak power generation and balances demand versus generation.

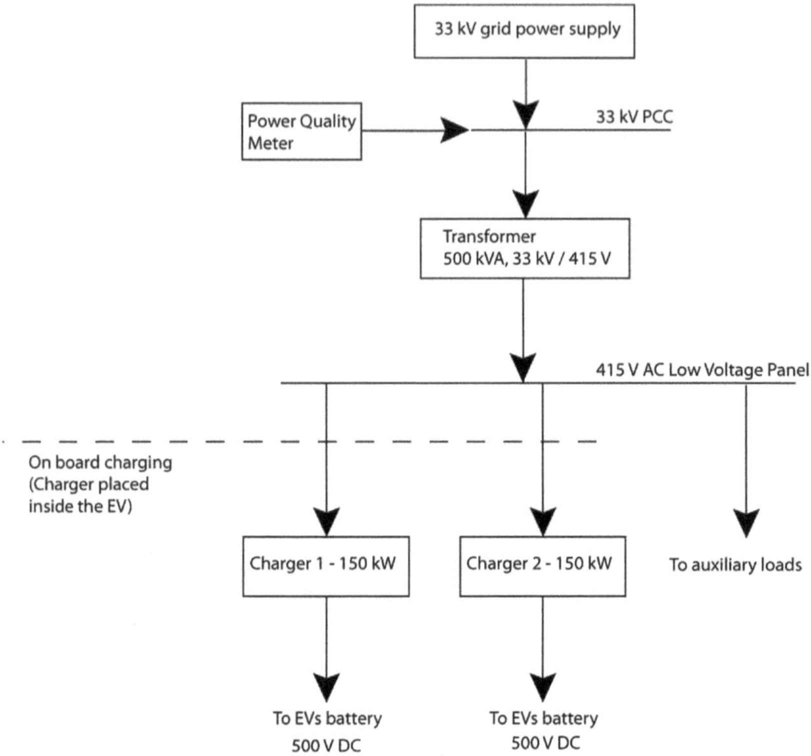

FIGURE 5.2 Typical single line diagram of AC charging or on-board charging. *EV*, electric vehicle; *PCC*, point of common coupling.

5.3 Power quality problems associated with EVSE and its impacts

PQ problems from EV charging stations are one of the major concerns for both power distribution companies and EV charging station owners. From a power distribution company perspective, power electronics-based converters used in EVSE inject harmonics, DC injection, and voltage flicker into the system [2]. Also, fast charging EVs will introduce rapid voltage changes into the system. From an EV charging station owner perspective, faults in transmission or distribution systems lead to voltage sag or interruption of supply voltage to EVSE and further result in disconnection from the grid [1].

5.3.1 Power quality problems from EV charging station into the grid

The charger in EVSE plays an important role in EV charging stations. These chargers are basically power electronics-based equipment or simply

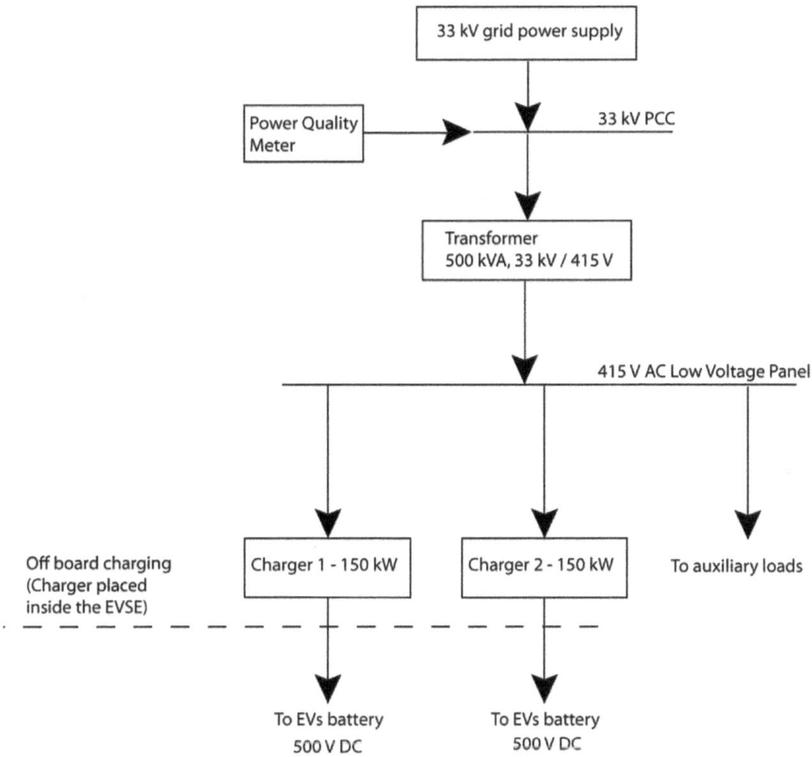

FIGURE 5.3 Typical single line diagram of DC charging or off-board charging. *EV*, Electric vehicle; *EVSE*, electric vehicle supply equipment; *PCC*, point of common coupling.

TABLE 5.1 Typical charger types for both fast and slow charging.

S. no.	Charger type	Charger connectors	Rated output voltage (V)	Charging vehicle type
1	Fast	Combined charging system (min 50 kW)	200–750 V or higher	4W
2		CHArge de Move (min 50 kW)	200–500 V or higher	4W
3		Type 2 AC (min 22 kW)	380–415 V	4W, 3W, 2W
4	Slow	Bharat DC-001 (15 kW)	48 V	4W, 3W, 2W
5		Bharat DC-001 (15 kW)	72 V or higher	4W
6		Bharat AC-001 (10 kW)	230 V	4W, 3W, 2W

W, Wheeler.

nonlinear loads. During operation, they draw nonlinear current from the given supply voltage. Due to the scarcity of fossil fuels and CO_2 emissions, many developed and/or developing countries are switching from conventional vehicles to EVs. To charge an EV, owners use a charging system/station. Due to this transition, the number of EVs connected to the distribution system is high and spread across the distribution system. The major concern is that if the current drawn by the EVs is highly nonlinear, then this may increase voltage distortions at the PoC and other users connected to the distribution system. The distribution system may pose a higher penetration or integration of EV charging infrastructure, therefore power quality problems created by EV charging infrastructure are going to be a major concern and distribution systems have to be well planned for this scenario [4]. Charging EVs during peak hours has an impact on load pattern, and creates higher losses, higher harmonic distortions, overloading of cables/lines and distribution transformers, etc. Also, distribution transformer life is reduced because of overheating caused by overload.

As described earlier, EV chargers are nonlinear loads and they draw nonsinusoidal current from the supply voltage, i.e., they inject harmonics into the supply system. EV chargers inject harmonic components (spectral content less than 2 kHz) and high-frequency components (spectral content between 2 and 150 kHz). This high-frequency component is also called supraharmonics [10]. The emission of high-frequency content (2–150 kHz) is due to switching frequency of the EV charger. It creates adverse impacts on the distribution system resulting in equipment malfunction, flickering of illumination devices, audible noise, nuisance tripping of protective devices (residual current circuit breaker/earth leakage circuit breaker), interference with communication equipment, errors in instrumentation and metering equipment readings, system resonance conditions, excessive heat losses, etc. [10].

The increased penetration of the EV charging infrastructure, especially fast charging, will severely affect the voltage profile of the distribution system. As explained earlier, fast chargers are capable of charging 80% of the SoC of a battery within 15 min and draw high power in a short time from the distribution grid [7]. Due to this, many end users' equipment experiences rapid voltage changes. IEC 61000-3-3 defines rapid voltage change as a voltage fluctuation and voltages vary between the two steady-state conditions [8]. During this time, voltage variation should not exceed the threshold limits. If the voltage exceeds the tolerance limits, then linear loads are operating at reduced performance and power electronics-based equipment is disconnected from the source.

Two-wheelers are widely used in most residential homes for transportation. Since the rating of a two-wheeler is considerably less, manufacturers offer single-phase chargers to charge the EV's battery. Single phase EVs charger is not a concern for lesser penetration into the distribution system. However, it is a major concern when large numbers of single-phase EV chargers are connected into the distribution system because they cause unbalanced loading into the three-phase/four-wire distribution system [5,6,9]. It is difficult to balance the connection of a single-phase EV charger equally among all the three phases. Due to unbalanced loading, the current flowing in all the three phases is different in a three-phase/four-wire system. This results in voltage unbalance in the distribution grid and overloading of the neutral conductor. The performance of other end users loads are severely affected by voltage unbalance [5].

5.3.2 Power quality problems from grid into EV charging station

Electrical faults in the transmission system result in voltage sag in the distribution system equipment. Sometimes, faults in the distribution system result in supply interruption on the line and voltage sag on neighboring lines. The propagation of grid voltage sag into end users' EVs will affect EV performance, i.e., reduction of voltage from EV charger tolerance limits will lead to activation of undervoltage protection and disconnection from the grid [1]. A typical EV charger name plate rating is shown in Fig. 5.4.

In Fig. 5.4, the input voltage of the charger is 380 V and the tolerance is ±10%. If the input voltage to the charger exceeds this tolerance limit, then the charger is disconnected from the grid.

5.3.3 Case study: harmonics analysis for a DC fast charger of 150 kW rating

Harmonic measurement is carried out at the input side of a 150 kW EV fast charging station in India. The SLD of the charging station is shown in Fig. 5.5.

FIGURE 5.4 Name plate rating of electric vehicle charger.

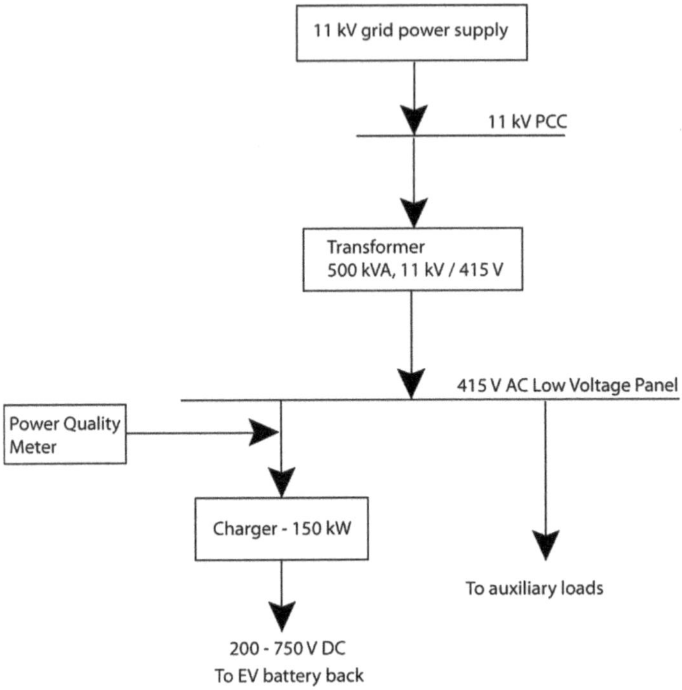

FIGURE 5.5 Single line diagram of a 150 kW electric vehicle (EV) fast charging station. *PCC*, Point of common coupling.

A Fluke 435-ii, class-A power quality analyzer as per IEC 61000-4-30 is used for field measurement. The harmonics are measured at the input side of the charger and are shown in Fig. 5.5. The actual measurements recorded at the EV fast charging station are shown in Fig. 5.6.

The harmonics injected by the 150 kW EV charger at 0%–99% and 100% of battery SoC are listed in Table 5.2.

From Table 5.2, current harmonics injected by the EV charger are less (around 3%) when battery SoC is 0%–99% and the current harmonics are drastically increased to 13% from 3% during 100% of battery SoC.

5.4 Mitigation of impact of higher penetration of EVs into distribution system

The higher penetration of EVs into the distribution system has various impacts as discussed earlier in this chapter. There are various methods available in the literature to mitigate or minimize the impacts due to EV charging [4,7]:

1. Energy storage system (ESS)
2. Distributed flexible alternating current transmission system (D-FACTS) devices
3. Demand response

FIGURE 5.6 Fluke 435-ii at the input side of the charger.

TABLE 5.2 Harmonics injected by the 150 kW EV charger at 0%–99% and 100% of battery SoC.

S. no.	Battery SoC (%)	Active power (kW)	Voltage THD (%)			Current THD (%)		
			R	Y	B	R	Y	B
1	0%–99%	129.3	1.5	1.5	1.6	2.9	3.1	3
2	100%	72	1.75	1.63	1.84	13.01	13.78	13.23

B, phase; *EV*, Electric vehicle; *R*, phase; *SoC*, state of charge; *THD*, total harmonic distortion; *Y*, phase.

5.4.1 Energy storage system

The ESS is charged at the time of reduced grid power demand and discharged at higher grid power demand. These systems supply both active and reactive power into the system. This will reduce the peak power demand to a certain extent. Whenever the EV charging infrastructure has renewable energy sources

(RES) like solar photovoltaic systems, the ESS can be charged at a higher RES generation time and discharged at a lower RES generation time [11].

5.4.2 Distributed FACTS devices

D-FACTS devices are used to reduce various power quality problems in distribution systems. The other salient features of D-FACTS devices are voltage stabilization or regulation, current harmonic mitigation, reactive power compensation, power factor improvement, reduction of active power losses, etc. D-FACTS devices used in modern distribution systems are distributed static compensator (D-STATCOM), distributed thyristor-controlled switched capacitor, distributed switched filter compensator, distributed static series compensator, dynamic voltage restoration (DVR), etc. Voltage source converter-based DSTATCOMs have the capability to enhance voltage, power quality improvement, and power factor improvement, and reduce losses in vehicle to house (V2G), vehicle to house, and grid to vehicle operating modes [12].

5.4.3 Demand response

Demand response enforces end users or customers to consume their normal electricity requirements in a different pattern based on electricity price over a period of time. The main motive of demand response is to reduce power consumption during peak demand, i.e., shifting power consumption from peak hours to off-peak hours. The various methodologies of demand response are peak pricing, time of use (ToU), day ahead pricing, and dynamic pricing. ToU is a widely adapted methodology that prefers the charging of EVs during a predefined or prior specified time when the price of electricity is less. This concept significantly reduces the peak demand created by EV charging [4,11], i.e., it encourages users to charge their EVs during times of reduced power demand.

5.5 Conclusion

EVs are an alternative to conventional vehicles and they are free from CO_2 emissions. Automobile manufacturers are producing EVs and encouraging consumers to switch to them from conventional vehicles. However, EV loads (during charging) vary greatly with respect to type of battery technology used and charging methodology (slow charging, fast charging, and flash charging) adopted. In flash charging mode, EVs are charged in less than a minute by consuming more power from the grid. Highly varying EV charging characteristics have adverse impacts on the distribution network such as overloading, stability problems, power quality problems, poor voltage profile with other users connected in the same distribution network, etc.

References

[1] Z. Wang, X. Guo, J. Li, X. Wang, Impact of voltage sag on electric vehicle charger and critical voltage sag determination, IEEE Trans. Power Deliv. 31 (No. 3) (June 2016) 1397−1399.

[2] CEA Technical Standards for Connectivity of the Distributed Generation Resources, 2019.

[3] Charging Infrastructure for Electric Vehicles (EV) − Revised Guidelines and Standards, Ministry of Power, Government of India, 2019.

[4] A. Dubey, S. Santoso, Electric vehicle charging on residential distribution systems: impacts and mitigations, IEEE Access 3 (September 2015) 1871−1893.

[5] P. Sivaraman, C. Sharmeela, Existing issues associated with electric distribution system, in: B. Kahn, H. Haes Alhelou, Ghassan (Eds.), Handbook of Research on New Solutions and Technologies in Electrical Distribution Networks, IGI Global, Hershey, PA, 2020, pp. 1−31.

[6] P. Sivaraman, C. Sharmeela, Introduction to electric distribution system, in: B. Kahn, H. Haes Alhelou, Ghassan (Eds.), Handbook of Research on New Solutions and Technologies in Electrical Distribution Networks, IGI Global, Hershey, PA, 2020, pp. 1−31.

[7] S. David, T. Le Anh, Impacts of fast charging of electric buses on electrical distribution systems, in: CIRED 24th International Conference on Electricity Distribution, Paper 0802, Glasgow, June 2017, pp. 12−15.

[8] V. Ravindran, selcuk Sakar, S. Ronnberg, H. Math, J. Bollen, Characterization of the impact of PV and EV induced voltage variations on LED lamps in a low voltage installation, Elec. Power Syst. Res. 185 (March 2020) 1−11.

[9] J. Meyer, S. Mueller, S. Ungethuem, X. Xiao, A. Collin, S. Djokic, Harmonic and supraharmonic emission of on-board electric vehicle chargers, in: IEEE PES Transmission & Distribution Conference and Exposition, Morelia, 2016.

[10] A. Espin-Delgado, S. Ronnberg, T. Busatto, V. Ravindran, M. Bollen, Summation law for supraharmonic currents (2−150 kHz) in low voltage installations, Elec. Power Syst. Res. 184 (March 2020) 1−12.

[11] Energy Storage: A Key Enabler for the Decarbonisation of the Transport Sector, European Association for Storage of Energy, Brussels, February 2019.

[12] F.H. Gandoman, A. Ahmadi, A.M. Sharaf, P. Siano, J. Pou, B. Hredzak, V.G. Agelidis, Review of FACTS technologies and applications for power quality in smart grids with renewable energy systems, Renew. Sustain. Energy Rev. 82 (September 2017) 502−514.

[13] P. Sivaraman, C. Sharmeela, IoT based Battery Management System for Hybrid Electric Vehicle, in: A. Chitra, P. Sanjeevikumar, Jens Bo (Eds.), Holm-Nielsen and Himavathi S. & (Eds.), *Artificial Intelligent Techniques for Electric and Hybrid Electric Vehicles*, Wiley-Scrivener publication, 2020.

Further reading

[1] C. Jiang, R. Toruato, Diogo Salles, W. Xu, Method to assess the power-quality impact of plug-in electric vehicles, IEEE Trans. Power Deliv. 29 (No 2) (April 2014) 958−965.

[2] B. Morris, F. Foiadelli, M. Longo, The exploitation of vehicle-to-grid function for power quality improvement in a smart grid, IEEE Trans. Intell. Transport. Syst. 15 (No.5) (October 2014) 2169−2177.

[3] M.K. Gray, W.G. Morsi, Power quality assessment in distribution systems embedded with plug-in hybrid and battery electric vehicles, IEEE Trans. Power Syst. 30 (No. 2) (March 2015) 663−671.

Chapter 6

Impact of power quality issues in residential systems

R. Senthil Kumar[1], I. Gerald Christopher Raj[2], S. Saravanan[1], P. Leninpugalhanthi[1], P. Pandiyan[1]

[1]*Department of Electrical and Electronics Engineering, Sri Krishna College of Technology, Coimbatore, Tamil Nadu, India;* [2]*Department of Electrical and Electronics Engineering, PSNA College of Engineering and Technology, Dindigul, Tamil Nadu, India*

Chapter outline

6.1 Introduction 164
6.2 Power quality disturbances in residential systems 165
6.2.1 Unpredictable disturbances 165
6.2.2 Residential customers 166
6.2.2.1 Overvoltage and undervoltage 167
6.2.2.2 Harmonics 167
6.2.2.3 Voltage flickers 167
6.2.2.4 Voltage sags 168
6.2.2.5 Voltage swells 168
6.2.2.6 Transients 168
6.2.2.7 Voltage unbalance (voltage imbalance) 168
6.2.2.8 Interruptions 168
6.2.2.9 Frequency deviation 168
6.2.3 Effects of power quality in residential systems 169
6.3 Power quality measurement 170
6.4 Study of power quality disturbances caused by a home appliance using MATLAB/Simulink 172
6.4.1 Modeling methodology 172
6.4.2 Power quality impacts due to single-phase nonlinear loads 172
6.4.3 Power quality impacts due to single-phase AC motors starting 174
6.5 Mitigation techniques 181
6.5.1 Power factor correction (compensation) 184
6.5.2 Harmonic filter 184
6.5.3 Power quality-based equipment 186
6.5.3.1 Chokes 186
6.5.3.2 Neutral blocking filter 186
6.5.3.3 Zigzag reactors 186
6.5.4 Surge protection devices 187
6.5.5 Programmable protection devices for household appliances 188
6.6 Conclusion 190
References 190

Power Quality in Modern Power Systems. https://doi.org/10.1016/B978-0-12-823346-7.00009-8

6.1 Introduction

The quality of power supply is critical in power system (PS) networks, specifically in residential systems. It is essential to maintain power quality (PQ) in residential systems to increase the lifespan of household appliances [1]. However, consumers are unaware when household loads create more total harmonic distortion (THD), which affects the PQ [2]. Linear loads, such as heaters and incandescent lamps, do not cause distortion and current is drawn in a sinusoidal waveform. However, nonlinear loads increase distortions and nonsinusoidal waveforms [3].

Today, a large number of nonlinear loads are preferred in home appliances. Harmonic current can interact with supply system impedance, which creates further distortions in the output current and voltage. Furthermore, it develops an effect on all other loads connected in residential systems. As a result, the performance of the system is affected, producing overheating and power losses. Varying residential load demands and disturbances produce voltage and frequency deviations from their normal characteristics [3]. These disturbances affect many home appliances like computers, printers, single-phase AC motors, compact fluorescent lamps (CFLs), and other appliances [4]. Recently, power electronics-based loads operated in residential systems have created supply variations.

Voltage and current waveforms are distorted due to the utilization of semiconductor-based devices, nonlinear loads, and adjustable speed drives [5], [7]. The effects of harmonics are:

1. Reduced performance of electrical appliances, e.g., changing the brightness of a picture.
2. Overheating of mercury arc lamps and fluorescent lamps.
3. Synchronizing problems in converter equipment.

So, it is necessary to maintain the precise quality of power, which increases reliability. To resolve the foregoing list of PQ problems in residential systems, various mitigation techniques are used.

While mitigating PQ problems, the design of the equipment plays a vital role to reduce harmonic currents. This leads to technological advancement such as ballasts in fluorescent lamps, variable speed drives, uninterrupted power supply (UPS) units, battery charges, etc. Also, by incorporating PQ improvement devices such as power factor capacitors, harmonic filters, and surge protection schemes, PQ issues in residential systems can be reduced [6].

In this chapter, the causes and effects of PQ disturbances in residential systems are described. Also investigated is the impact of PQ disturbance by using MATLAB/Simulink. Finally, mitigation techniques for PQ problems in residential systems are addressed.

6.2 Power quality disturbances in residential systems

PQ is commonly defined as the power grid's ability to supply a constant, clean, and stable power supply. The power flow should have a pure sinusoidal waveform and should remain within specified voltage and frequency limits. Fig. 6.1 shows the consequences of harmonic currents generated by nonlinear loads on an ideal PS.

The roots of PQ problems are classified into three types: predictable and small (e.g., harmonics originated by residential customers), incidental and large (e.g., flicker and voltage fluctuation), and predictable and large (e.g., uncharacteristic harmonics and harmonic instability).

6.2.1 Unpredictable disturbances

More than 60% [7] of PQ problems are initiated by natural and unpredictable events by both residential users and electrical services (e.g., lighting surge production, faults, geomagnetically induced currents, resonance, and ferroresonance).

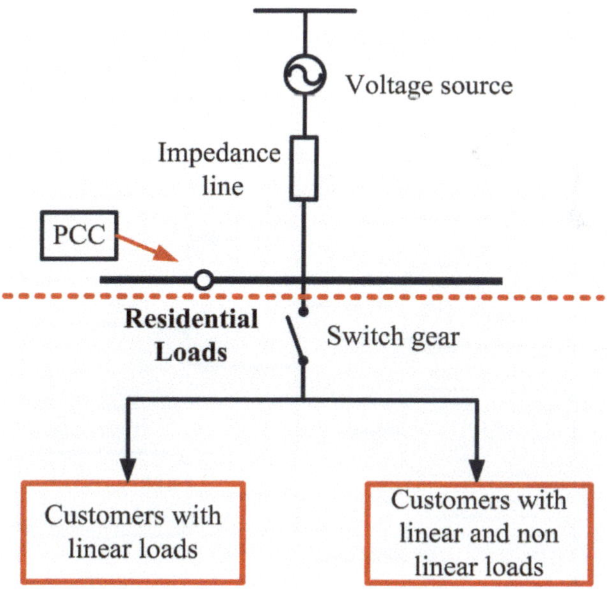

FIGURE 6.1 Generation of harmonics in nonlinear load. *CFL*, Compact fluorescent lamp; *PCC*, point of common coupling; *SMPS*, switch-mode power supply.

6.2.2 Residential customers

In today's PS, a considerable portion of PQ problems is caused by customer loads. Harmonics are one of the main PQ problems created by various loads in residential systems like laser printers, CFLs, laptops, power electronics equipment, and variable speed drives. Other PQ problems are transients (arcing, electrostatic discharge), improper grounding, and poor power factor (inductive loads like induction motors [IMs], air conditioning units). So, it is required that manufactures, technicians, engineers, and PS operators are well informed about the consequence of PQ disturbances. Fig. 6.2 shows an outline of the major categories of PQ problems from a customer's viewpoint.

To summarize the various PQ issues in PS, problems like transients, voltage sags and swells, interruptions, over- and undervoltage, harmonics, flickers, and frequency variations are considered [8] and discussed in the following sections.

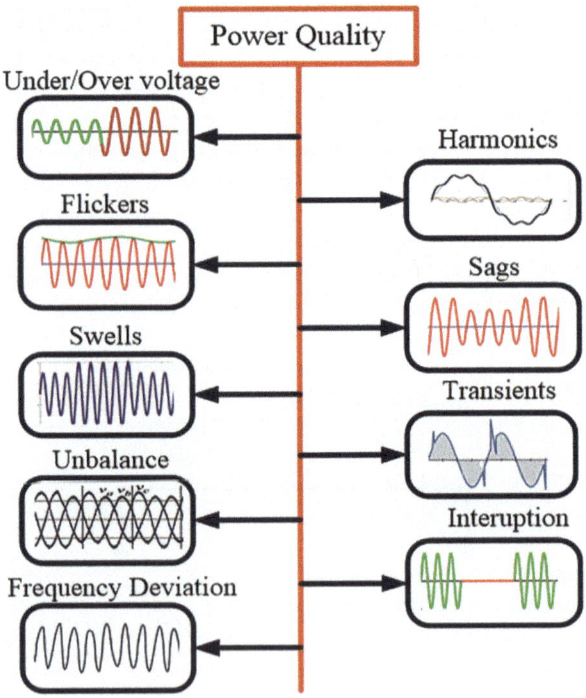

FIGURE 6.2 Typical power quality problems in residential systems.

6.2.2.1 Overvoltage and undervoltage

Electrical home appliances are made to perform at a specific voltage with predefined tolerances. Most of the time, the nominal voltage within acceptable tolerance limits is delivered by the electric utility to residential systems. But it can differ under several uncertain conditions, for example, overvoltage or undervoltage problems appearing in the customer side. The main sources of overvoltage are transmission equipment defects, lightning strikes, accidental tripping of transmission circuits, inaccurate tap settings on transformers, capacitor bank energization, and sudden switching off of bulky loads. This can have a drastic impact on electrical devices used in homes and creates various stresses on the insulation of electrical equipment.

An undervoltage occurs during a period of extensive electrical demands like an increase in air conditioning units running at the same time and the starting of high-power motors. Undervoltage causes voltage uncertainty and leads to increased system losses and poor performance of the equipment. For example, it can make a fan stop or cause a humming noise when running, whereas overvoltage causes sparking in the electric motor. If the voltage falls 10% or 20% below the rated voltage, the motor works less efficiently and if the load connected to it is maximum, then the motor winding will suffer considerable wear and tear as not enough torque is being produced to the load.

6.2.2.2 Harmonics

Due to the heavy utilization of power electronics devices, most loads are nonlinear in residential systems, which is a major reason for generating harmonic currents. Harmonic currents reduce the performance, reliability, and safety of the system.

Prevalent harmonics sources in residential systems are used for power electronics converters, electronic ballast in fluorescent lamps and fan regulators, UPS units, and switch-mode power supplies.

6.2.2.3 Voltage flickers

Flickers normally originate from the use of heavy loads that have abruptly fluctuated in real and reactive power demand. However, with the growth in number of customers and power installations, the flicker issue has risen quickly [9]. Flicker occurs due to heavy inductive loads frequently occurring at the same time, such as IMs and refrigerators. Voltage fluctuation rapidly reduces the performance of household appliances like refrigerators, air conditioning units, and microwave ovens.

6.2.2.4 Voltage sags

Voltage sags are one of the most significant PQ disturbances faced by customers. Voltage sags have an effect on delicate electrical apparatus like communication equipment and laptops. If the number of sag events is high it may lead to loss of data and improper operation of protection equipment. Sag originates from the energization of large loads during initial current carried by refrigerators, air conditioning motors, and compressors.

6.2.2.5 Voltage swells

A voltage swell produces catastrophic effects on the operation of an electrical system and induces flickering. Besides, the age of electrical network connections, insulation degradation of the equipment, and malfunction in power electronic devices are the effects of voltage swell. PQ meters are needed to collect complete information about voltage sags and swells. These meters are used to obtain the root causes of machine problems, capture high-resolution waveforms, and provide an exact detailed event profile.

6.2.2.6 Transients

Transients are generated due to faulty connections in wiring and periodic switching equipment like air conditioning units and compressors. Transients can remove or alter computer data and cause electrical equipment failure, which are all crucial to identify computational errors. In severe cases, transients may destroy electrical equipment and damage electronic circuitry.

6.2.2.7 Voltage unbalance (voltage imbalance)

Voltage imbalance happens due to unequal distribution of single-phase loads from three-phase circuits, which leads to load unbalance. It may also occur when one phase is not working in three-phase motors; in this case the remaining phases meet the high currents. Voltage unbalance creates overheating in windings and enhances additional power losses of three-phase induction machines.

6.2.2.8 Interruptions

The primary sources of electrical interruptions are malfunction in controlling system devices, breaker opening, fuse blowing, and PS faults. It can be classified according to standards and time.

Interruptions can be divided into two types according to European EN 50160 standards [10], namely accidental and preorganized interruptions. Accidental interruptions happen when transient or rapid equipment failure takes place, and restoring the electrical supply can be a lengthy process. In the case of preorganized or planned interruptions, customers are notified before the arrival of a fault.

6.2.2.9 Frequency deviation

Frequency deviation happens between several cycles and for several hours. It has more of an effect in classical electronics equipment used in residential systems, e.g., clocks run faster (or slower), or motor speed increases (or decreases) by a few revolutions.

6.2.3 Effects of power quality in residential systems

The consequences of PQ issues can differ from one device to another due to their age. The quality of electric power plays a predominant role for both customers and electric utilities. The leading issues faced by residential customers are described in Table 6.1.

In general, PQ issues affect equipment used in computers and data communication devices, produce overheating of AC motors, incorrect readings of power meters, and the risk of critical electrical shocks.

TABLE 6.1 Consequences of power quality disturbances in the home.

Sl no.	Power quality disturbances	Consequences in the home
1	Overvoltage/ undervoltage	Damage to lighting filaments and reduced lifespan of motors
2	Harmonics	High electricity bills, blown fuses, maloperation of relays, and overheating in AC motors
3	Flickers	Variations in luminance from illumination devices, and TV screens displaying flicker and data loss
4	Sag	Malfunctioning communication equipment, computer failures, and overheating in induction motors
5	Swell	Computer failures, lights and screens displaying flicker, and sensitive equipment may not perform or be permanently damaged
6	Transients	Damage to highly sensitive components such as microprocessors, insulating coils, TV regulators, and computer chips
7	Voltage unbalance	Negative sequence may be generated into the system and overheating in the windings of induction motors
8	Interruptions	Sensitive electric equipment may cease to function, inefficient data processing or loss of data, protection devices will trip, and if there are long interruptions, all electrical equipment will not perform
9	Frequency deviations	Disturbances in communication and electronic equipment (not a serious issue in residential systems)

6.3 Power quality measurement

Numerous different instruments can be used to measure PQ issues like harmonic analyzers, spectrum analyzers, flicker meters, and energy monitors. To measure PQ disturbances, many indices have been developed and a few are listed in Table 6.2. The most extensively used PQ index is THD, which measures the intensity of current and voltage harmonics to the fundamental component [11].

Table 6.3 shows the current THD of various nonlinear loads in residential systems. From the table, laptops, personal computers, and printers provide more current THD, higher-order current harmonic distortion, and reduced power factor [12].

Fig. 6.3 shows a single line diagram of an electrical wiring system for a 230 V, single-phase, 63 A distribution consumer unit with residual current devices (RCDs). The various nonlinear loads are connected through a miniature circuit breaker to RCDs for the purpose of protection in residential systems.

TABLE 6.2 Power quality indices.

Index	Definition	Main applications
Total harmonic distortion	$\left(\sqrt{\sum_{i=2}^{\infty} I_i}\right) / I_1$	General purpose; standards
Power factor	$P_{total}/\lvert V_{rms}\rvert\lvert I_{rms}\rvert$	Potentially in revenue metering
Telephone influence factor	$\left(\sqrt{\sum_{i=2}^{\infty} w_i^2 I_i}\right) / I_{rms}$	Interference in audio circuits
C message index	$\left(\sqrt{\sum_{i=2}^{\infty} c_i^2 I_i}\right) / I_{rms}$	Interference in communication devices
IT products	$\sqrt{\sum_{i=2}^{\infty} w_i^2 I_i^2}$	Interference in audio circuits, shunt capacitor stress
VT products	$\sqrt{\sum_{i=2}^{\infty} w_i^2 V_i^2}$	Voltage distortion index
Crest factor	V_{peak}/V_{rms}	Dielectric stress
Unbalance factor	$\lvert V_-\rvert/\lvert V_+\rvert$	Three-phase circuit balance
Flicker factor	$\Delta V/\lvert V\rvert$	Incandescent lamp operation

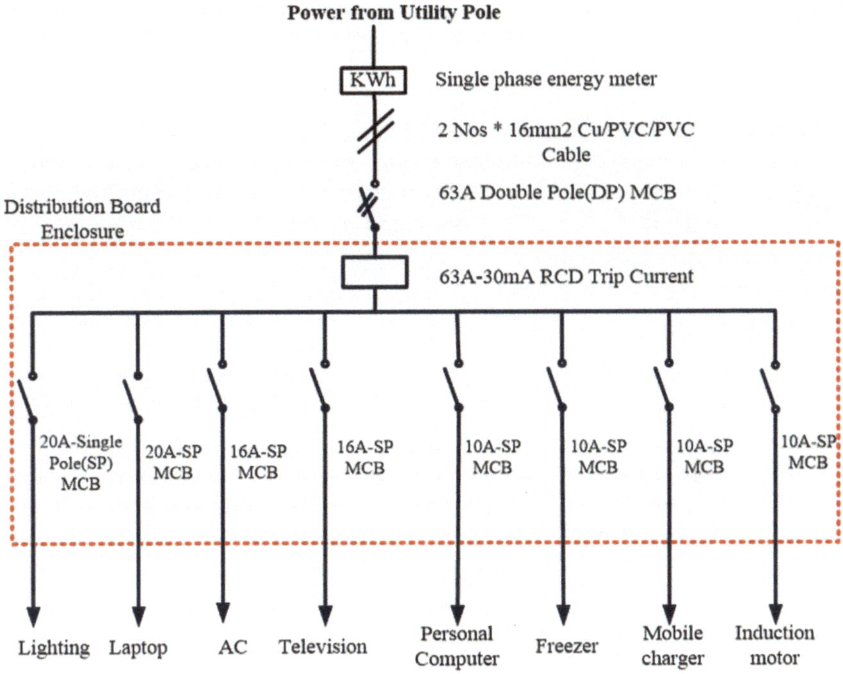

FIGURE 6.3 Single line diagram of residential systems. *MCB*, Miniature circuit breaker; *PVC*, polyvinyl chloride.

TABLE 6.3 Comparison of total harmonic distortion (THD) values in various nonlinear loads.

S. no.	Appliance	Power factor	THD (%)	3rd (%)	5th (%)	7th (%)	9th (%)	11th (%)
1	Laptop	0.52	86.9	49.5	44.1	36.2	27.5	17.7
2	Personal computer	0.53	80.2	52.8	43.5	31.5	19.1	8.5
3	Printer	0.49	83.7	46.8	41.2	36.1	28.5	21.4
4	Television	0.66	72.3	55.1	36.8	20.3	11.5	10.9
5	Freezer	0.44	61.8	11.1	4.8	11.2	7.1	7.2
6	Fluorescent lamp	0.57	11.3	10.7	2.1	1.8	0.96	0.6
7	Amplifier	0.71	48.1	32.2	30.7	14.2	7.6	2.7

6.4 Study of power quality disturbances caused by a home appliance using MATLAB/Simulink

This section demonstrates a widespread set of MATLAB/Simulink models used to study different PQ issues caused by home appliances. The simulation models presented include nonlinear loads (home appliances like a personal computer, laptop, power bank, mobile charger, CFL, light-emitting diode [LED], smart LED TV, etc.). The presented models are used to simulate and study different PQ disturbances.

6.4.1 Modeling methodology

MATLAB has a Simscape platform, which has a collection of specific block libraries and distinctive simulation attributes for simulation of a physical systems model. In this chapter, simulation is performed on MATLAB/Simulink. Developed modeling is used to simulate different PQ issues and identify how the disturbances affect the pure sinusoidal waveforms in different nonlinear loads [13]. Models have been designed with fewer blocks in mind and use default settings where possible for reader reproducibility purposes.

6.4.2 Power quality impacts due to single-phase nonlinear loads

The single phase with power electronics components (nonlinear load model) has been developed using Simulink, which is represented in Fig. 6.4. This model is used to identify triplen harmonics problems in nominal voltage, which are produced by a single-phase AC to DC converter with capacitance filters. Also, it is typically found in home appliances. The parameters used in the single-phase nonlinear loads are listed in Table 6.4.

The simulation response of the harmonic waveforms in a high-voltage (11 kV) and low-voltage (0.4 kV) bus is shown in Fig. 6.5. Simulation runtime is set for 0.04 s to visualize the two cycles of source and load side waveforms.

From the results, at 0.4 kV the harmonic distortions are considerable at 60 and 240 degrees of every cycle. However, harmonic distortion at the 11 kV bus has been critically suppressed using a delta/wye transformer.

THD is injected with the help of the nonlinear loads at all phases, and the simulation time is applied up to 0.2 s to examine 10 cycles. The power graphical user interface block with a fast Fourier transform analysis is used to analyze the harmonic content. The harmonic analysis of the high-voltage and low-voltage bus with full load conditions is shown in Figs. 6.6. and 6.7. The voltage and current THD value of the high-voltage bus is very low when compared to the low-voltage bus.

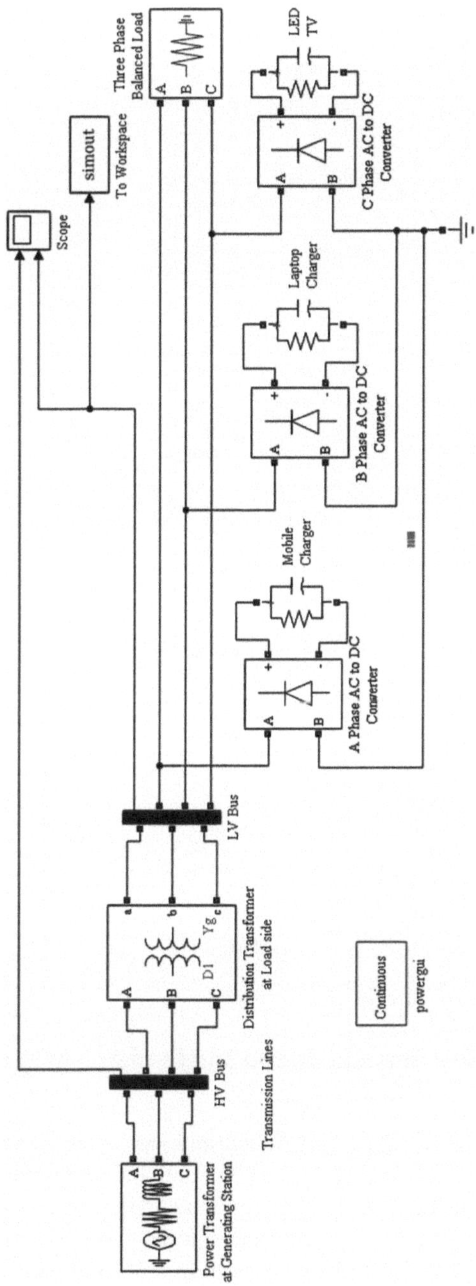

FIGURE 6.4 Simulink model of single-phase nonlinear loads. *HV*, High voltage; *LED*, light-emitting diode; *LV*, low voltage.

TABLE 6.4 Simulink block parameters of single-phase nonlinear loads.

S. no.	Power system component	Simulink block	Block parameters
1	Power transformer at generating station	Three-phase source	30 MVA, 11 kV, and 50 Hz [14]
2	Distribution transformer	Delta/wye transformer	1 MVA, 1 kV/ 0.4 kV,
3	Mobile charger connected in A-phase	Resistive loads, single-phase uncontrolled bridge block and capacitor	2000 μF, 10 Ω
4	Laptop charger connected in B-phase		220 μF, 1 kΩ
5	LED TV charger connected in C-phase		22 μF, 1 kΩ

LED, Light-emitting diode.

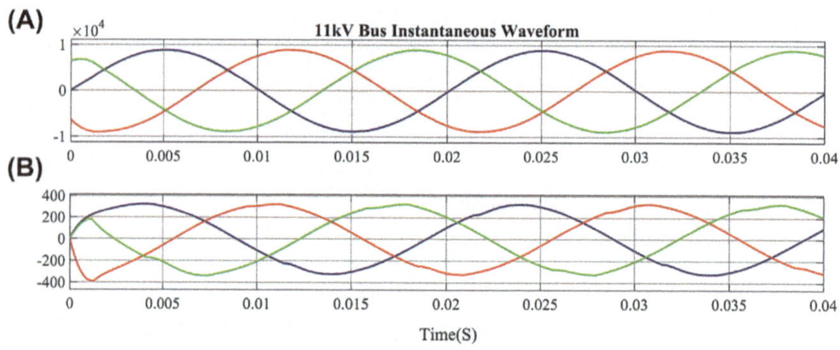

FIGURE 6.5 Simulated output of instantaneous waveforms: (A) 11 kV bus and (B) distorted waveforms low voltage (0.4 kV).

6.4.3 Power quality impacts due to single-phase AC motors starting

The starting model of an IM is developed and simulated to study the voltage transients and sag. The starting of heavy rated induction machines and parameters is shown in Table 6.5. The developed Simulink model is shown in Fig. 6.8. In this simulation, a three-phase squirrel cage IM and the following single phase motors like split phase, capacitor start, and capacitor-run IM are considered for residential loads. Performance of the machines is validated from 0 to 0.4 s.

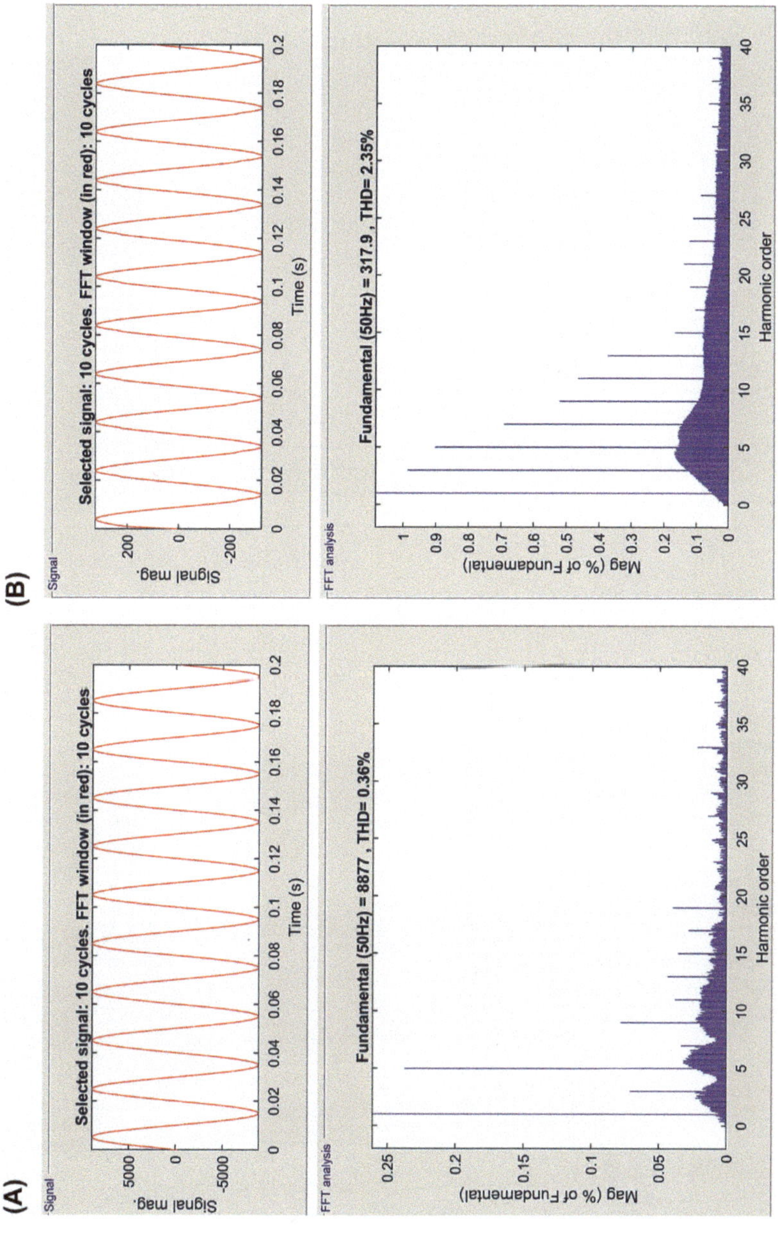

FIGURE 6.6 Voltage total harmonic distortion (THD) when all loads were running: (A) Phase "A" THD on high-voltage side, (B) Phase "A" THD on low-voltage side, (C) Phase "B" THD on high-voltage side, (D) Phase "B" THD on low-voltage side, (E) Phase "C" THD on high-voltage side, (F) Phase "C" THD on low-voltage side. *FFT*, Fast Fourier transform.

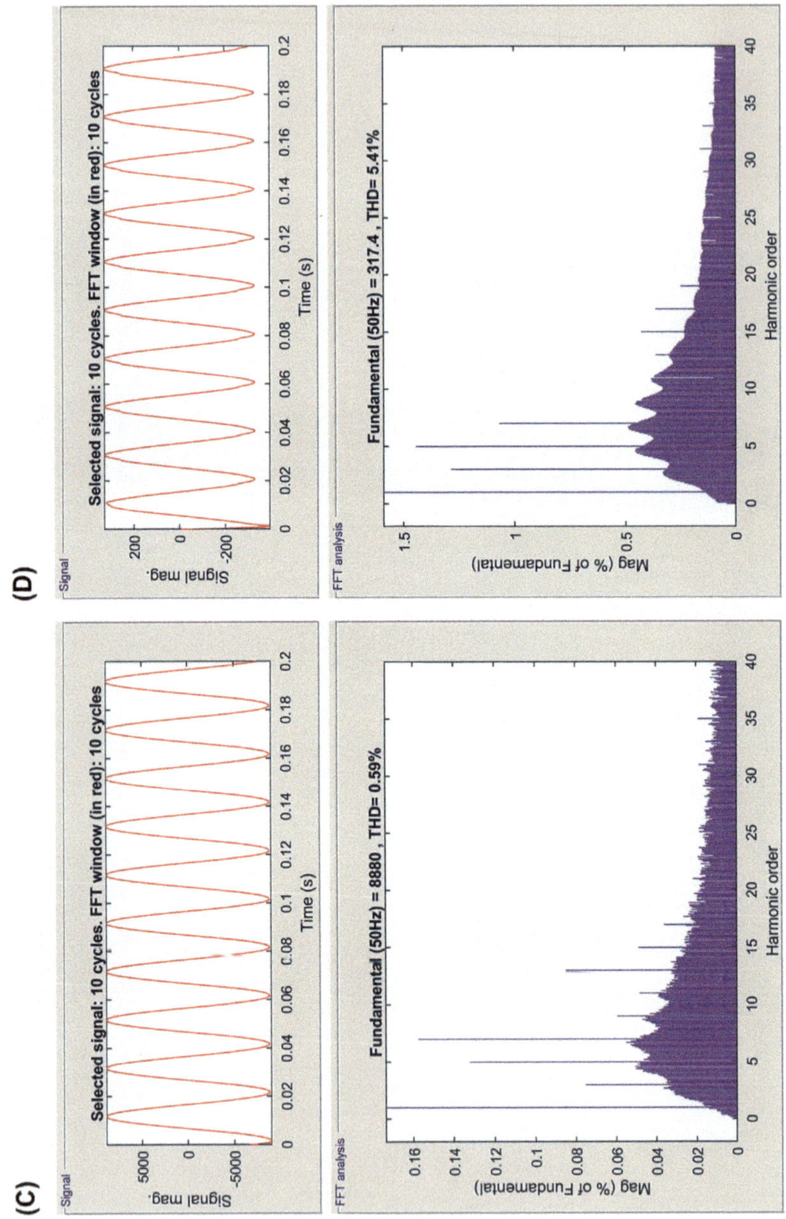

FIGURE 6.6 cont'd

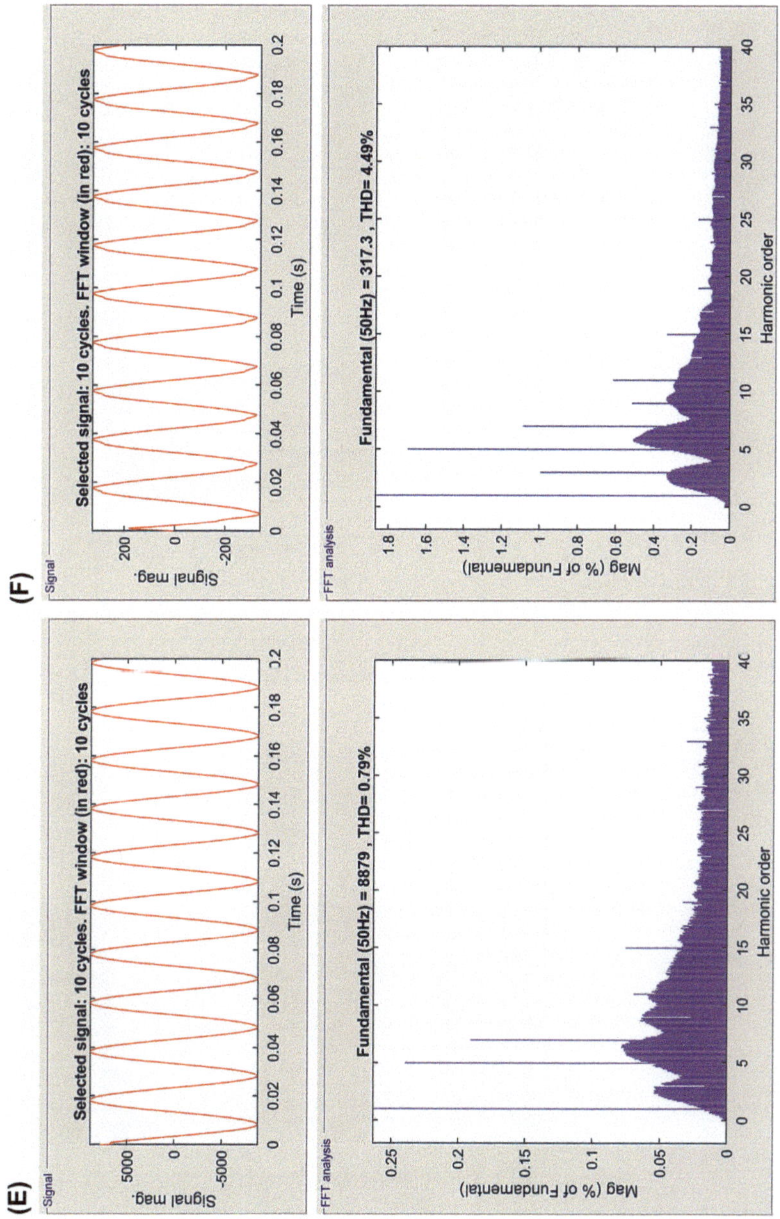

FIGURE 6.6 cont'd

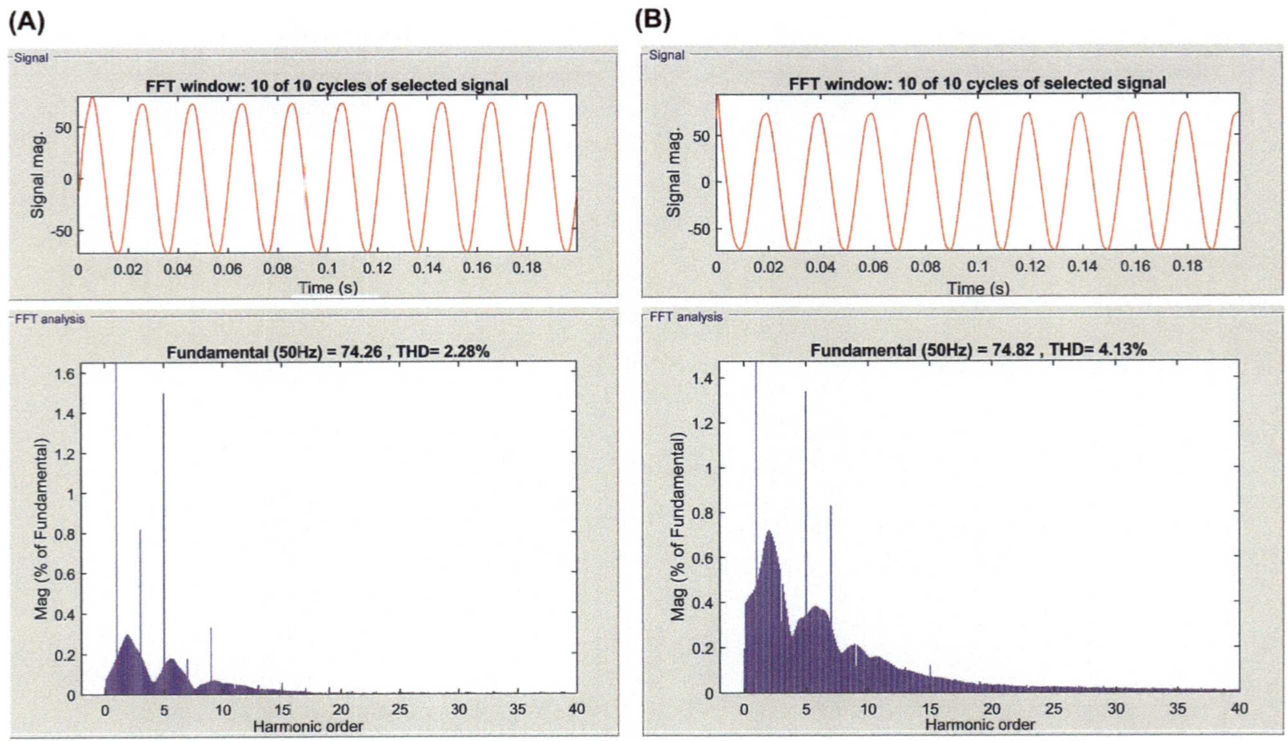

FIGURE 6.7 Current total harmonic distortion (THD) when all loads were running (A) Phase "A" THD on high-voltage side (B) Phase "A" THD on low-voltage side (C) Phase "B" THD on high-voltage side (D) Phase "B" THD on low-voltage side, (E) Phase "C" THD on high-voltage side, (F) Phase "C" THD on low-voltage side. *FFT*, Fast Fourier transform.

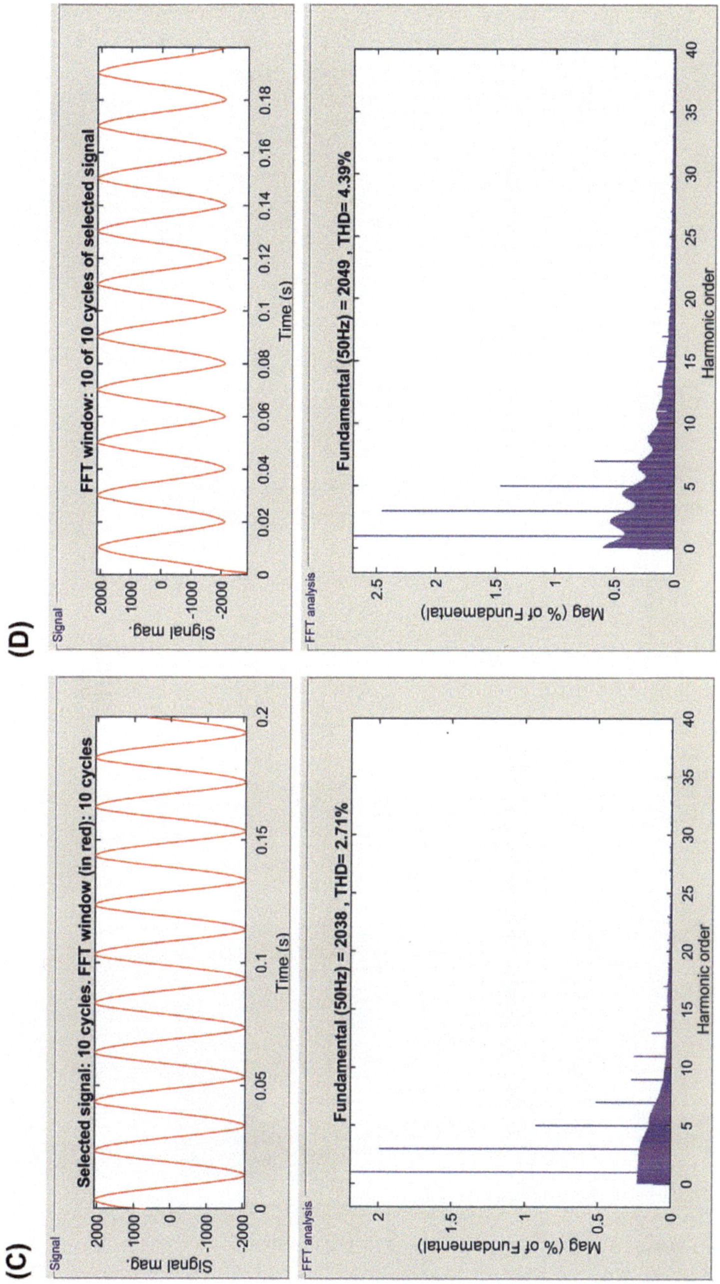

FIGURE 6.7 cont'd

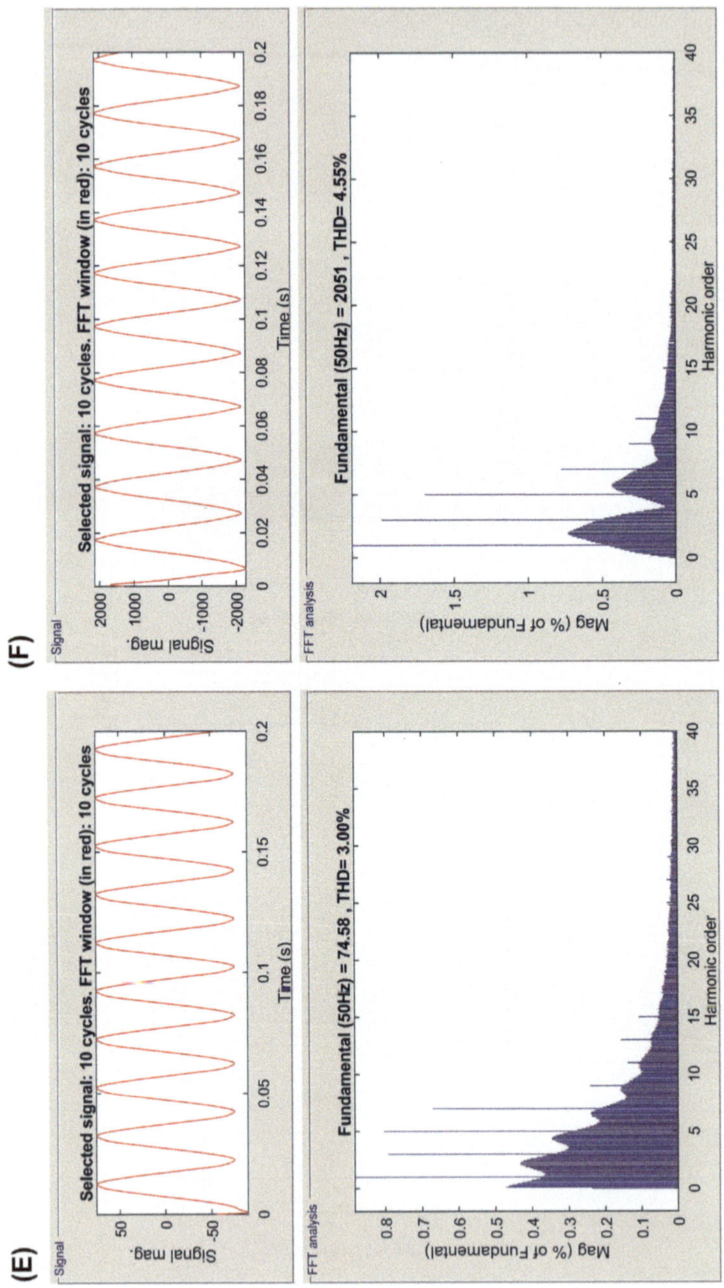

FIGURE 6.7 cont'd

TABLE 6.5 Simulink block parameters of an induction motor (IM) during starting/loading.

S. no.	Component	Simulink block	Block parameters
1	Power transformer at generating station	Three-phase source	30 MVA, 11 kV, and 50 Hz [15]
2	Distribution transformer	Delta/wye transformer	1 MVA and 11 kV/0.4 kV
3	Circuit breaker or switch	Three-phase breaker	Initial status = open Switching times = 0.1 s
4	3-Phase induction motor (industrial)	Asynchronous machine	Squirrel-cage preset model 19: 100 hp
5	1-Phase induction motor Residential applications motor	Asynchronous machine (single phase)	Single-phase, split-phase, capacitor start, and capacitor-run IM

Fig. 6.9A and B shows the voltage sag instantaneous waveform in the starting of a three-phase IM, which is generated by a preset model 75 kW (100 hp). During starting, the speed is set at 1000 rpm and output voltage is analyzed between 0.2 and 0.4 s. From 0.2 s, the voltage transients/sag are observed at the 0.4 kV bus compared to the 11 kV bus. The voltage sag reduces in magnitude and propagates poorly in an 11 kV feeder.

The starting of a three-phase IM voltage sag is generally stabilized and has a shallow decrease up to 34 V in magnitude. The voltage sag in RMS waveform is clearly visualized and demonstrated in Fig. 6.10A and B. Fig. 6.10A shows the RMS voltage waveform of a low-voltage bus where all three- and single-phase motors are running. When the machine is started, a voltage sag is developed between 0.2 and 0.3 s. Fig. 6.10B shows the RMS voltage waveform of a low-voltage bus where only a three-phase motor is running. When the motor is started, a sag is developed at 0.27 s. The magnitude of the sag is much reduced when compared to all loads in simultaneous running conditions. Based on the power rating of machines, the magnitude of sag is different in IMs. A higher power rating IM leads to less sag in magnitude.

6.5 Mitigation techniques

If residential applications cause PQ issues, then this raises utility and operating costs. For a three-phase PS, PQ problems mainly occur because of disturbance to parameters like frequency, magnitude, or waveform. PQ problems can cause equipment failures and unexpected and expensive outages. PQ in residential

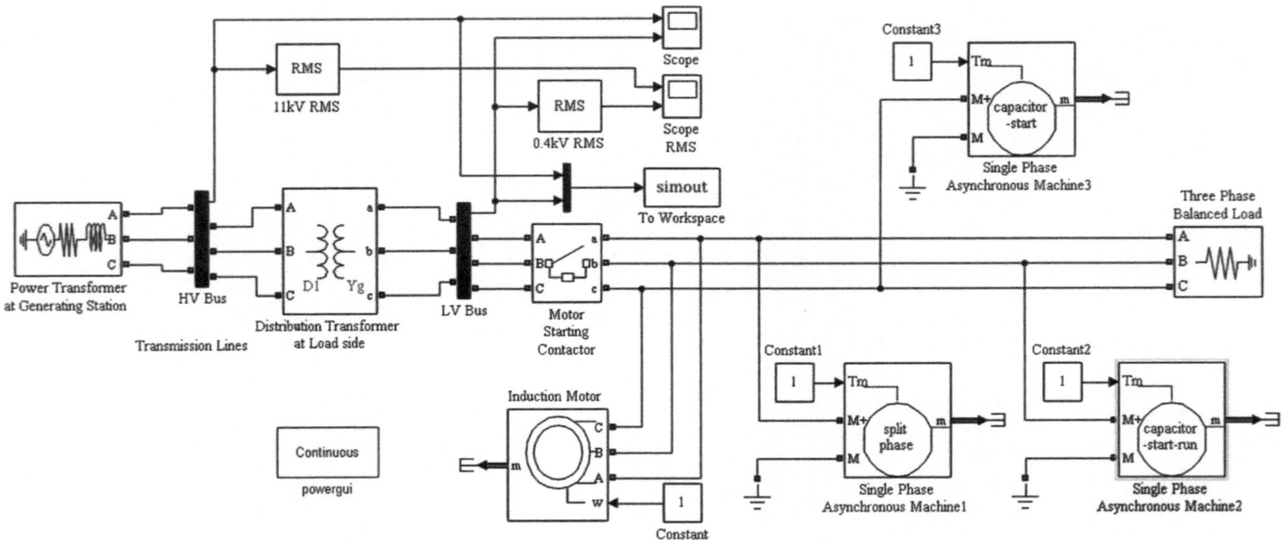

FIGURE 6.8 Simulink model for the starting of industrial and residential induction motors. *HV*, High voltage; *LV*, low voltage; *RMS*, root mean square.

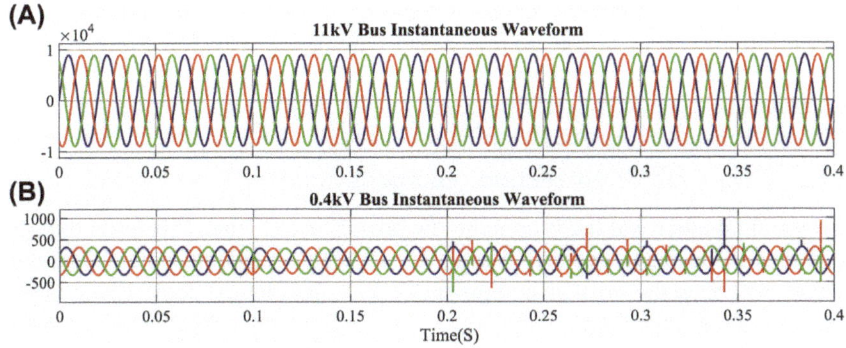

FIGURE 6.9 Instantaneous waveform of line-to-line voltages with transients: (A) 11 kV and (B) 0.4 kV bus instantaneous waveform.

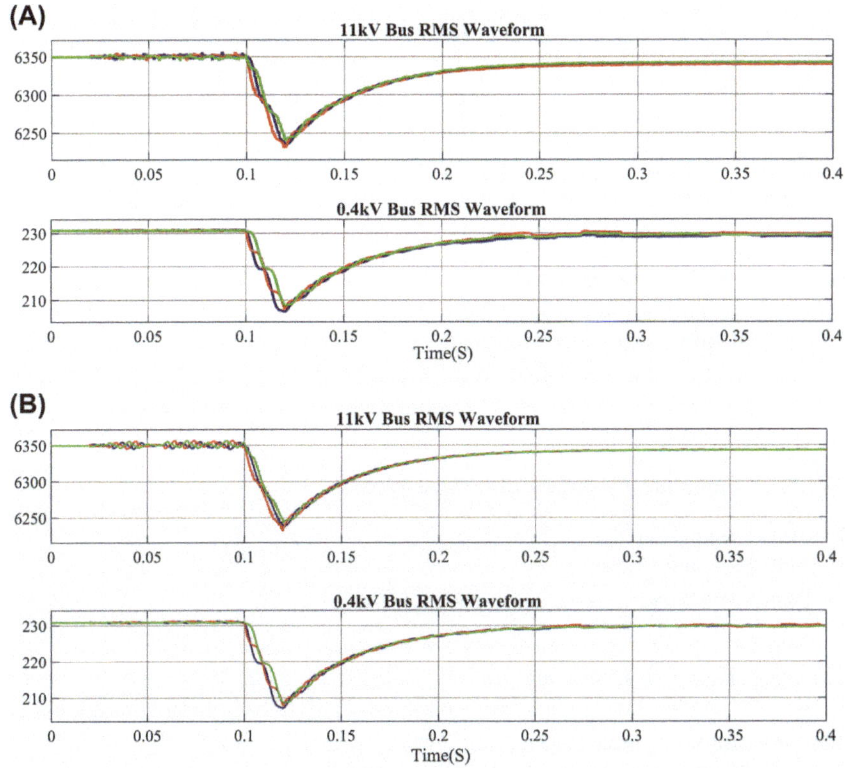

FIGURE 6.10 Root mean square (RMS) waveform of (line-to-line) voltage sag generated by the starting of an induction motor: (A) when all three- and single-phase motors are running and (B) when the three-phase motor is running and single-phase motors are not running.

systems can be improved using mitigation techniques like power factor correction and other power quality-based equipment. The foregoing two solutions are affirmed for maintaining acceptable energy costs. As a result, the reduction in low power factor penalties and energy savings can be achieved.

6.5.1 Power factor correction (compensation)

If residential electrical loads contain low power factors, then this leads to huge increases in power to perform a task. Thus energy consumption is increased, which increases the cost of equipment or installations. A poor power factor creates high voltage drops in the power distribution network. Generally, power suppliers penalize industries that operate power factors below the corresponding value.

Power factor correction reduces power demand by consuming equipment and improves total power factor. A good power factor is achieved by using a compensation network maintained between 0.95 and 0.98. If a power factor operates at 0.85 or less, it is commonly considered to be poor [16].

6.5.2 Harmonic filter

Commonly, shunt-type harmonic filters are used in PSs because they provide low impedance paths for currents to ground on one or more harmonic frequencies. Compared to series filters, shunt filters are preferable based on the economic aspects of PS applications.

1. A series filter that is connected in series requires high component sizes, which is costly. Series filter components are chosen based on their full current rating and power frequency.
2. A shunt filter that is connected in parallel requires small component sizes, which is economical. Shunt filter components are chosen based on system voltage.

Shunt filters are classified into three types:

1. Single-tuned filters
2. Multiple-tuned filters
3. Damped filters

Single- and double-tuned filters are commonly used to filter specific frequencies. Damped filters are mostly utilized to filter extensive frequency ranges. If residential applications contain small harmonic-generated loads, they are filtered by employing single-tuned filters to remove the problematic harmonic currents. For bulky applications, multituned and damped filters are mostly utilized. Fig. 6.11 shows single- and double-tuned filter circuit diagrams and Fig. 6.12 illustrates a damped filter-type circuit diagram.

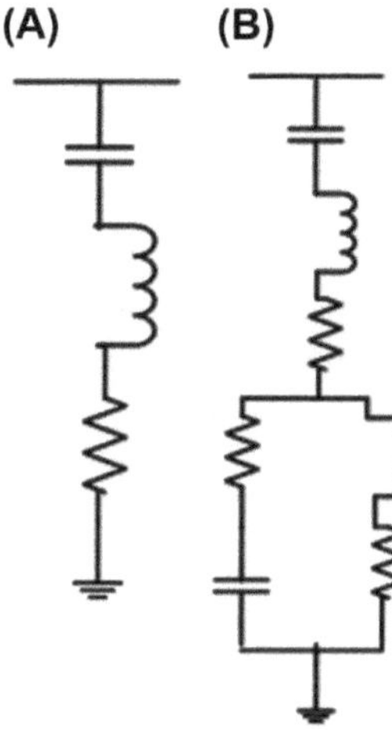

FIGURE 6.11 Harmonic filters: (A) single tuned and (B) double tuned.

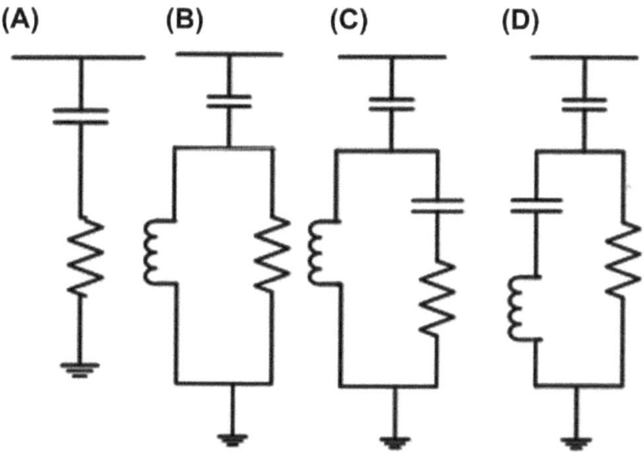

FIGURE 6.12 Damped filter: (A) first order; (B) second order; (C) third order; (D) C-type damped filter.

The damped filter type is mostly used to control higher-order harmonics in the network. It contains higher resistance than single- and double-tuned filters, so this type of filter is not used to remove harmonics near a power frequency. Commonly, damped filters are used to reduce the 11th and 13th, 17th, 19th, etc. pairs of harmonics.

6.5.3 Power quality-based equipment

The following solution can enhance PQ but does not deal with normal energy saving.

6.5.3.1 Chokes

To improve PQ, it is not necessary to use costly devices; instead, a three-phase choke connected in series to the network for the purpose of harmonic trap and phase balance. Compared to digital power conditioning systems, the choke operates at a low level but satisfies a large number of consumers without incurring extra cost.

6.5.3.2 Neutral blocking filter

This type of filter is used to block triple frequency harmonics by connecting the neutral conductor between the transformer and the three-phase load, as shown in Fig. 6.13. Instead of an individual phase filter, the neutral blocking filter is used to block triple zero-sequence harmonics in each phase and it is also more economical [17].

6.5.3.3 Zigzag reactors

Zigzag reactors are based on incorporating phase shifting into a single- or multiphase transformer through a very low zero-sequence impedance, which considerably eliminates third, fifth, and seventh order harmonics. This method

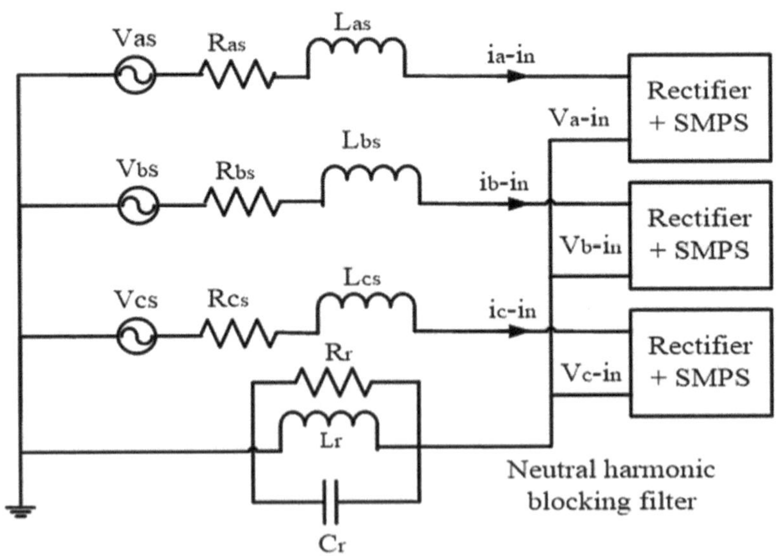

FIGURE 6.13 Neutral blocking filter circuit diagram. *SMPS*, Switch-mode power supply.

is another way to safeguard the transformer neutral conductor from third order harmonics by removing these harmonics near the load. An autotransformer is shunt connected to the supply, which gives a zero-sequence current path to block and eliminate third order harmonics, as illustrated in Fig. 6.14 [17].

6.5.4 Surge protection devices

Surge protective devices (SPDs) are used to safeguard electrical devices from overvoltage, as shown in Fig. 6.15. SPDs are fixed in three locations in residential applications:

- Point of entry protection—meter board devices
- Panel mount suppressors
- Plug-in surge suppressors

The purpose of the SPD is to reduce overvoltage that occurs in normal electrical power, which is supplied to the load. A general source of overvoltages produced inside a building are devices sensed by using a thermostat operating a switch to power on and off. In total, 60%–80% of overvoltages are produced inside a facility. These surges have limited energy but can cause damage to household equipment. Surges that occur from outside buildings are

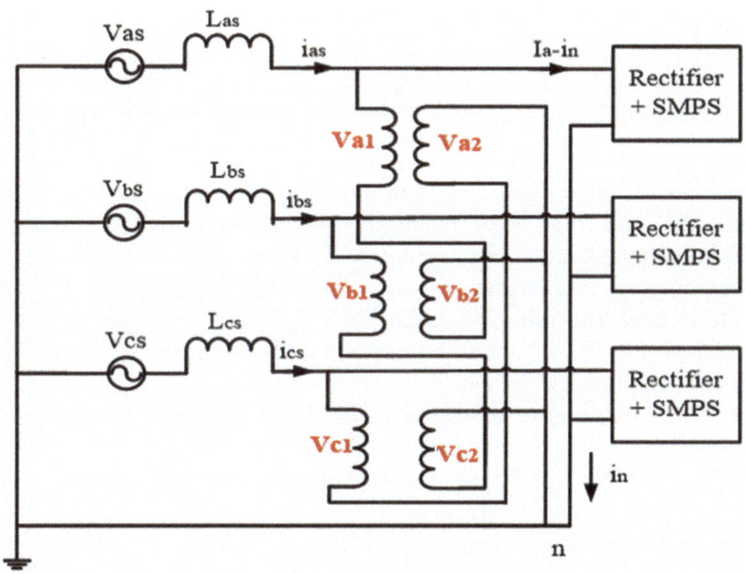

FIGURE 6.14 Zigzag reactor circuit diagram. *SMPS*, Switch-mode power supply.

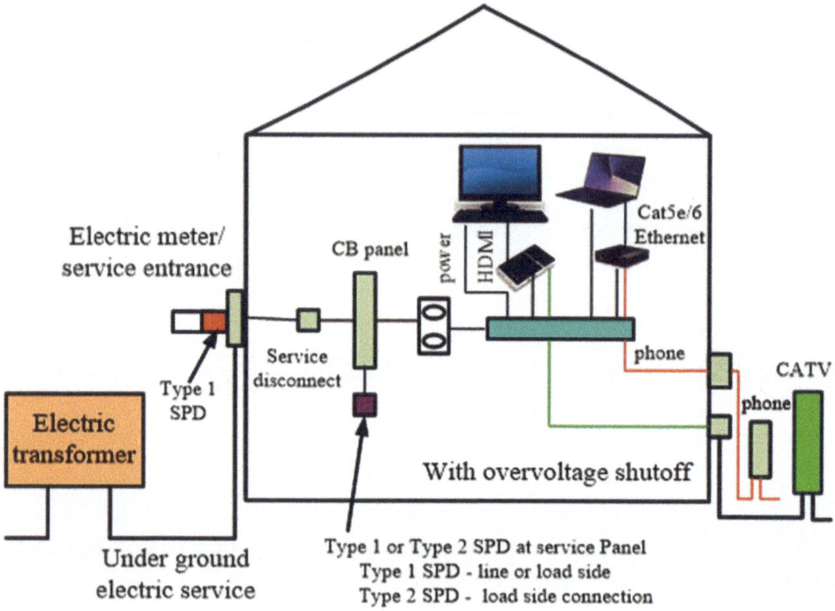

FIGURE 6.15 Surge protective device (SPD)-based household applications. *CATV*, Cable television; *CB*, Circuit breaker; *HDMI*, high definition multimedia interface.

caused by the switching of utility grids and lightning. These external overvoltages rarely arise, but can cause severe damage to equipment compared to internal sources.

6.5.5 Programmable protection devices for household appliances

A programmable protection device is used to measure voltage deviation and maintain the set point depending upon the low- and high-voltage level, which prevents the load from being damaged. Fluctuation in the AC mains itself occurs in various domestic and commercial systems.

The tripping mechanism protects household appliances such as refrigerators, fans, and other electrical appliances from damage, as shown in Fig. 6.16. It consists of a four-comparator integrated circuit. In addition, a couple of comparators are also used as window comparators. If the system produces an error due to deviation in the input voltage above the window range, it triggers the relay to cut off the load (fan, lamp, etc.). Due to this operation, high PQ will be ensured and it also improves the power factor. Obviously, power usage is maximized.

The protection circuit depicted in Fig. 6.16 offers protection against frequently occurring phenomena such as overvoltage, undervoltage, and

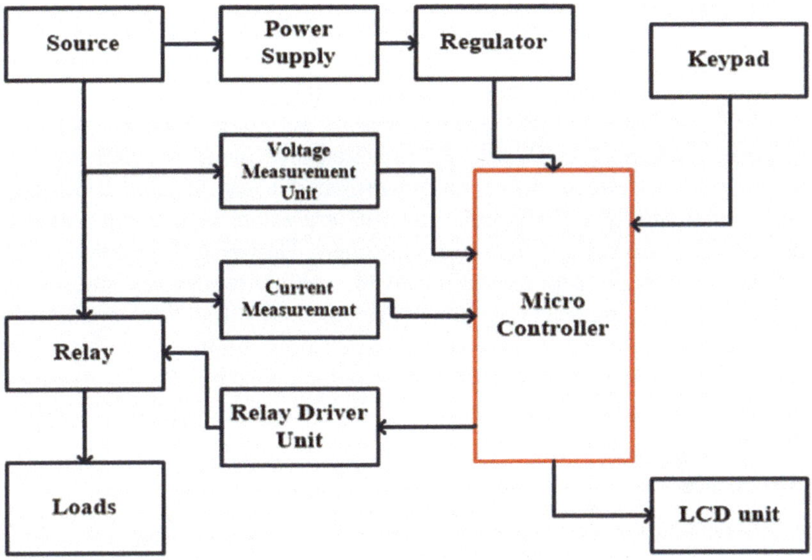

FIGURE 6.16 Block diagram of a programmable protection scheme. *LCD*, Liquid crystal display.

overcurrent for single-phase systems. In general, the single-phase operating voltage is considered to be in the range 220–230 V with ±10% tolerance. This AC voltage is converted into DC voltage by means of a rectifier with an RLC filter. The filter employed with the rectifier is used to suppress the ripples available in the DC output voltage. The output of the filter is given to the regulator in which the output voltage varies up to 440 V. The regulator supplies the power to the microcontroller and relay even though there is no variation in input voltage. The output voltage of the regulator is 12 V. If overvoltage/current fault occurs at the supply side, it is converted into a digital signal that is recognized by the microcontroller through a potential/current transformer as well as analogue-to-digital converter. The reference voltage/current is fed to the microcontroller through keypads. The microcontroller checks whether the voltage/current measured concurrently through the potential/current transformer is smaller or greater than the reference voltage/current; it then provides a decision signal to actuate the relay. The liquid crystal display utilized in this circuit shows readings of voltage and current. In addition, the PQ meter monitoring system is needed to protect residential appliances from unbalanced loads or harmonics.

Hence, these mitigation techniques and protection schemes are used to maintain a high quality of power and improve the power factor, thus the power is fully utilized for residential applications.

6.6 Conclusion

In this chapter, different PQ disturbances in residential systems were described, and provided simple solutions to safeguard household loads. To understand the impact of PQ disturbances in the home, various loads were considered, and their THD, as well as voltage variation of both the high-voltage and low-voltage sides, was investigated using MATLAB/Simulink. Specific standards like IEEE, EN, and IEC are established to minimize PQ issues and propose various filters for customers. However, the main obstacles faced by residential customers to install PQ mitigation devices are complex control, high cost, increasingly larger devices, and difficulty realizing huge power levels. Therefore considerable research opportunities in this field and accompanying concerns need to be considered to achieve pollution-free power and zero interruption, which is necessary for today's residential systems.

References

[1] A. Grevener, et al., Survey of supraharmonic emission of household appliances, CIRED-Open Access Proc. J. 1 (2017) 870–874.

[2] G. Ye, et al., Stochastic residential harmonic source modeling for grid impact studies, Energies 10 (3) (2017) 372.

[3] S. Jain, Power Quality: An Introduction. Modeling and Control of Power Electronics Converter System for Power Quality Improvements, Academic Press, 2018, pp. 1–29.

[4] C.K. Neha, et al., Effect of home appliances on power quality of conventional grid, in: International Conference on Circuits, Controls, Communications and Computing (I4C). IEEE, 2016, 2016.

[5] A. Ghosh, G. Ledwich, Power Quality Enhancement Using Custom Power Devices, Springer science & business media, 2012.

[6] S.H.E.A. Aleem, Z. Ahmed F, M.E. Balci, Optimal resonance-free third-order high-pass filters based on minimization of the total cost of the filters using Crow Search Algorithm, Elec. Power Syst. Res. 151 (2017) 381–394.

[7] E. Fuchs, M.A.S. Masoum, Power Quality in Power Systems and Electrical Machines, Academic press, 2011.

[8] A.F. Zobaa, A., Shady, E.B. Murat, in: Power System Harmonics: Analysis, Effects and Mitigation Solutions for Power Quality Improvement, BoD—Books on Demand, 2018.

[9] D. Geiger, M. Arechavaleta, S. Mark Halpin, Evaluating voltage fluctuations and flicker using half-cycle RMS values, in: IEEE International Conference on Environment and Electrical Engineering and 2017 IEEE Industrial and Commercial Power Systems Europe (EEEIC/I&CPS Europe). IEEE, 2017, 2017.

[10] 50160: 2010+ A1: 2015 B.S. EN, Voltage Characteristics of Electricity Supplied by Public Electricity Networks, BSI, UK, 2010.

[11] J. Stones, A. Collinson, Power quality, Power Eng. J. 15 (2) (2001) 58–64.

[12] J. Pinzón, A. Pedraza, F. Santamaría, Harmonic impact of non-linear loads in a power distribution system, in: Conferência Brasileira sobre Qualidade da Energia Elétrica (CBQEE). São Paulo. Brasil, 2013.

[13] J. Meyer, S. Haehle, P. Schegner, Impact of Higher Frequency Emission above 2 kHz on Electronic Mass-Market Equipment, 2013, 0999–0999.

[14] R.H.G. Tan, V.K. Ramachandaramurthy, H.G. Rodney, A Comprehensive Modeling and Simulation of Power Quality Disturbances Using MATLAB/Simulink. Power Quality Issues in Distributed Generation, InTech, Rijeka, 2015, pp. 83–107.
[15] M.H.J. Bollen, L.D. Zhang, Different methods for classification of three-phase unbalanced voltage dips due to faults, Elec. Power Syst. Res. 66 (1) (2003) 59–69.
[16] https://passive-components.eu/an-introduction-to-capacitor-based-power-factor-correction-circuits/.
[17] H.A. Kazem, Harmonic mitigation techniques applied to power distribution networks, Advan. Power Electron. 2013 (2013).

Chapter 7

Voltage control in active distribution networks

Fco. Javier Zarco-Soto, José L. Martínez-Ramos, Pedro J. Zarco-Periñán
Department of Electrical Engineering, University of Seville, Seville, Spain

Chapter outline

7.1 Introduction **193**
7.2 Voltage quality requirements **194**
7.3 Traditional voltage control strategies **196**
7.4 Voltage control strategies in active distribution networks **199**
7.4.1 Transmission and distribution systems: two different scenarios 201
7.4.2 Novel voltage control strategies 203
7.4.3 Case study 206
7.4.3.1 Base case: passive network 207
7.4.3.2 High distributed generation and low demand 208
7.4.3.3 High demand due to electric vehicles and low generation 212
7.5 Conclusions **214**
References **215**

7.1 Introduction

Electric power systems were conceived more than a century ago to transport energy over long distances. From large power plants that may produce gigawatts of power, electricity traveled to reach end users. Around this concept, massive infrastructures that have stood the test of time were developed to satisfy increasing electricity demands. These huge systems are usually divided into transmission and distribution networks according to the amount of energy to be transmitted. The aim of the former is to shorten distances between generators and end users, whereas the latter is in charge of delivering the end product to consumers at adequate voltage.

However, the whole electric system is evolving to successfully accomplish new requirements regarding resource and energy efficiency. Renewable energy sources have shown to be one of the most sustainable ways of generating

Power Quality in Modern Power Systems. https://doi.org/10.1016/B978-0-12-823346-7.00003-7

electrical energy and distribution networks are facing large-scale penetration of distributed generation. In addition, urban transportation is pivoting to replace combustion vehicles with electric vehicles that may be charged through distribution systems. These mobile loads are particularly heavy and concentrated in short periods of time.

This new paradigm contrasts with traditional distribution networks that were initially developed to supply energy only to passive loads. The relatively low energy managed in conjunction with the large amount of equipment and number of clients led to simplifying the operation that was mainly solved during the planning stage. However, the increasing penetration of distributed generation and electric vehicles complicates the operation since the network becomes active. Thus novel control algorithms are required that consider both traditional equipment and the new devices introduced to achieve effective integration of new energy sources while avoiding costly investments. Along these lines, this chapter analyzes the ongoing transition to modern distribution systems within the scope of voltage control.

7.2 Voltage quality requirements

Electric power system planning is focused on defining and sizing the necessary facilities to reach all users according to certain quality standards. During real-time operation, all the electrical magnitudes must be maintained close to their nominal values despite the effects of demand evolution and any unpredictable event [1]. Since electric systems are dynamic by nature, upper and lower bounds are defined for the essential parameters and thus both planning and operation are focused on maintaining them within limits as well as correcting any deviation.

Among all the parameters that model energy systems, the most important magnitude is the voltage of the system. International and national standards regulate the characteristics of the wave within the network and that supplied to customers, which include frequency, amplitude, and harmonic distortion. Two of the most widespread standards to define voltage characteristics are EN 50160 [2], mainly in Europe, and ANSI C84.1 [3], in America. Both set main features during normal operation along with the expected bounds within which voltage must remain; however, the main difference is the nominal frequency of the system. The former is developed for 50 Hz and the latter is focused on 60 Hz. Other differences also arise from considering the idiosyncrasy of each system, for instance, the nominal low voltage of the standard EN 50160 is 230 V, whereas the most common values given by ANSI C84.1 are 120 and 480 V.

On the other hand, the events that affect the network could result in transient disturbances or modify the steady state according to their nature, e.g., lightning surge is considered an impulse since the frequency of the resulting wave ranges between 10 kHz and 3 MHz [4], whereas load reduction lasts for hours. In this way, the objective of voltage control is to correct any deviation with respect to the amplitude bounds in steady state.

Standard EN 50160 limits voltage variation around the nominal value in percentage. The same maximum voltage variation is predefined for medium and low voltage that must not exceed the range ±10%. However, these bounds are relaxed for distribution networks that are not connected to the transmission system or for special remote users. Standard ANSI C84.1 distinguishes between service voltage and utilization voltage depending on the element connected at the corresponding node. The former is the voltage at the point where the electrical systems of the supplier and the user are connected, whereas the latter is defined as the voltage at the line terminals of utilization equipment. Two ranges are defined to limit deviations with respect to their nominal value. Roughly speaking, range A sets the normal operating values of electric supply systems and user systems so that out-of-limit voltages are infrequent. Range B determines a wider range than A to consider practical design and conditions on supply or user systems, or both. When these values occur, corrective measurements need to be undertaken to meet range A again.

In contrast to EN 50160, the limits provided by ANSI C84.1 are given as absolute values depending on the nominal voltage. Since range A determines the normal operating limits, the values given for the most common medium-voltage and low-voltage systems can be rewritten as percentages to compare the resulting ranges with standard EN 50160, as shown in Table 7.1. It is worth mentioning that, although both standards define low voltage for system nominal voltages less than 1 kV and medium voltage larger than 1 kV, EN 50160 considers medium voltage for nominal voltages up to 36 kV and ANSI C84.1 up to 100 kV.

TABLE 7.1 Maximum and minimum voltage range of standards ANSI C84.1 and EN 50160 in percentage, with respect to the nominal system voltage.

	EN 50160			ANSI C84.1		
	Maximum	Minimum	Minimum for special remote users	Maximum utilization and service voltage	Minimum service voltage	Minimum utilization voltage
Low voltage	+10	−10	−15	+5	−5	−10
Medium voltage	+10	−10	−15	+5	−2.5	−10

7.3 Traditional voltage control strategies

The vast number of users connected to distribution networks and the substantial amount of equipment necessary to reach all customers make it impracticable to achieve similar feasibility to transmission systems. The large amount of energy managed has made these assets of crucial importance to transmit electricity in contrast to distribution networks. Due to the relatively low and intermittent power consumption of clients at low voltage, distribution companies needed to develop their facilities while tightening their budget. Even so, by taking advantage of the intrinsic characteristics of traditional systems, high-quality services were provided to end users.

Given that only passive loads were connected to distribution networks and operation was radial, power flows could be considered unidirectional, flowing from the substations that fed the network to downstream loads. According to this assumption, the classical operation of distribution systems was considerably simplified. Voltage decreased through power lines depending on power consumption: as demand increased, so did the flowing current through lines that led to higher voltage drop.

Considering this particular profile, the voltage of the whole system is mainly controlled in passive networks at the upstream substation by means of the transformers that feed the distribution network and connect to the rest of the electric system. Several taps are provided within the primary and/or secondary winding that allow one to add or subtract extra turns and thus vary the ratio. To avoid supply interruptions that may affect several consumers, the substation transformer is usually equipped with on-load tap changers to modify the turns ratio during normal operation. It is normally installed in the primary winding since the current is lower than in the secondary side and therefore wear and tear is reduced.

In this way, voltage regulation is performed considering the current that flows to loads to correct the voltage drop. As depicted in Fig. 7.1A, voltage decreases along the feeder due to the impedance of lines. Should the tap changer remain fixed at its nominal value, voltage would be within bounds in case power consumption is not too high. On the contrary, increasing demands result in out-of-bound voltages that surpass the lower limit at the end of the line. The on-load tap changer is used to elevate the voltage at the secondary side of the transformer according to the magnitude of the current supplied and therefore the tail end of the feeder accomplishes with the required limits (Fig. 7.1B). Although it can be manually operated, the operation is usually automated by a line drop compensator that measures the current to select the adequate tap position.

It is worth noticing that the voltage provided to the low-voltage grids or medium-voltage consumers depends on their relative position within the feeder: the closer to the substation, the higher the voltage. In this way, medium-voltage/low-voltage transformers are also equipped with tap

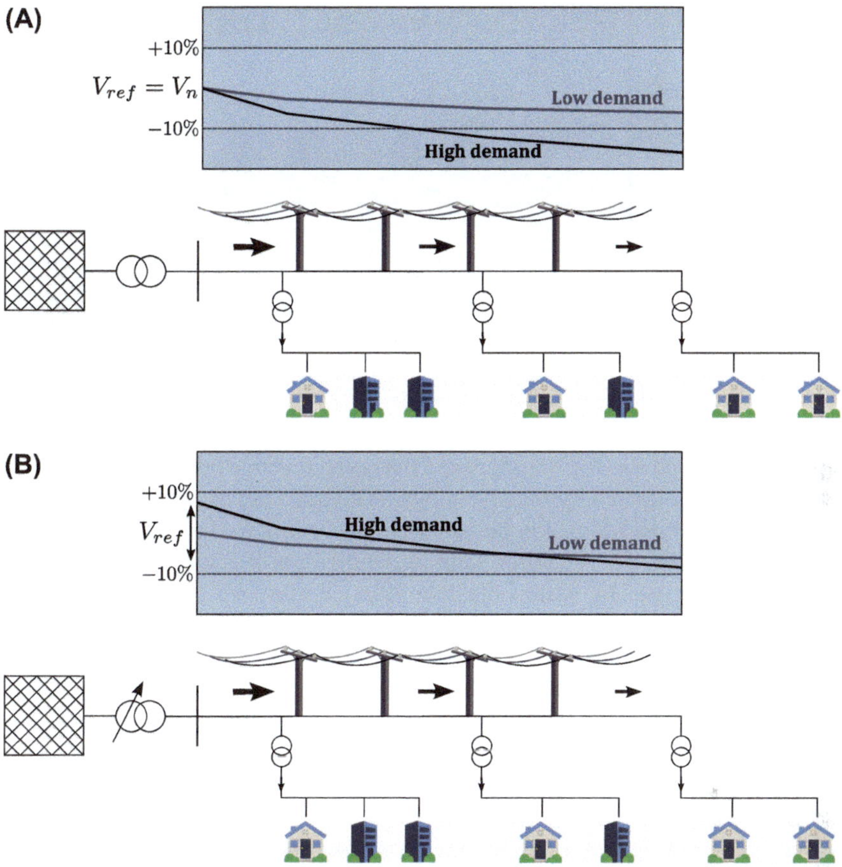

FIGURE 7.1 Traditional voltage profile with (A) fixed taps position and (B) on-load tap changer.

changers; however, these are simpler due to the large number of branches and need to be operated deenergized. The turns ratio is thus set during installation according to the position along the feeder and the length of the subsystem to deliver electricity at adequate voltage. Even so, it must be considered to select the appropriate tap that the topology of the system could change if meshed and the closest nodes to the substation might become the farthest ones in case a fault occurs near the substation, as shown in Fig. 7.2.

However, tap changers could be insufficient to compensate for excessive voltage drop in long lines, which are common in rural areas. A voltage regulator is typically employed that consists of an autotransformer with an on-load tap changer to elevate the voltage at its secondary side and to maintain it around the nominal voltage [5]. As with the substation transformer, the tap changer is also automated by a line drop compensator that estimates the voltage drop between the regulator and the load center.

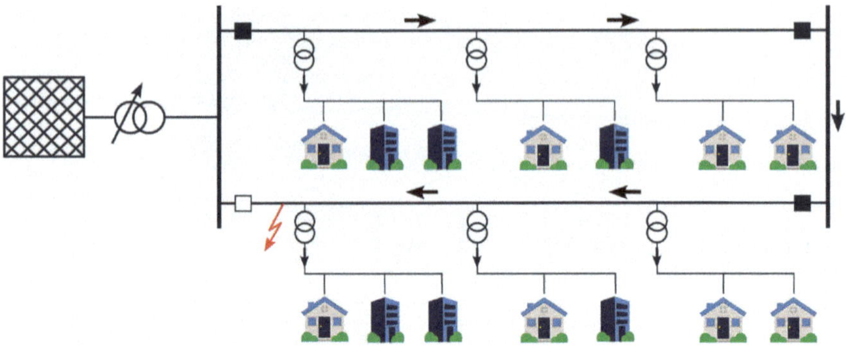

FIGURE 7.2 Topology change due to a fault close to the substation.

The voltage profile is also improved thanks to reactive power compensation since voltage is affected by reactive power. The main device for this purpose is the shunt capacitor, which can be classified according to the position where it is connected. First, the capacitor bank in the medium-voltage substation is the most common type. Its main objective is to compensate the reactive power consumption of the distribution system and avoid importing it from sub-transmission or transmission, which should be paid for. The capacitor bank in the substation typically allows one to connect or disconnect several steps to balance the reactive power injected with the consumption. However, it must be considered that such demand not only comes from grid users, but also from the transformer, which is indeed significantly higher.

In general, most users are residential loads that do not need to correct their power factor and directly consume the necessary reactive power from the grid. On the contrary, industrial loads are usually larger and have lower power factors. A distribution system operator usually collects reactive power, which may even cost as much as real power, because it needs to be transmitted within the facilities leading to "useless" increased power losses and voltage drop. Therefore industries also install capacitor banks to compensate their power factor to avoid extra costs. On the other hand, a distribution operator could also provide reactive power close to end users in long feeders to avoid excessive voltage drop and extra losses. In this way, capacitors are connected to overhead lines, that is, pole-mounted capacitor banks.

Traditional distribution networks were characterized by achieving high-quality standards monitoring and controlling a relatively low number of nodes. The operation was typically solved at the planning stage, which led to the so-called "fit-and-forget" approach. Conductors and equipment were initially sized considering the current power consumption and forecasting future loads in conjunction with certain economic criteria. Then, the equipment was properly tuned to work under the expected situations and it would only be revised in case the device presented any problem, e.g., consumers complained about over- or undervoltage that may be caused by an incorrect tap position of a transformer. Likewise, preventive maintenance is uncommon in low-voltage levels and the equipment is only repaired or substituted in case it fails.

7.4 Voltage control strategies in active distribution networks

The evolution of distribution networks is closely linked to the rest of the electric system. The increasing penetration of distributed generation close to end users and mainly based on renewable technologies challenges the classical operation. As explained in the previous section, power flows were traditionally assumed to be unidirectional, from the upstream substation to downstream loads. Since the system becomes active, depending on the balance between generation and load consumption, the electric power circulates upstream or downstream and the whole network may even inject power back to the subtransmission system. Therefore power flows become bidirectional and the traditional voltage profile changes, as shown in Fig. 7.3.

Power injection increases the voltage at the node where the generator is connected and its surroundings as demonstrated in Section 7.4.1; however, the impact depends on the relative position with respect to loads, the state of the system, and the location within the feeder. In case close loads account larger demand than the energy production, the presence of distributed generation can be considered as net demand reduction, provided no congestion arises. On the contrary, if generation is larger than power consumption, the voltage will increase according to net generation and the profile will be more or less affected depending on the location within the feeder.

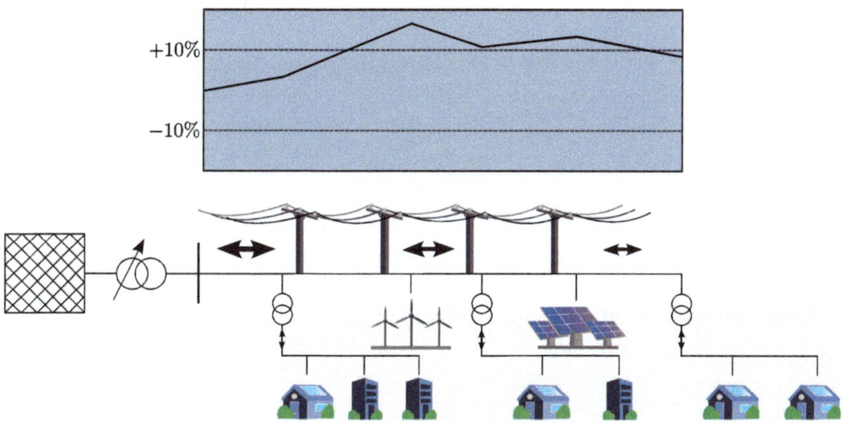

FIGURE 7.3 Example of voltage profile in active distribution networks.

The position where distributed generation is connected within the feeder is also important since it could blind automatic voltage controllers [6]. Line drop compensation is based on measuring the current to estimate the voltage drop according to the model of downstream lines. If a generator is connected just downstream of an on-load tap changer transformer or a voltage regulator, it will contribute to the upstream grid to feed loads but also make the current measured decrease. The automatic control will thus detect reduced demand and the selected tap will not correspond with the real current that circulates. Therefore the setting voltage will be lower than is necessary to maintain the voltage within bounds and it may result in excessive voltage drop at the tail end of the feeder.

In addition, renewable energy sources are intrinsically intermittent and nondispatchable if no storage system is available. Increased or decreased generation due to external factors affect the voltage profile of the network; however, this effect can be transient or permanent, e.g., scattered clouds affect photovoltaic generation in the short term, but energy production is reduced in the long term as night falls. In that sense, voltage control is oriented to correct any deviation in steady state.

On the other hand, the current trend in urban transportation toward electrification also affects distribution systems. Plug-in electric vehicles need to be connected to the grid to store energy and then work, such as battery-based vehicles, or to increase their total efficiency rate by charging the battery, such as hybrid vehicles that reduce their fuel consumption. In this way, the electric system faces a novel load whose power consumption depends on the charging mode, namely, slow and fast [7]. Slow charging mode is characterized by a small rate of charging and takes a long serving time; consequently, its use is mostly limited to overnight charging at home. The main advantage is the possibility of directly using home electrical installation due to its low power consumption. Fast charging mode is almost the opposite, it takes a relatively short amount of time but needs for considerably higher currents and voltages.

The growing tendency to reduce charging times makes electric vehicles behave as a heavy mobile load, which was not considered when distribution networks where designed. Electric vehicles have thus the opposite effect on the voltage profile in comparison with distributed generation since they increase power demand considerably. What is more, this fact is worsened due to their random space–time characteristic that complicates any prediction, and, if left uncontrolled, charging may cause excessive voltage drop within the system depending on electric vehicle concentration [7].

It is thus evident that traditional operation is no longer valid. Voltages at some nodes within the network may be higher than the voltage at the substation due to distributed generation. In fact, the system operator may not be conscious of such a situation since real-time measurements are uncommon. On the contrary, electric vehicles may cause excessive voltage drop that could result in undervoltage.

7.4.1 Transmission and distribution systems: two different scenarios

The key importance of transmission networks in the energy supply process has typically led to substantial investments to enhance reliability and efficiency. Cutting-edge technology was developed to control the system and to cope with large generation and demand. Since distribution networks are also facing the presence of both generation and demand, similar problems arise. Along these lines, many of the approaches to overcome the new challenges are based on previous techniques but tuned to fit the current idiosyncrasy of distribution.

Among all the methodologies employed in transmission, optimal power flow (OPF) is the predominant tool to attain a favorable state. The formulation of the problem allows one to optimize an objective function by rescheduling the available controls while satisfying system constraints. It is used for both planning and operation and, depending on the objective function, costs are minimized, reliability is improved, etc. Several techniques are used to provide robustness and seek a feasible solution; however, the current limitations of OPF [8] complicate its use to solve constraint violations, e.g., the solution provided to correct voltage deviations typically consists of several control actions, which makes the implementation difficult [1]. A possible solution is to perform subsequent iterations reducing the number of available controls at the expense of increasing computing time. Likewise, defining a proper objective function to reduce the number of elements also becomes complex due to the wide variety of equipment and devices of a different nature existing in the electric system, e.g., tap changers, capacitor banks, and voltage setpoint of generators.

A common approach to maintain voltages within bounds is to provide an operator in the control center with the necessary tools to monitor the system and to control the equipment remotely. In case any deviation is detected, the operator should consider the available control variables and the state of the network to decide the best corrective action. Alternatively, heuristic algorithms can be used to make the right decisions automatically as if an operator were in the control center. Such techniques are typically based on defining certain rules to select the corrective action considering the characteristics of the system in combination with sensitivities that provide the relation between the available controls and voltages [9,10]. However, the actions determined by the operator or the heuristic tool are focused on correcting voltage violations effectively, but these will certainly not be the optimal actions. In this way, solving periodically the OPF problem complements any of these corrective methodologies. The equipment is thus rescheduled considering the demand variations or any other event during the last interval to achieve a more favorable state by reducing global losses or according to any other objective.

Voltage is intimately linked to reactive power in transmission systems; consequently, voltage control is typically simplified by considering only reactive power actions and it is usually denoted as volt/var control. The two main reasons that result in weak coupling between active power and voltage amplitudes in high-voltage levels are the small phase angle differences between adjacent buses, on the one hand, and the inductive nature of transmission lines, on the other [11]. For illustration purposes, the circuit given at Fig. 7.4 is considered. A generator supplies energy to a load that consumes certain complex power through a single line. Kirchhoff's second law of this small system can be written as:

$$\mathcal{U}_g = \mathcal{I}(R + jX) + \mathcal{U}_L \tag{7.1}$$

where $\mathcal{U}_g$ and $\mathcal{U}_L$ are the phasor voltage of the generator and at the load, respectively, R and X are the series resistance and inductance of the line, respectively, and $\mathcal{I}$ is the phasor current, which depends on the load consumption as given by the subsequent expression.

$$\mathcal{I} = \left(\frac{\mathcal{S}_L}{\mathcal{U}_L}\right)^* = \frac{P_L - Q_L}{\mathcal{U}_L^*} \tag{7.2}$$

where $\mathcal{S}_L$ is the complex power demanded by the load, which is equal to the complex sum of the active power, P_L, and reactive power, Q_L.

Eq. (7.1) can then be rewritten as follows using Eq. (7.2) and operating:

$$\mathcal{U}_g\mathcal{U}_L^* - \mathcal{U}_L\mathcal{U}_L^* = (P_L - jQ_L)(R + jX) \tag{7.3}$$

Expanding this expression and considering separately the real and the imaginary part, the following two equations are obtained:

$$V_g V_L \cos(\theta_g - \theta_L) - V_L^2 = RP_L + XQ_L \tag{7.4}$$

$$V_g V_L \sin(\theta_g - \theta_L) = XP_L - RQ_L \tag{7.5}$$

where V_g and V_L denote the voltage amplitude of the generator and the load, respectively.

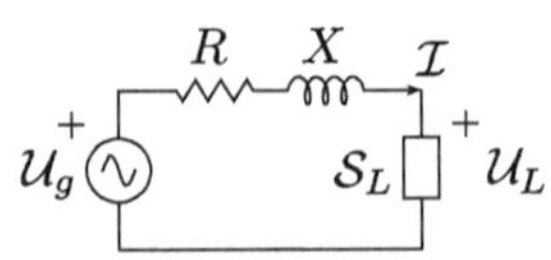

FIGURE 7.4 Example of a load supplied through a single line.

As mentioned, the phase angle differences are small between adjacent nodes in transmission systems, which means $\cos(\theta_g - \theta_L) \approx 1$ and $\sin(\theta_g - \theta_L) \approx \theta_g - \theta_L$. Besides, the resistance is negligible in comparison with the inductance, $R/X \ll 1$. Therefore Eqs. (7.4) and (7.5) are rewritten as:

$$V_g - V_L \approx \frac{XQ_L}{V_L} \tag{7.6}$$

$$\theta_g - \theta_L \approx \frac{XP_L}{V_g V_L} \tag{7.7}$$

The fact that electric systems are operated close to their nominal values allows one to take the denominators of both equations as constants. It is thus evident that reactive power is coupled with voltages and active power is linked with the phase angle difference, which is ultimately related to the frequency. In this way, voltages can be efficiently controlled through reactive power in high-voltage systems. These assumptions are not valid in distribution networks, however. The ratio R/X of lines [11] in 400 kV systems is $1/10$ and $1/5$ for 220 kV, leading to $R/X \ll 1$, but lines become increasingly resistive as voltage level decreases. The ratio is approximately unity in 50 kV lines and the resistance is clearly greater than the inductance in 20 kV lines.

Therefore Eqs. (7.4) and (7.5) can only be simplified for distribution systems considering the small phase angle difference between adjacent nodes in case lines are not too loaded:

$$V_g - V_L \approx \frac{RP_L + XQ_L}{V_L} \tag{7.8}$$

$$\theta_g - \theta_L \approx \frac{XP_L - RQ_L}{V_g V_L} \tag{7.9}$$

It is clear that voltage drop is affected by both active and reactive power. Indeed, resistance is considerably greater than inductance in low-voltage systems, where many distributed generation and electric vehicles are being connected. Distribution voltage controls will thus need to consider not only reactive power as did volt/var controls, but also active power injections or consumption.

7.4.2 Novel voltage control strategies

Distribution system operators need to develop their networks to face the new scenario and to allow increasing distributed generation and plug-in electric vehicle penetration. The traditional "fit-and-forget" approach suggests increasing investments during the planning stage to reinforce the network to cope with the associated problems; however, the reality is rather different. The intermittency of renewable energy sources, which are not dispatchable, in conjunction with the mobile characteristic of electric vehicles make this task extremely difficult and, if coordinated actions are not considered, it cannot be efficiently solved.

Voltage control used to rely on a few items of equipment in traditional distribution networks that simplified substantially the system operation, but the presence of generation and heavy loads disturbs the initial planning. The converter interface nature of these elements, however, becomes the solution to meet voltage requirements. Depending on the implemented control, power converters allow one to determine the active and reactive power injected, to support the system in voltage control, and to define the voltage wave characteristics at the connection node, that is, the amplitude and the frequency [12]. Similarly, the so-called custom power devices apply power electronic technologies to system quality enhancement. Among all the solutions included in this concept, the compensating devices are used for voltage regulation, reactive power compensation, and power factor correction in addition to other services [13]. Despite the large number of topologies and control strategies, there are three main categories according to the connection and functioning principle [14]. The distribution static compensator (DSTATCOM) consists of shunt-connected devices that provide the required services by means of injecting certain current according to the implemented control. In contrast, the dynamic voltage restorer (DVR) is its dual device since it is series connected and provides the required services by injecting series voltage into the system. These two structures are combined into the unified power quality conditioner (UPQC), leading to a more versatile device that allows one to inject both shunt current and series voltage.

On the other hand, the rapid advances in the field of energy storage systems [15], which continuously increase storage capacity and lifetime while reducing costs, allow one to manage certain amounts of energy to improve the voltage profile and system efficiency as well. In general, energy excess is stored to be used during peak load hours; however, since storage devices are also based on power converters, it is possible to use reactive power to correct voltage deviations according to previous considerations. Similarly, plug-in electric vehicles can also provide the system with flexibility thanks to vehicle-to-grid technology [7]. Coordinated charging or discharging results in bidirectional power exchange between the vehicle and the network to help with system operation, provided the vehicle achieves a certain state of charge when the user requires it.

However, due to the large number of distributed generators and plug-in electric vehicles along with their relatively small size, operation becomes considerably complex. The role of the aggregator arises to undertake the representation of these new actors in the ancillary service market, which includes voltage regulation, and to create a competitive bidding strategy [16]. On the one hand, the aggregator allows many low and medium size generators and electric vehicles to participate in the market that would otherwise not be able to.

On the other hand, the aggregator handles more information than individuals separately and can make better use of it to increase clients' profits. These issues are analyzed for batteries in [17]. Therefore the tasks of the system operator are simplified since the aggregator is in charge of dealing with generators, plug-in electric vehicles, and storage devices. In addition, all these elements can be aggregated as a microgrid or a virtual power plant considering their parameters and constraints in conjunction with those of the corresponding portion of the network [18,19].

To maintain high service quality, new control methodologies that consider all these possibilities should be developed to maintain voltages within the predefined bounds. The specialized literature includes a wide variety of proposals that range from centralized controls based on modern communication infrastructures to distributed controls that rely on local measurements. The simplest approach to deal with distributed generation, which typically causes overvoltage, is to define a droop for every generator to curtail its active power. This technique is discussed for photovoltaic inverters in [20] along with the design of droop coefficients. However, despite providing an effective solution that allows one to increase the presence of renewable energies, active power curtailment is essentially nonefficient.

One of the most appreciated approaches to solve out-of-limit voltages is to compute sensitivities to determine the relation between the available controls and voltages. The centralized tool proposed in [21] corrects any voltage deviation through a set of heuristic rules that considers the current state of the system, the network constraints, the capacity of the available controls according to their limitations, and the cost associated to each possible action. This methodology stands out by providing a framework that compares a wide range of control variables and merges technical aspects with economic criteria. Similarly, a local heuristic controller is presented in [22] to regulate the voltage profile at the connection node of wind turbines. It begins by trying to correct overvoltage by means of the reactive power injected and, if the capability curve limits this initial action, the active power is curtailed to solve the problem.

A combination of local and centralized controllers in a two-level architecture is presented in [23]. Whereas the lower level provides fast response to local voltage deviations through a linear static volt/var characteristic, the upper level adjusts such a characteristic. This centralized control develops a model predictive control that minimizes future reactive power corrections using sensitivities to predict system behavior. However, in contrast with other methods, the closed-loop nature of the model predictive control avoids updating sensitivities continuously and thus the computational burden remains low.

The network can also be divided into areas [24,25] that may be governed by decentralized controllers to achieve an intermediate solution between centralized and local. For instance, in [26] a representative bus is selected for every control area whose voltage is controlled through reactive power injections of photovoltaic generators within the corresponding area. As a matter of fact, photovoltaics has been particularly studied due to its increasing penetration within low-voltage distribution networks. Such advances are mainly encouraged thanks to their modularity that allows one to install them on rooftops in residential areas. A local voltage regulation control for photovoltaic inverters is developed in [27], whereas coordination between energy storage systems in the presence of photovoltaics is studied in [28] to regulate voltages while considering the state of charge of batteries to effectively use storage capacities.

However, the fact that traditional distribution networks do not usually have real-time measurements limits the use of most of the centralized approaches. To avoid, or at least reduce, substantial investments, some authors have developed methodologies to deal with insufficient information. For instance, fuzzy-based voltage regulators are proposed in [29] to correct voltage problems while avoiding excessive wear and tear of on-load tap changers and limiting power curtailment. Likewise, a perturb-and-observe algorithm is presented in [30] to estimate sensitivities heuristically instead of analytically, which certainly requires much information from the network. In that sense, many of the existent sensitivity-based algorithms could be adapted to reduce their communication requirements.

Despite the efficacy of these methodologies to correct voltage deviations, most of them lead to nonoptimal states in the long term where operational constraints are fulfilled but operation is far from efficient. Following the same approach as in transmission systems, quick response techniques should be used to solve any out-of-limit voltage value, while the voltage profile may be optimized during normal operation by rescheduling control variables after solving an OPF. However, this approach is limited to distribution networks with modern communication infrastructure to provide enough measurements and a distribution management system to filter them.

7.4.3 Case study

The aim of this section is to illustrate the issues addressed within this chapter. A low-voltage distribution network is proposed to study how classical voltage profiles are modified due to the presence of distributed generation and plug-in electric vehicles, and the importance of using controls able to manage different types of elements to tackle voltage problems effectively. First, the base case that is a passive network with classical voltage profile is analyzed. Then, distributed generation, energy storage, and electric vehicles are successively included to reach two extreme scenarios where voltage controls are particularly useful: low demand with high generation, which results in overvoltage, and high demand due to electric vehicles with low generation, which leads to undervoltage.

To correct any voltage deviation, sensitivity matrixes will be employed in conjunction with intuition to determine the necessary control actions to meet voltage requirements. Sensitivities are calculated analytically for each scenario proposed as summarized in [21]; however, commercial tools could also be used, e.g., DIgSILENT PowerFactory. Note that sensitivities are estimated for certain operating states and the accuracy to forecast a new state relies on the size of the considered control action.

7.4.3.1 Base case: passive network

The proposed low-voltage subsystem is a three-phase network whose nominal voltage is 400 V. As depicted in Fig. 7.5, it is radial and consists of five nodes, four residential loads, one transformer, and three overhead lines (Tables 7.2 and 7.3).

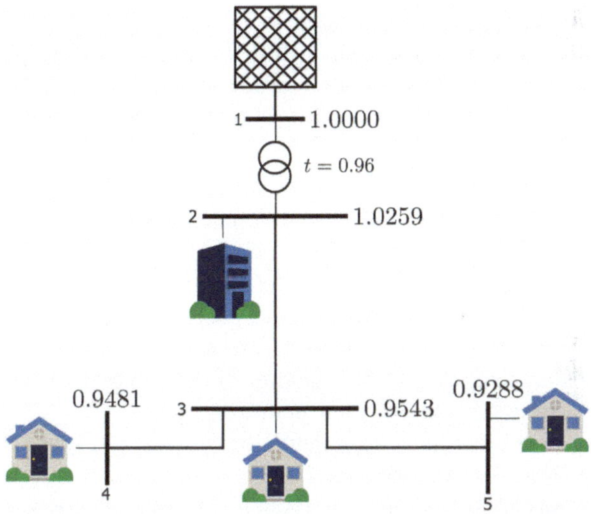

FIGURE 7.5 Proposed low-voltage network and resulting voltages considering the maximum load.

TABLE 7.2 Maximum expected loads.

Node	S (kVA)	Power factor
2	150	0.95
3	100	0.95
4	30	0.95
5	60	0.95

TABLE 7.3 Line and transformer parameters.

Initial node	Final node	*R* (ohms)	*X* (ohms)
1	2	0.0032	0.0128
2	3	0.05	0.03
3	4	0.03	0.01
3	5	0.06	0.02

The subsystem is connected within the feeder of a 32 kV medium-voltage system through a transformer that has a fixed tap position. As addressed in this chapter, the tap position is selected during installation according to the "fit-and-forget" approach. Since EN 50160 limits are considered, the turns ratio of the transformer, t, is set equal to 0.96 to maintain all voltages between 1.10 and 0.90 under any load condition. In Fig. 7.5 the voltage amplitude is presented in per unit next to each node considering the maximum load. On the other hand, note that no shunt capacitor is typically available in low voltage.

7.4.3.2 High distributed generation and low demand

The first scenario considers high photovoltaic penetration within the low-voltage system that could even surpass nominal loads (Table 7.4). Since all loads are associated with residential consumers, power consumption will decrease during the morning while most people are at work. In contrast, photovoltaic generation will increase as the sun reaches its zenith. In this scenario, loads are supposed to be a quarter of their maximum value, whereas photovoltaic generators achieve their nominal values.

The initial state presents voltage deviations at nodes 3, 4, and 5 as depicted in Fig. 7.6. Since the transformer has no on-load tap changer, the only available controls to solve the contingencies are the photovoltaic inverters, which allow one to reduce active power generation and to inject or consume

TABLE 7.4 Nominal photovoltaic generation.

Node	*P* (kW)
2	200
3	100
4	120
5	50

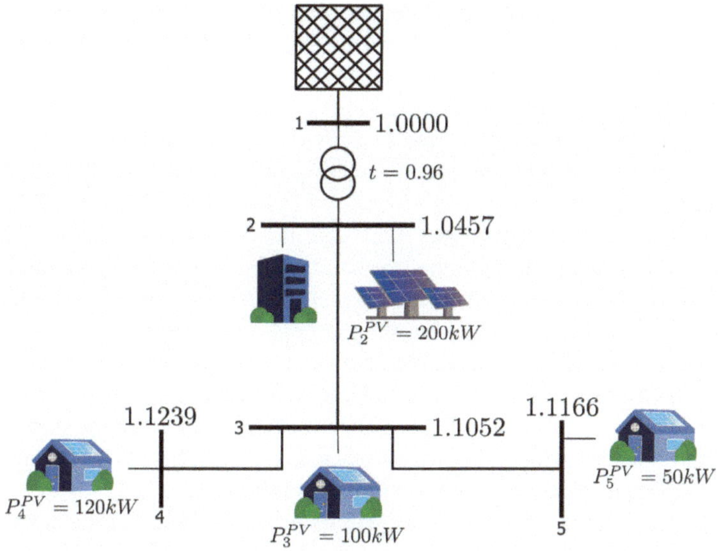

FIGURE 7.6 High distributed generation and low demand: initial state. *PV*, photovoltaics.

reactive power. However, the latter is limited by the inverters' capability curve that depends on the active power. Such constraints will be approximated considering that inverters are only limited by the apparent power, which is taken equal to the nominal values given in Table 7.4. In addition, it will be considered that photovoltaic generators allow one to determine the active and reactive power setpoints.

To decide the best control actions that should be taken, the sensitivity matrixes that provide the relation between the available controls and voltages are computed. The rows refer to voltages at demand nodes (2−5) where active and reactive power injections are known, whereas the columns apply to power injections at demand nodes:

$$S_{V_D,Q_D} = \begin{bmatrix} 7.69 & 7.77 & 7.77 & 7.78 \\ 7.32 & 24.61 & 24.64 & 24.71 \\ 7.20 & 24.21 & 29.81 & 24.31 \\ 7.24 & 24.36 & 24.39 & 35.71 \end{bmatrix} \cdot 10^{-5}\text{kvar}^{-1} \quad (7.10)$$

$$S_{V_D,P} = \begin{bmatrix} 1.69 & 1.02 & 0.90 & 0.94 \\ 1.61 & 27.20 & 26.04 & 26.45 \\ 1.59 & 26.76 & 41.99 & 26.02 \\ 1.60 & 26.93 & 25.77 & 59.37 \end{bmatrix} \cdot 10^{-5}\text{kW}^{-1} \quad (7.11)$$

where S_{V_D,Q_D} denotes the sensitivity matrix that relates the changes in the voltage at demand nodes with respect to the reactive power injected at demand nodes, and $S_{V_D,P}$ the sensitivity matrix that relates the changes in the voltage at demand nodes with respect to the active power injected.

The first row is not considered because it provides sensitivities with respect to the voltage at node 2, which is within bounds. Having analyzed the first column, it is clear that any action taken at node 2 will hardly improve the contingencies; therefore corrective actions should be focused on nodes 3–5. Besides, the inverters are not able to provide reactive power since they are working at full load and the output active power is equal to the admissible apparent power.

There are several approaches that can be developed combining power curtailment between nodes. On the one hand, each photovoltaic generator could be focused on the contingency at its connection node or a common curtailment might be estimated to be shared between generators. In both cases, it should be considered that curtailment is coupled between nodes to avoid excessive loss of generation. On the other hand, an initial corrective action may be taken to solve the most severe out-of-limit voltage. Given that all sensitivities are positive, the other contingencies will be improved or even solved. Should any contingency remain, subsequent actions should be taken similarly.

In this way, the last methodology will be employed to meet the constraints. The highest overvoltage is 1.1239 at node 4, as depicted in Fig. 7.6, and the highest sensitivity for node 4 is provided by $S_{V_D,P}(3,3)$, which refers to curtailing generation at such node. To effectively redirect the system to a new safe state and to consider the intrinsic error associated to sensitivity matrixes, the objective is that voltage at node 4 reaches 1.0950 instead of 1.10. Therefore the required control action is calculated as:

$$\Delta P_4^{PV} = \frac{1.0950 - 1.1239}{41.99 \cdot 10^{-5}} = -68.8\text{kW} \tag{7.12}$$

Having rescheduled the photovoltaic generator, all voltages are within bounds as presented in Fig. 7.7. Note that the voltage at node 4 is not 1.0950 because sensitivities are the result of a linear approximation. Even so, the error is negligible as it is 0.36 V.

Although the contingencies have been solved, the new state is not efficient since only power curtailment has been applied. Reactive power should also have been taken into account to maximize generation; however, sensitivities calculation is based on considering that control actions are taken solely. An iterative process could be developed to recalculate sensitivities after varying the active or reactive power, but the computational burden would increase considerably. Since the subsystem remains close to the initial operating point after corrective actions, it will be considered that sensitivities are constant at the expense of increasing uncertainty about the expected state. Therefore the new control action is curtailing 20.8 kW from the photovoltaic generator at node 4 in conjunction with consuming 67.5 kvar, which results in the state shown in Fig. 7.8. By also considering reactive power, curtailing 48 kW is avoided. Similarly, photovoltaic inverters could have been oversized to provide reactive power under nominal conditions to avoid derating generation.

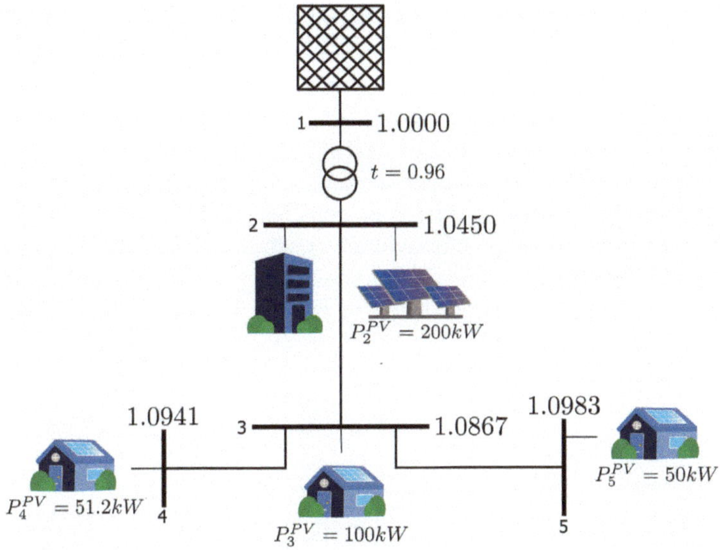

FIGURE 7.7 New state after curtailing 68.8 kW at node 4. *PV*, photovoltaics.

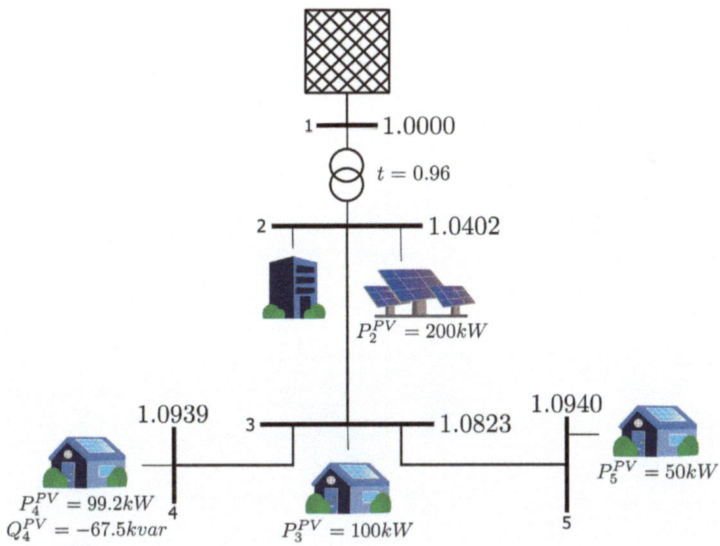

FIGURE 7.8 New state after curtailing 20.8 kW and consuming 67.5 kvar at node 4. *PV*, photovoltaics.

On the other hand, storage systems can also be used to store the energy excess that could be used later during peak load hours. Should electrochemical batteries be connected at node 3 and be able to consume 110.9 kW, voltage limits would be met as depicted in Fig. 7.9. Again, reactive power could be considered to reduce the use of active power.

7.4.3.3 High demand due to electric vehicles and low generation

The proposed example supposes plug-in electric vehicle penetration, which produces excessive voltage drop, in conjunction with previous photovoltaic generators according to Tables 7.4 and 7.5. It is considered an evening when customers are back to home from work and thus consumption rises. In addition, electric vehicles are plugged in and start charging since there is no coordination. The scenario is created by considering that load rate is 75% with respect to the nominal value, photovoltaic generation is null due to sunset, and electric vehicles are charging at full load. As a result, voltage at node 5 surpasses the lower limit as presented in Fig. 7.10.

TABLE 7.5 Maximum plug-in electric vehicle consumption.

Node	*P* (kW)
2	100
3	75
5	50

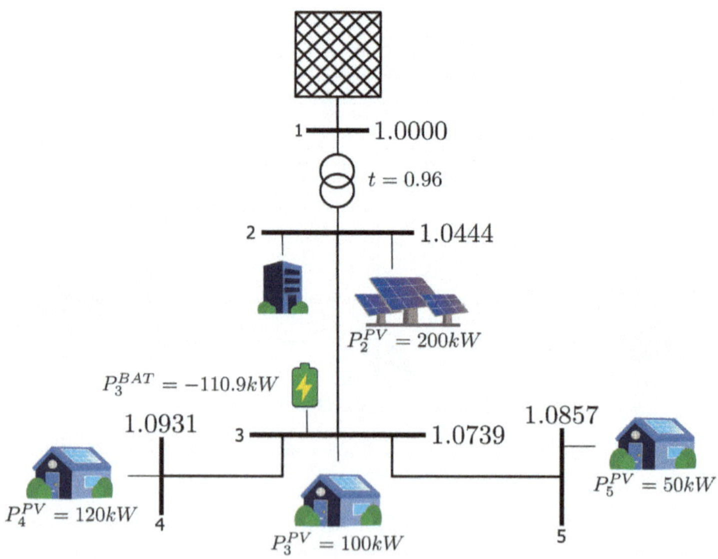

FIGURE 7.9 New state considering storage devices at node 3. *PV*, photovoltaics.

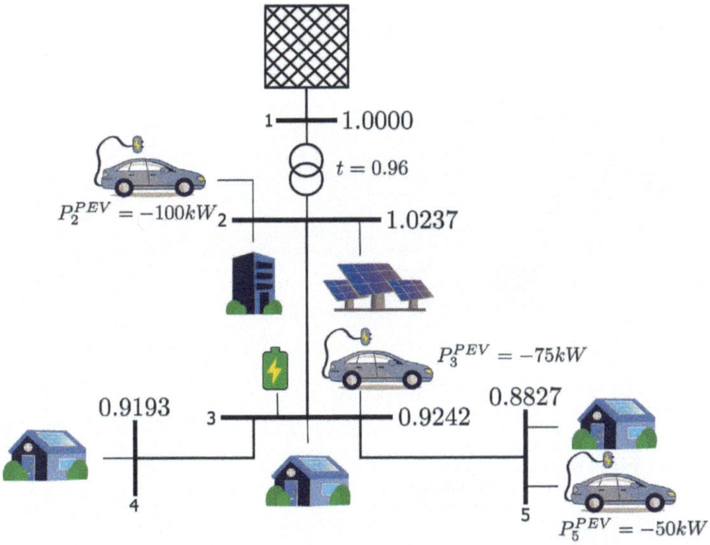

FIGURE 7.10 High demand due to plug-in electric vehicles and low generation: initial state. *PV*, photovoltaics; *PEV*, plug-in electric vehicle.

Sensitivity matrixes are again computed to obtain the relation between voltages and the available controls:

$$S_{V_D,Q_D} = \begin{bmatrix} 7.93 & 8.27 & 8.29 & 8.37 \\ 8.94 & 30.77 & 30.94 & 31.57 \\ 8.99 & 30.93 & 37.93 & 31.73 \\ 9.38 & 32.29 & 32.46 & 47.64 \end{bmatrix} \cdot 10^{-5}\text{kvar}^{-1} \tag{7.13}$$

$$S_{V_D,P} = \begin{bmatrix} 2.28 & 3.83 & 3.89 & 4.45 \\ 2.58 & 43.28 & 43.80 & 48.29 \\ 2.59 & 43.51 & 64.53 & 48.55 \\ 2.70 & 45.42 & 45.95 & 95.36 \end{bmatrix} \cdot 10^{-5}\text{kW}^{-1} \tag{7.14}$$

In this case, photovoltaic inverters can provide as much reactive power as their nominal power and thus power curtailment is avoided. The most sensitive reactive control to voltage at node 5 is the generator connected at such node, which has 120 kvar available. In that sense, 46.7 kvar are injected that result in all voltages within bounds, as depicted in Fig. 7.11.

In this scenario the deviation is solved thanks to idle photovoltaic inverters, which are unloaded due to the lack of sun. The presence of elements of different nature allows one to increase the number of available control

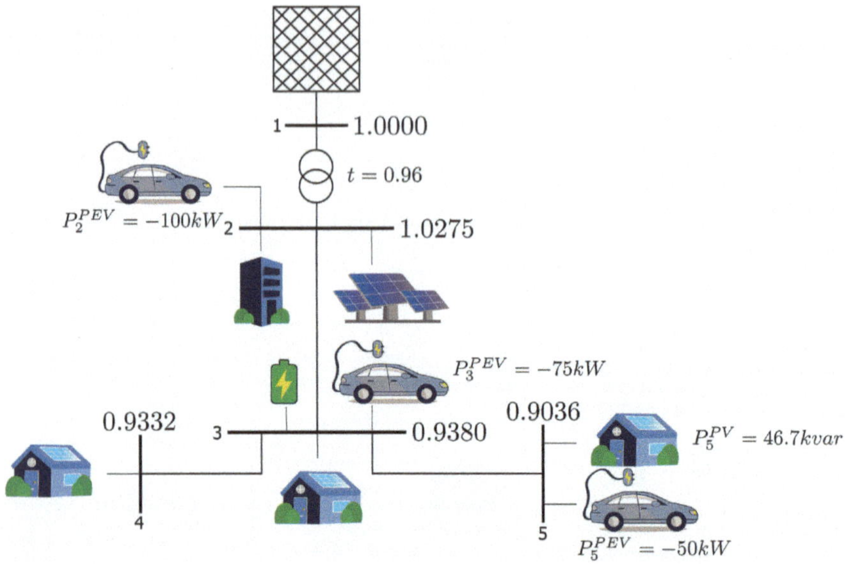

FIGURE 7.11 New state after injecting 46.7 kvar at node 5. *PV*, photovoltaics; *PEV*, plug-in electric vehicle.

variables, e.g., photovoltaics, wind turbines, battery storage, electric vehicles, DSTATCOMs, etc. Consequently, more efficient states are reached that avoid curtailing generation or limiting vehicle charging. Besides, it is evident that distributed generation and electric vehicles complement each other since generation excess can be used to feed electric vehicles and thus reduce the impact of these new actors on the grid.

7.5 Conclusions

Voltage control in traditional distribution networks was developed to provide high-quality services while avoiding considerable investments. By assuming that power flows were unidirectional, a few items of equipment were enough to maintain voltages within bounds. However, the increasing penetration of distributed generation and plug-in electric vehicles results in bidirectional power flows that modify the classical voltage profile, on the one hand, and excessive voltage drop due to vehicle charging, on the other.

The integration of these new actors to meet voltage quality standards is effectively achieved by developing techniques to reschedule the available control variables. In addition to distributed generators and electric vehicles, other solutions such as battery storage and custom power devices provide the system with flexibility to improve the voltage profile. On the other hand, as analyzed in the proposed examples, distributed generation and electric vehicles complement each other. Generation excess can be used to reduce electric vehicle demand and thus reduce their impact on the grid.

To sum up, the ongoing transition to active distribution networks challenges the classical operation. The traditional "fit-and-forget" approach suggests increasing investments during the planning stage to reinforce the network; however, if coordinated actions are not considered, it cannot be efficiently solved. In that sense, novel control strategies that consider the wide variety of available control variables need to be used.

References

[1] J.L. Martínez-Ramos, A.J. Conejo, Optimal and secure operation of transmission systems, in: A. Gomez-Exposito, A.J. Conejo, C. Canizares (Eds.), Electric Energy Systems: Analysis and Operation, second ed., CRC Press, Boca Raton, 2018, pp. 237–277.

[2] EN 50160. Voltage Characteristics of Electricity Supplied by Public Electricity Networks, British Standards Institute, 2010.

[3] ANSI C84.1, Electric Power Systems and Equipment - Voltage Ratings (60 Hertz), 2016.

[4] CIGRE WG 33.02, Guidelines for representation of network elements when calculating transients, CIGRE Tech. Brochure 39 (1990).

[5] W.H. Kersting, Distribution System Modeling and Analysis, third ed., CRC Press, Boca Raton, 2012.

[6] P.P. Barker, R.W. de Mello, Determining the impact of distributed generation on power systems: Part 1 - radial distribution systems, in: Proc: 2000 Power Engineering Society Summer Meeting (Cat. No. 00CH37134), vol. 3, 2000, pp. 1645–1656, https://doi.org/10.1109/PESS.2000.868775.

[7] Y. Zheng, S. Niu, Y. Shang, Z. Shao, L. Jian, Integrating plug-in electric vehicles into power grids: a comprehensive review on power interaction mode, scheduling methodology and mathematical foundation, Renew. Sustain. Energy Rev. 112 (2019) 424–439, https://doi.org/10.1016/j.rser.2019.05.059.

[8] F. Capitanescu, Critical review of recent advances and further developments needed in AC optimal power flow, Elec. Power Syst. Res. 136 (2016) 57–68, https://doi.org/10.1016/j.epsr.2016.02.008.

[9] A.G. Expósito, J.L. Martinez Ramos, J.L. Ruiz Macías, Y.C. Salinas, Sensitivity-based reactive power control for voltage profile improvement, IEEE Trans. Power Syst. 8 (3) (1993) 937–945, https://doi.org/10.1109/59.260908.

[10] J.L.M. Ramos, A.G. Expósito, J.C. Cerezo, E.M. Ruiz, Y.C. Salinas, A hybrid tool to assist the operator in reactive power/voltage control and optimization, IEEE Trans. Power Syst. 10 (2) (1995) 760–768, https://doi.org/10.1109/59.387914.

[11] A. Gómez-Expósito, F.L. Alvarado, Load flow, in: A. Gómez-Expósito, A.J. Conejo, C.A. Cañizares (Eds.), Electric Energy Systems: Analysis and Operation, second ed., CRC Press, Boca Raton, 2018, pp. 99–132.

[12] J. Rocabert, A. Luna, F. Blaabjerg, P. Rodríguez, Control of power converters in AC microgrids, IEEE Trans. Power Electron. 27 (11) (2012) 4734–4749, https://doi.org/10.1109/TPEL.2012.2199334.

[13] Y. Pal, A. Swarup, B. Singh, A review of compensating type custom power devices for power quality improvement, in: Proc: 2008 Joint International Conference on Power System Technology and IEEE Power India Conference, 2008, pp. 1–8, https://doi.org/10.1109/ICPST.2008.4745338.

[14] A. Ghosh, G. Ledwich, Power Quality Enhacement Using Custom Power Devices, first ed., Springer, 2002.

[15] D.O. Akinyele, R.K. Rayudu, Review of energy storage technologies for sustainable power networks, Sustain. Energy Technol. Assess. 8 (2014) 74–91, https://doi.org/10.1016/j.seta.2014.07.004.

[16] K. Oureilidis, et al., Ancillary services market design in distribution networks: review and identification of barriers, Energies 13 (4) (2020) 917, https://doi.org/10.3390/en13040917.

[17] K. Liu, Q. Chen, C. Kang, W. Su, G. Zhong, Optimal operation strategy for distributed battery aggregator providing energy and ancillary services, J. Modern Power Syst. Clean Energy 6 (4) (2018) 722–732, https://doi.org/10.1007/s40565-017-0325-9.

[18] P. Asmus, Microgrids, virtual power plants and our distributed energy future, Electr. J. 23 (10) (2010) 72–82, https://doi.org/10.1016/j.tej.2010.11.001.

[19] D. Pudjianto, C. Ramsay, G. Strbac, Virtual power plant and system integration of distributed energy resources, IET Renew. Power Gener. 1 (1) (2007) 10–16, https://doi.org/10.1049/iet-rpg:20060023.

[20] R. Tonkoski, L.A.C. Lopes, T.H.M. El-Fouly, Coordinated active power curtailment of grid connected PV inverters for overvoltage prevention, IEEE Trans. Sustain. Energy 2 (2) (2011) 139–147, https://doi.org/10.1109/TSTE.2010.2098483.

[21] J.L. Martinez-Ramos, F.J. Zarco-Soto, A. Marano-Marcolini, P.J. Zarco-Periñan, A.L. Trigo-Garcia, Coordination of distributed energy resources to solve voltage problems in distribution networks, in: Proc: 2018 International Conference on Smart Energy Systems and Technologies, 2018, pp. 1–6, https://doi.org/10.1109/SEST.2018.8495842.

[22] V. Calderaro, V. Galdi, G. Massa, A. Piccolo, Distributed Generation and local voltage regulation: an approach based on sensitivity analysis, in: Proc: 2nd IEEE PES International Conference and Exhibition on Innovative Smart Grid Technologies, 2011, pp. 1–8, https://doi.org/10.1109/ISGTEurope.2011.6162803.

[23] H.S. Bidgoli, T. Van Cutsem, Combined local and centralized voltage control in active distribution networks, IEEE Trans. Power Syst. 33 (2) (2018) 1374–1384, https://doi.org/10.1109/TPWRS.2017.2716407.

[24] W.M. Villa, J.L. Rueda, S. Torres, W.H. Peralta, Identification of voltage control areas in power systems with large scale wind power integration, in: Proc: 2012 Sixth IEEE/PES Transmission and Distribution: Latin America Conference and Exposition (T&D-LA), 2012, pp. 1–7, https://doi.org/10.1109/TDC-LA.2012.6319092.

[25] C.A. Aumuller, T.K. Saha, Determination of power system coherent bus groups by novel sensitivity-based method for voltage stability assessment, IEEE Trans. Power Syst. 18 (3) (2003) 1157–1164, https://doi.org/10.1109/TPWRS.2003.814886.

[26] A. Cagnano, E. De Tuglie, A decentralized voltage controller involving PV generators based on Lyapunov theory, Renew. Energy 86 (2016) 664–674, https://doi.org/10.1016/j.renene.2015.08.072.

[27] S. Ghosh, S. Rahman, M. Pipattanasomporn, Distribution voltage regulation through active power curtailment with PV inverters and solar generation forecasts, IEEE Trans. Sustain. Energy 8 (1) (2017) 13–22, https://doi.org/10.1109/TSTE.2016.2577559.

[28] Y. Wang, K.T. Tan, X.Y. Peng, P.L. So, Coordinated control of distributed energy-storage systems for voltage regulation in distribution networks, IEEE Trans. Power Deliv. 31 (3) (2016) 1132–1141, https://doi.org/10.1109/TPWRD.2015.2462723.

[29] M.A. Azzouz, H.E. Farag, E.F. El-Saadany, Real-time fuzzy voltage regulation for distribution networks incorporating high penetration of renewable sources, IEEE Systems J. 11 (3) (2017) 1702–1711, https://doi.org/10.1109/JSYST.2014.2330606.

[30] F. Tamp, P. Ciufo, A sensitivity analysis toolkit for the simplification of MV distribution network voltage management, IEEE Trans. Smart Grid 5 (2) (2014) 559–568, https://doi.org/10.1109/TSG.2014.2300146.

Chapter 8

Voltage dips caused by faults in a transmission system: a monitoring case study of a sensitive industrial consumer

Roger Alves de Oliveira, Bibiana Petry Ferraz, Ruth Agustini, Renato Ferraz, Roberto Chouhy Leborgne
Universidade Federal do Rio Grande do Sul (UFRGS), Porto Alegre, Rio Grande do Sul, Brazil

Chapter outline

8.1 Introduction **219**
8.2 Methodology **223**
8.2.1 Proposed approach for voltage dip quantification 223
8.2.2 Proposed approach for grouping the faults and voltage dips 228
8.3 Case study and results **229**
8.3.1 Quantification of voltage dips and comparison of IEEE 1564 and brazilian grid code 230
8.3.2 Grouping the main influence variables 237
8.3.3 Assessment of the voltage dip vulnerability area 241
8.4 Conclusions **242**
References **243**
Further reading **244**

8.1 Introduction

Among all the power quality disturbances, voltage dips still represent a challenge to be faced by energy utilities, electrical equipment suppliers, and consumers. Occurrences of voltage dips, combined with the sensitivity of modern equipment, have resulted in significant financial losses, mainly due to the expressive number of industrial shutdowns. Typically, voltage dips are associated with: (1) faults in power systems, (2) transformer energizing, (3) starting of heavy loads (such as large electrical machines), and (4) switching operations to isolate the faulted sections. In electrical power systems, faults can be caused mainly by storms, lightning, snow, freezing rain, and insulation breakdown. Thus voltage dips can be characterized as random events. Then, in

Power Quality in Modern Power Systems. https://doi.org/10.1016/B978-0-12-823346-7.00006-2

general, the most severe voltage dips are caused by solid faults or low-impedance faults, which propagate throughout the power system and can affect a sensitive consumer even when it is connected far from the fault location.

To avoid load shutdowns, the end-user's electrical equipment must feature minimum low-voltage ride-through capability. Nevertheless, the equipment might still be susceptible to a range of voltage dips. In this context, the quantification and mapping of voltage dips are needed by the energy utility and the end-users, considering that the improvement of power quality provides economic competitiveness.

The standardization of the sensitivity of electronic equipment concerning power quality was introduced by IEEE Standard 1346 [1]. Since then, several studies have analyzed the influence of magnitude, duration, and frequency of occurrence of events. For instance, methods for detecting electromagnetic interference in the operation of equipment were developed in [2], sensitivity regions based on semiconductors were estimated in [3], equipment tolerance ranges to short-term voltage variations were studied in [4], and stratification of the events into categories based on occurrence limits was proposed in [5].

Considering that a voltage dip with a residual voltage close to 0.9 pu and wide duration may have a greater impact when compared to another voltage dip of shorter duration, it is necessary to propose standards to measure the impact and severity considering both the magnitude and the duration of the events. To characterize voltage dips, voltage dip indices were established by [6]. Then a working group (task force P1564) developed methods for quantifying the severity of individual events, which were published in the IEEE Guide for Voltage Sag Indices. Such severity allows one to evaluate both local (final consumer) and global (distribution and transmission utilities) performance.

In Brazil, the Agência Nacional de Energia Elétrica (ANEEL) opened Public Consultation no. 05/2011 [7] to improve the short-term voltage variations regulatory procedures, updating the distribution systems' grid codes [8]. However, due to the complexity of the topic, the discussion was expanded with Public Consultation no. 18/2014 [9] to establish a procedure for measuring short-term voltage variations, as well as to implement performance indexes and apply reference limits for phenomena involving power quality disturbances. In the standardization of [8], a categorization of events similar to the one developed in [5] was proposed, and from there a method for calculating a single index was presented. Although the methodology was developed for the distribution networks, the Brazilian regulation presents weighting for levels up to 230 kV.

In this framework, the IEEE 1564 update was handled by [10], where an overview of standardization was presented. That work summarizes the methods for the individual, local, and global quantification of voltage dips, but does not include a case study. Additionally, [11] presented the quantification of the impact of the dips in an industrial distribution circuit through the severity index. It was concluded by [11] that the average monthly or annual severity was a more important parameter than the number of dips in the same period. Similarly, [12] carried out a comparative analysis between the Brazilian grid code update proposal and South Africa's NRS 048 [5] using real measurement data in the distribution circuit. It was concluded that although the Brazilian grid code was based on the South African grid code, when applying the same dataset to the two standards, the obtained diagnosis was different. Thus the main conclusion refers to the question of the period of analysis, since NRS 048 applies to 1 year, while the ANEEL applies to only 1 month. For this reason an update to the Brazilian standard was recommended to indicate permanent monitoring. Finally, [13] presented a summary of contributions and indications for improving ANEEL's public consultation, emphasizing the need for installing meters at strategic points in the system, aiming to capture a larger number of events and to improve the weighs of the impact factor (FI), in accordance with weather parameters of each state.

The vulnerability area is a concept used to assess the probability of a sensitive consumer being affected by voltage dips due to faults in the power system. Its mapping plays an important role in transmission utility and sensitive consumers. Traditionally, the area of vulnerability is obtained by fault simulations in all buses of the system, which allow one to define an area around the point of common coupling, where the consumer and its sensitive equipment or processes are located.

Several studies have established the correlation between voltage dips and electrical faults in a power system. Generally, disturbances in the transmission system are more severe for industrial equipment when compared to voltage dips registered within the distribution system. In this context, [14] demonstrated the dependence of voltage dips on the characteristics of the faults at different voltage levels of a power system. Due to the meshed characteristic of transmission systems, voltage dips caused by transmission line faults spread over long distances.

Additionally, the studies presented by [14,15] have contributed to explaining the propagation and causes of power system faults. Similarly, [16] presented some advances based on load variation and undelivered energy in a transmission system due to voltage dips. In this case, the power flow variation was used to assess the severity of the voltage dips. Moreover, [16] established important conclusions by using actual power system faults.

The study presented in [17] dealt with the evaluation of the origin of voltage dips due to faults in real networks by using predictive methods, where measures of voltage dips only were used. In that work, the data from the protection systems were not used and the electrical faults were estimated for each voltage dip record. Thus most studies contemplate the assessment of voltage dips by predictive methods.

The work presented in[16] assessed the characteristics of electrical faults through real data of operation of the power system. In relation to the survey of the vulnerability area, all works that establish this delimitation used only the magnitude of the voltage dips as an indicator.

Therefore this chapter provides a study of the vulnerability area based on sensitivity classes defined by the Brazilian grid code [8] and the severity index defined by [18]. It brings relevant information to assess industrial plants' vulnerability considering transmission system fault characteristics, such as location, type, and fault cause.

Even though there are many papers about voltage dip mitigation, the importance of considering the event correlation to establish measures for preventive maintenance in the power system is presented in [17]. Consequently, the proposal for mitigation measures in the power system to reduce the incidence of voltage dips in the industry was also justified.

In the context thus far described, this chapter deals with the quantification of voltage dips utilizing standardized indexes, as well as the mapping of their causes by a monitoring case study of a Brazilian paper and pulp industry. The most sensitive production areas for voltage dips are sheet formation and drying because both need accurate speed control. If one or more of the machines in the drying or sheet formation process suddenly stops due to a voltage dip, the paper will break inside the machine [19]. Due to the size of these machines, it might take hours to clean the equipment and to restart the whole process.

Each shutdown in the paper industry due to voltage dips is estimated to cost U$30,000 [20]. Thus to compensate for the poor power quality of the utility in question, continuous power quality monitoring is needed.

In this case study, the sensitive consumer is connected to the transmission level (230 kV) by its 230/34.5/13.8 kV substation. The three voltage levels were monitored for 24 months by power quality analyzers (Schneider ION8650). Also, a real dataset of transmission system faults was obtained for 18 months of this 2-year monitoring period. The voltage dips were assessed according to the Brazilian grid codes and the IEEE 1564 standard. Also, simultaneous voltage dips and recorded faults allowed one to analyze the impact of the events, to study the vulnerability area, and to propose mitigations in the power system under study.

The main parts of the chapter are: (1) assessment of voltage dip number and of severity and FI on the industrial consumer, based on a 2-year monitoring period and on the most recent voltage dip standardization; (2) analysis of the impact of (a) the correlation between monitored voltage dips and recorded faults in the region of analysis, (b) the particularities of the anomaly in the power system, and (c) the influence of the fault variables on the monitored voltage dips; (3) evaluation of voltage dip propagation in power systems, based on the vulnerability area influence and on the relationship between the characteristics of recorded faults and monitored voltage dips; and (4) proposition of a technique to mitigate the occurrence of voltage dips in a sensitive consumer.

After this introduction, Section 8.2 presents the proposed approaches to assess voltage dips and map their origin. Then, Section 8.3 describes the case study and discusses the results. Finally, Section 8.4 presents the main conclusions as well as recommendations to apply the methodology in other power systems.

8.2 Methodology

In this section, two approaches are presented for the analysis of voltage dips in a sensitive consumer. The first approach deals with the quantification of voltage dip impact for sensitive loads. Two standards are presented for this quantification: IEEE 1564 and the Brazilian grid code.

The steps to obtain the standardized indexes are detailed and the differences between the two are analyzed. The second approach presents the mapping for voltage dip origin obtained through the correlation between the faults in the power system and the voltage dips monitored at the end-user. From the grouping of the dips and faults, the area of vulnerability is assessed. Voltage dip magnitude, severity, and load sensitivity classes are used to find solutions in the power system to mitigate the effect of voltage dips in sensitive consumers.

8.2.1 Proposed approach for voltage dip quantification

The proposed methodology for the quantification of dips and the comparison between IEEE 1564 and the Brazilian grid code is illustrated in Fig. 8.1. Subsequently, all the steps used to establish this comparison are explained: data input, phase aggregation, dip aggregation, dip stratification, and dip indexes determination, for further assessment of voltage dips and comparative analysis.

A. Data input

The data input must contain the results of the monitoring of voltage dips for the analyzed period. Each voltage dip must contain the date and hour of the occurrence, as well as its magnitude, duration, and busbar.

B. Phase aggregation

During an event, monitoring instruments may record dips in one phase or detect dips in more than one phase. In the second case, IEEE 1564 recommends representing multiphase measurements with only one measurement, using the phase with the largest deviation from the nominal voltage. IEEE 1564 also mentions the use of spatial aggregation to find the worst voltage dip at more than one monitoring point.

Similarly, the Brazilian grid code establishes the aggregation of the parameters of amplitude and duration of each event. The aggregation of phases occurs with the use of a single event at the monitoring point, in which, for the voltage dip, the smallest amplitude is used during the event. The duration is

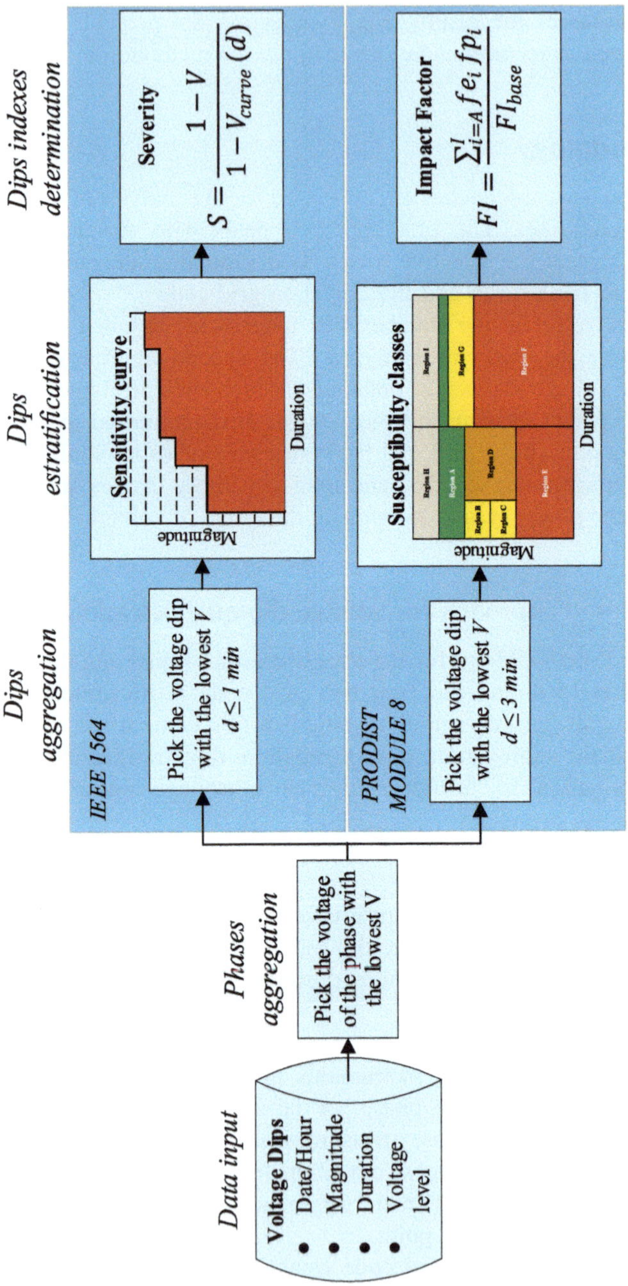

FIGURE 8.1 Proposed approach for the quantification of voltage dips.

defined as the time interval between the instant when the first event crosses a certain limit and the instant when the last event returns to that limit, not exceeding 3 min.

Thus the Brazilian grid code presents two procedures for carrying out the aggregation, based on a criterion of the union of phases. The first procedure consists of aggregation by critical parameters, where the duration of the event is defined as the maximum duration between the three phases and the magnitude value that distances itself the most from the reference voltage. The second procedure proposes aggregation by the critical phase, in which the duration of the event is defined as the duration of the event of critical amplitude, that is, the phase event with minimum amplitude.

C. Dip aggregation

According to the IEEE 1564 standard, if there is a succession of events in the same location, the aggregation is performed considering a single event within a time limit of 1 min. So, the event with lowest magnitude is used to represent the set of events. Similarly, the Brazilian grid code aggregates successive events in the period of 3 min.

D. Dip stratification

The severity index of the voltage dip proposed by IEEE 1564 stratifies the events through equipment sensitivity curves such as CBEMA [21] and SEMI F47 [22]. Those curves characterize an event taking into account magnitude and duration as well as arbitrating intervals according to the probability of causing a load malfunction.

Fundamentally, the calculation is based on the comparison between the stresses obtained by a sensitivity curve and the amplitude and duration of a monitored event. As illustrated in Fig. 8.2, the region above the curve is the immunity area, with severity values less than 1, and events located below the curve (red (gray in print version) area), represent the conditions for possible equipment failures, with severity values greater than 1.

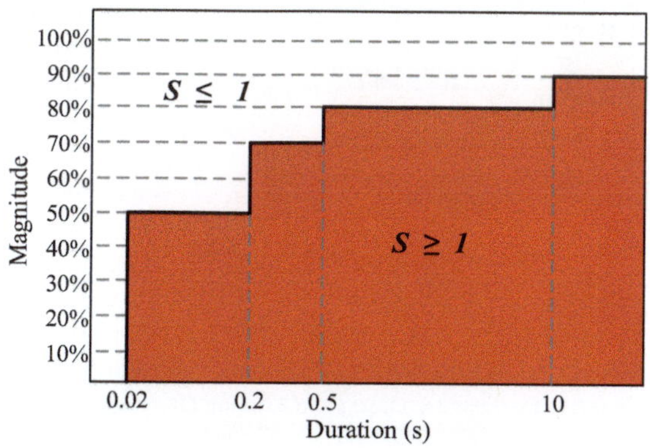

FIGURE 8.2 Voltage dip immunity curve SEMI F47.

According to the Brazilian grid code, the stratification method used for the calculation of the voltage dips' FI is based on the indicators of voltage dip tables. These tables are divided into duration and magnitude intervals, as shown in Table 8.1.

E. Dip indexes determination

To assess voltage dips, each method classifies the event with a severity index or impact factor, depending on the standardization applied to the quantification. In this scenario, the severity index of a voltage dip, proposed by IEEE 1564, characterizes the level of impact that a certain event caused according to:

$$S = \frac{1 - V}{1 - V_{curve}(d)}. \tag{8.1}$$

where $V_{curve}(d)$ is the voltage dip magnitude of the tolerance curve for a voltage dip duration and d and V are the voltage dip magnitude of the monitored event.

Thus voltage dips with a severity index less than 1 are considered within the area of operation of the equipment, but this does not imply that the equipment or the process will not be affected. Conversely, voltage dips with a severity index greater than 1 may produce process failures. Events with magnitude above 0.9 pu will have a severity index of 0. As for the monitoring period, IEEE 1564 indicates 1 year, and the index can be normalized for each monthly case. Note that the total voltage dip severity corresponds to the sum of

TABLE 8.1 Sensitivity classes proposed in the Brazilian grid code.

<table>
<tr><th rowspan="2">Magnitude [pu]</th><th colspan="7">Duration</th></tr>
<tr><th>D1</th><th>D2</th><th>D3</th><th>D4</th><th>D5</th><th>D6</th><th>D7</th></tr>
<tr><td>> 1.15</td><td colspan="3" rowspan="2">Region H</td><td colspan="4" rowspan="2">Region I</td></tr>
<tr><td>(1.10 - 1.15]</td></tr>
<tr><td>(0.85 - 0.9]</td><td colspan="3" rowspan="2">Region A</td><td colspan="4"></td></tr>
<tr><td>(0.80 - 0.85]</td><td colspan="4" rowspan="2">Region G</td></tr>
<tr><td>(0.70 - 0.80]</td><td rowspan="2">Region B</td><td colspan="2" rowspan="4">Region D</td></tr>
<tr><td>(0.60 - 0.70]</td><td colspan="4" rowspan="7">Region F</td></tr>
<tr><td>(0.50 - 0.60]</td><td rowspan="2">Region C</td></tr>
<tr><td>(0.40 - 0.50]</td></tr>
<tr><td>(0.30 - 0.40]</td><td colspan="3" rowspan="4">Region E</td></tr>
<tr><td>(0.20 - 0.30]</td></tr>
<tr><td>(0.10 - 0.20]</td></tr>
<tr><td>< 0.10</td></tr>
</table>

where D1: 16.67 ms $< d <$ 100 ms, D2: 100 ms $\leq d <$ 300 ms, D3: 300 ms $\leq d <$ 600 ms, D4: 600 ms $\leq d <$ 1 s, D5: 1 s $\leq d <$ 3 s, D6: 3 s $\leq d <$ 1 min, D7: 1 min $\leq d <$ 3 min.

TABLE 8.2 Weighting factor for each sensitivity class.

Weighting factor								
A	B	C	D	E	F	G	H	I
0.00	0.04	0.1	0.2	0.3	0.4	0.1	0	0.04

the severity indexes for all voltage dip events computed at a given location over a given time period, and the average voltage dip severity is the average severity index of all voltage dip events computed at a given location for a given period of time.

Similarly, to quantify the intensity of the events of both voltage dips and voltage swells, the Brazilian grid code proposed the FI. The FI calculation considers a weighting factor for each sensitivity region, with weights defined according to the intervals with the highest incidence of problems in processes when submitted to the disturbances. Using a base value, which analyzes the voltage level where the event occurred, the FI is calculated according to:

$$FI = \frac{\sum_{i=A}^{I} fe_i \, fp_i}{FI_{base}}, \tag{8.2}$$

where fe_i is the number of occurrences of events in a determined sensitivity class (A, B, C, D, F, G, H, and I), fp_i is the weighting factor for each sensitivity class, and FI_{base} is the base FI, obtained from the sum of the products of the weighting factors by the maximum allowed number of occurrences for each class. The values of fp_i for each sensitivity class are shown in Table 8.2 and values of FI_{base} for different voltage levels are shown in Table 8.3.

The reference value for the FI index is 1.0 pu. Regarding the minimum monitoring period, even when based on NRS 048, which establishes a 1-year monitoring period, the Brazilian grid code establishes a minimum period of 30 consecutive days of measurement. The purpose of monitoring in a short period of time aims to ensure that the industrial consumer has a faster response with respect to the quality of energy supplied. On the other hand, short monitoring periods could yield a very random result of FI due the stochastic and weather-dependent characteristics of voltage dips.

TABLE 8.3 FI_{base} for distinct voltage levels.

FI_{base}	
$1 < V_N < 69$ kV	$69 < V_N < 230$ kV
2.13	1.42

After the quantification is completed with the severity index and FI, the results of both indexes are compared for the same periods, aiming to highlight the classification differences between the standardizations. In this way, it will be possible to compare the national and international recommendations, raising the particularities of each one for the different cases of voltage dips.

8.2.2 Proposed approach for grouping the faults and voltage dips

As there is a cause/effect relationship between faults and voltage dips, it is of utmost importance to analyze the incidence relationship between the fault characteristics and the voltage dips. With the purpose of verifying the association between both, MATLAB is used to group data from voltage dips monitoring and fault recording. Fig. 8.3 illustrates the methodology applied to find the groups between faults and clustering, to assess the vulnerability area, and to recommend mitigation actions. Each step of the knowledge discovery process from the databases is detailed.

A. Data Entry

The input consists of two datasets of historical records: one containing the data of the voltage dips in the consumer under analysis (the same used in the previous proposition), and another containing the electrical faults of the system in which the consumer is connected. The data for the voltage dips are date/time of the disturbance, magnitude, duration, severity, and sensitivity class. The data of the electrical faults are given by date/time of the fault, faulted equipment, and cause of the fault.

B. Simultaneity of events

The correlation between faults and voltage dips can be identified through the simultaneity of the events. If both fault and voltage dip records correspond to the same date/time, the electrical fault at that moment is considered to be the cause of the voltage dip. Thus the records of the two phenomena are related, creating a new database that presents all simultaneous events of faults with voltage dip.

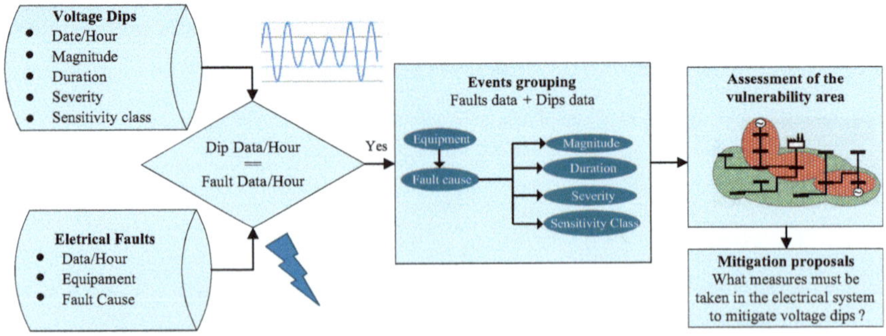

FIGURE 8.3 Proposed approach for grouping the main influence variables.

C. Events grouping

Once the new database was established, data processing and visualizing are carried out by MATLAB. Through scatter histograms, it is possible to generate data visualization and understand the relationship between the fault characteristics and the magnitudes of the voltage dips.

D. Assessment of the vulnerability area

From the relations obtained in the previous step, it is possible to draw the area of vulnerability of voltage dips of the sensitive consumer. The area of vulnerability corresponds to the area of the power system associated with a monitoring point, where the occurrence of electrical faults can cause severe voltage dips in the analyzed busbar.

In addition to the vulnerability according to the dip magnitude as proposed by [23], the area of vulnerability can also be drawn according to the event severity (IEEE) and the dip sensitivity classes (Brazilian grid code). Consequently, such a survey will allow a comparison of the faulted area that produces critical voltage dips according to the severity index and the FI. These results will show accuracy only when the voltage dip magnitude is considered to characterize the effect of an electrical fault on sensitive loads.

E. Mitigation recommendations

In addition to assisting in the assessment of the area of vulnerability, the relationship between electrical faults and voltage dips allows for the establishment of measures to mitigate the occurrence of load shutdowns in sensitive consumers. In this research, the proposals for mitigation are based on records of faults that caused severe voltage dips, taking into consideration their magnitude and magnitude duration classes with higher weighting (E or F). Thus, in this step, measures are sought to reduce the occurrence of these faults in the power system and, consequently, reduce the occurrence of the corresponding critical voltage dips.

8.3 Case study and results

The simplified single-line diagram of the sensitive consumer connection to the power system is illustrated in Fig. 8.4. The sensitive consumer is connected at 230 kV to bus 12 and has a stepdown substation to 34.5 kV and 13.8 kV. Additionally, the industry contains embedded generation composed of five turbogenerators connected at 13.8 kV, which have a nominal power equal to 251 MW.

Power quality monitoring was carried out at the three voltage levels using Schneider power quality analyzers model ION8650 [24] for 24 months. The months, which correspond to the period May 2015 to May 2017, are represented in numbers from 1 to 24. These analyzers perform measurements in accordance with IEC 61000-4-34 [25] class A/S, recording date/time,

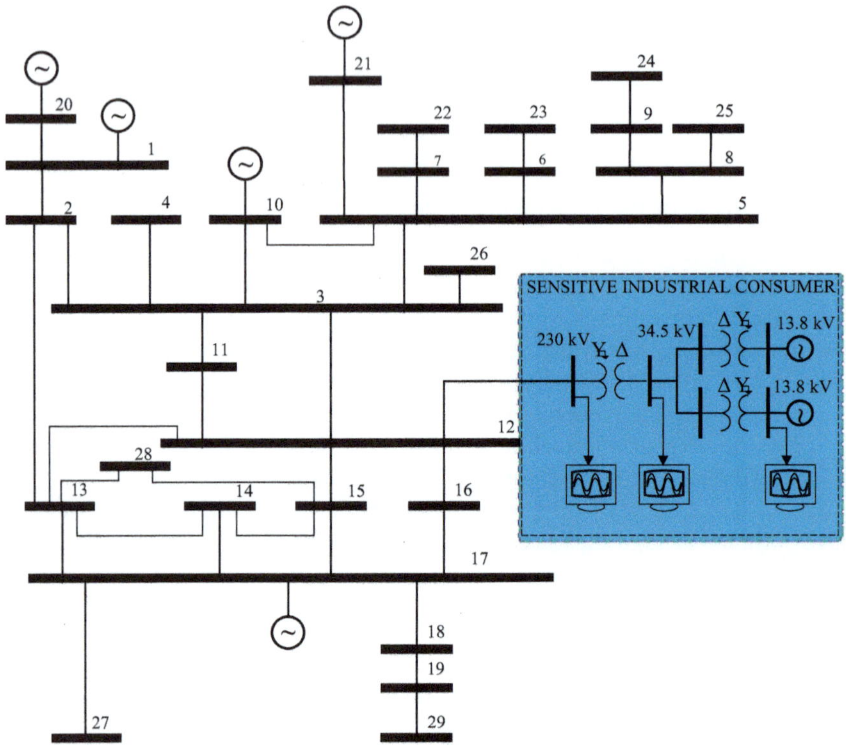

FIGURE 8.4 Simplified single-line diagram of the sensitive consumer connection to the power system.

magnitude, and duration of events. Similarly, the history of electrical faults was obtained from the fault records of the 230 kV power transmission equipment for 17 months of the 2-year period of voltage dip monitoring.

8.3.1 Quantification of voltage dips and comparison of IEEE 1564 and brazilian grid code

The number of occurrences by voltage level was determined by the aggregation of events. Fig. 8.5 shows the distribution of the number of voltage dips during the 24 months at each monitored busbar, where the blue (light gray in print version) bars indicate the quantification by IEEE 1564 and the red (dark gray in print version) bars indicate the quantification by the Brazilian grid code.

As seen in Fig. 8.5, it is possible to evaluate the months with the highest occurrence of voltage dips for each monitoring level. For the 13.8 kV level, the seventh month was the most affected, with nine events and, subsequently, the 11th month with seven occurrences. For 34.5 kV, the month with the highest incidence of voltage dips was the 20th with five voltage dips and then the 18th

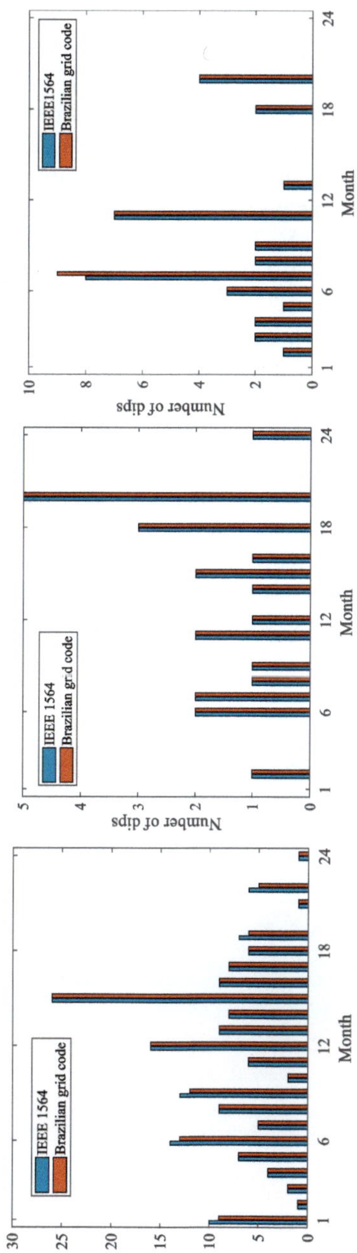

FIGURE 8.5 Number of voltage dips per month quantified by IEEE 1564 and the Brazilian grid in three voltage levels: 230 kV (*left*), 34.5 kV (*center*), and 13.8 kV (*right*).

with three events. For 230 kV, the highest number of events was the 15th month with 26 voltage dips, followed by the 12th month with 16 voltage dips, and the sixth and ninth months both with 13 voltage dips. Due to the random fault occurrence, the monthly number of voltage dips was different for periods 1−12 (year 1) and 13−24 (year 2).

In total, the IEEE 1564 computed 228 voltage dips in 2 years, distributed over the three busbars, as follows: 35 dips at 13.8 kV, 23 dips at 34.5 kV, and 170 dips at 230 kV. In contrast, according to the Brazilian grid code there were 224 voltage dips in the same period, with 36 dips at 13.8 kV, 23 dips at 34.5 kV, and 165 dips at 230 kV. Therefore the quantification by IEEE 1564 was slightly higher than that by the Brazilian grid code at the levels of 230 and 13.8 kV. For the voltage level of 230 kV, months 1, 6, 9, 19, and 22 computed different numbers of voltage dips between IEEE 1564 and the Brazilian grid code. For the voltage level of 13.8 kV, month 7 was the one that showed differences in quantification. Table 8.4 presents a description of these events, which were differently quantified by each standard.

With respect to the 13.8 kV level, Table 8.2 shows that the difference in month 7 is due to the record of a voltage dip whose duration is 135 s. The divergent quantification of the recommendations results from the maximum duration of voltage dips; since IEEE 1564 considers events lasting up to 1 min, and the Brazilian grid code considers those lasting up to 3 min. As the duration of the event is longer than 60 s, IEEE 1564 did not quantify such event as a voltage dip.

TABLE 8.4 Stratification of the voltage dips over the 24 months by the Brazilian grid code.

<table>
<tr><th rowspan="2">Magnitude [pu]</th><th colspan="7">Duration</th></tr>
<tr><th>D1</th><th>D2</th><th>D3</th><th>D4</th><th>D5</th><th>D6</th><th>D7</th></tr>
<tr><td>>1.15</td><td colspan="3" rowspan="2">H=[0, 0, 0]</td><td colspan="4" rowspan="2">I=[0, 0, 0]</td></tr>
<tr><td>(1.10 - 1.15]</td></tr>
<tr><td>(0.85 - 0.9]</td><td colspan="3" rowspan="2">A=[125, 16, 21]</td><td colspan="4"></td></tr>
<tr><td>(0.80 - 0.85]</td><td colspan="4" rowspan="2">G=[3, 0, 1]</td></tr>
<tr><td>(0.70 - 0.80]</td><td rowspan="2">B=[25, 1,5]</td><td colspan="2" rowspan="4">D=[4, 0, 1]</td></tr>
<tr><td>(0.60 - 0.70]</td><td colspan="4" rowspan="8">F=[0, 0, 0]</td></tr>
<tr><td>(0.50 - 0.60]</td><td rowspan="2">C=[3,0,3]</td></tr>
<tr><td>(0.40 - 0.50]</td></tr>
<tr><td>(0.30 - 0.40]</td><td colspan="3" rowspan="4">E=[5, 5, 5]</td></tr>
<tr><td>(0.20 - 0.30]</td></tr>
<tr><td>(0.10 - 0.20]</td></tr>
<tr><td><0.10</td></tr>
</table>

For the 230 kV busbar, the difference in quantification between the standardizations resulted from the voltage dip aggregation window. The events of months 1, 6, 9, 19, and 22 at 230 kV shown in Table 8.2 were distinctly aggregated by the two standards. In month 1, the difference was the result of two voltage dip records in 2 min. As the aggregation of the Brazilian grid code is carried with a time window of 3 min and the aggregation window of IEEE 1564 is 1 min, the Brazilian standard registered only one event for this interval, while the IEEE standard registered two voltage dips. In month 6, the aggregation by the IEEE 1564 again computed one more event, due to the occurrence of two voltage dips in a period of 110 s. In month 9, the occurrence of two events over a 93 s period also led IEEE 1564 to quantify one more voltage dip than the Brazilian grid code. In months 19 and 22, the justification for the difference was also the same, since two events were recorded one each month in periods of 178 s and 85 s, respectively.

The events were classified according to the distribution of magnitude and duration. Fig. 8.6 shows the stratification through the curve SEMI F47 as proposed in IEEE 1564 for the three voltage levels. Similarly, Table 8.4 shows the stratification proposed by the Brazilian grid code in the three voltage levels. The number of voltage dips in each sensitivity class is presented by the following format: dips in 230 kV, dips in 34.5 kV, and dips in 13.8 kV. For both Fig. 8.6 and Table 8.4, all the voltage dips over the 24-month period were glutted in the same plot. The foregoing is just to exemplify how to perform the stratification and to show an overview of the voltage dips for the whole period. To quantify the severity index and FI, this stratification should be made monthly.

Most of the voltage dips are classified as nonsevere because they are above the sensitivity curve as shown in Fig. 8.6. In the 230 kV busbar most of the severe voltage dips presented a magnitude lower than 50% and a duration lower than 0.2 s; but also, severe voltage dips presented a magnitude higher than 70% and a duration longer than 0.7 s. On the 34.5 kV busbar, the severe voltage dips presented a magnitude lower than 40% and a duration lower than 0.2 s; this busbar also suffered severe voltage dips with magnitude higher than 70% and duration longer than 0.8 s. The 13.8 kV busbar presented severe voltage dips with magnitude lower than 40% and duration shorter than 1 s.

Table 8.4 shows that most of the voltage dips are classified in the sensitivity class A (green (grey in print version)) and therefore does not contribute to the FI. A lower number of voltage dips is classified into the sensitivity classes E and F, which have the highest weighting factor when compared to the number of severe voltage dips classified by IEEE 1564. The 230 kV busbar presented seven severe voltage dips, but only five dips in class E and none in class F. The 34.5 kV busbar presented six severe voltage dips, five dips in class E and none in class F. The 13.8 kV busbar presented eight severe voltage dips, five dips in class E and none in class F.

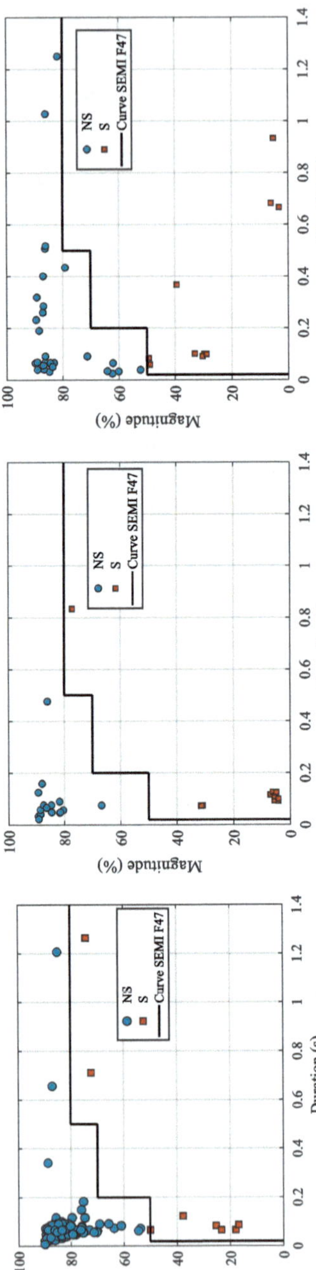

FIGURE 8.6 Magnitude versus duration of the voltage dips over the 24 months in three busbars: 230 kV (*left*), 34.5 kV (*center*), and 13.8 kV (*right*).

Table 8.5 details the months that presented differences in the stratification between both standards. The stratification in sensitivity classes did not classify the severe voltage dips into regions E and F for months 6, 9, 15, and 17 at the 230 kV busbar. By using Table 8.5 it is possible to analyze those voltage dips that were classified by the Brazilian grid code in classes D and G. In the 34.5 kV busbar, month 18 presented a single severe voltage dip that was classified in class G. The severe events at the 13.8 kV busbar were in class G during months 4 and 20.

A severity index and FI were obtained for each busbar according to the monthly stratification shown in Fig. 8.7. The severity index was higher than 1 for the last month of monitoring and the FI was below unity for the 230 kV busbar. During the monthly analysis, a single event whose severity index is larger than 1 resulted in a severe month classification, regardless of it being a single occurrence. According to the Brazilian grid code, this event classification is E. One event alone at class E is not enough to create an FI larger than 1 according to Eq. (8.2). For this voltage level, the number of voltage dips in class E must be at least nine to obtain an FI larger than 1.

For the 34.5 kV level, only month 20 was classified as severe, but the FI was also less than the limit value. Month 20 presented a severity index greater than 1 because the five voltage dips that occurred in the month presented $S \geq 1$. Thus when calculating the monthly severity index, the average among the five dips was higher than unity. The five dips were classified into class E according to the Brazilian grid code. The weight used in Eq. (8.2) for voltage dips in class E is 0.25. Therefore, when applying Eq. (8.2), the FI is less than 1.

TABLE 8.5 Monthly stratification of voltage dips.

Voltage level	Month	IEEE 1564		Brazilian grid code								
		NS	*S*	*A*	*B*	*C*	*D*	*E*	*F*	*G*	*H*	*I*
230 kV	6	12	2	4	0	0	2	0	0	0	0	0
	9	12	1	8	1	0	2	0	0	1	0	0
	15	25	1	23	0	1	1	0	0	1	0	0
	17	7	1	6	1	0	0	0	0	1	0	0
34.5 kV	18	2	1	2	0	0	0	0	0	1	0	0
13.8 kV	4	1	1	0	0	2	0	0	0	0	0	0
	20	3	1	3	0	1	0	0	0	0	0	0

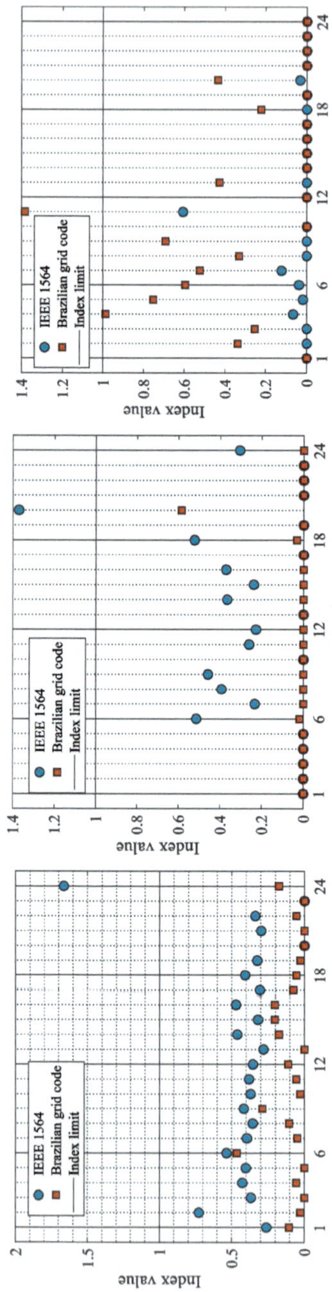

FIGURE 8.7 Severity index and FI for 24 months in three busbars: 230 kV (*top left*), 34.5 kV (*top right*), and 13.8 kV (*bottom*).

For the 13.8 kV busbar, the 11th month was the only month that presented a severity index greater than unity; however, the FI was lower than unity. Month 11 presented a severity index greater than 1 because from the seven events that occurred in the month, five presented $S > 1$. Thus when calculating the monthly index, the average among the seven events was greater than 1. On the other hand, the FI calculation depends on the classification of monthly events, which in this case, according to Table 8.5, is: A = 1, B = 1, E = 5. The weighting factors of these regions are 0, 0.04, and 0.25, respectively, and the FI_{base} is 2.13. Thus, according to Eq. (8.2), the FI is less than 1.

8.3.2 Grouping the main influence variables

The methodology presented in Fig. 8.3 was applied to the datasets of voltage dips and electrical faults in the power system to which the industry is connected. As the occurrences in the power system were obtained only from the fourth to the 21st month, the analysis was performed during these 18 months. Fig. 8.8 shows the total number of dips in the three voltage levels and the number of electrical faults recorded in the power system for each month. The Pearson correlation index (ρ) obtained between the two events is 0.6874. According to [26], a ρ between 0.5 and 0.7 indicates a moderate correlation between the two variables. Considering all the voltage levels, 140 voltage dips out of a total of 203 were simultaneous to electrical faults.

Thus it was possible to relate 51% of the voltage dips at the 230 kV busbar with electrical faults, 29% at the 34.5 kV busbar, and only 12.5% at the 13.8 kV busbar. The lower percentage of relationships in the levels of 13.8 and 34.5 kV occurs because those events correspond to the propagation of voltage dips from 230 kV to the lower levels. Fig. 8.9 exemplifies this propagation.

A fault in the transmission line 12–16 (red (dark gray in print version)), which is close to the industry consumer, generated a voltage dip with a magnitude of 53.73%. Due to the low-voltage magnitude in 230 kV, the

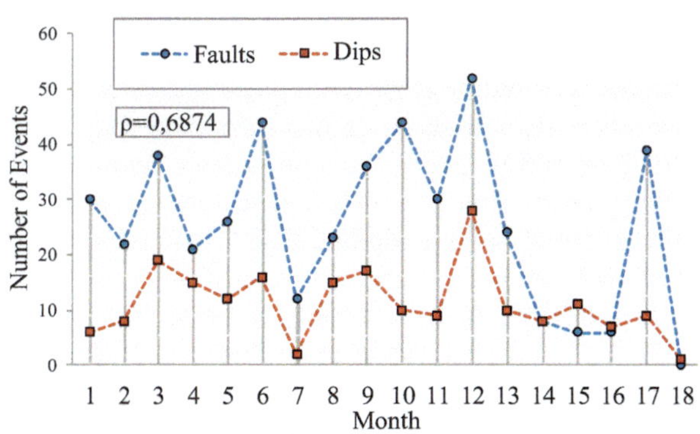

FIGURE 8.8 Number of faults and voltage dips over 18 months.

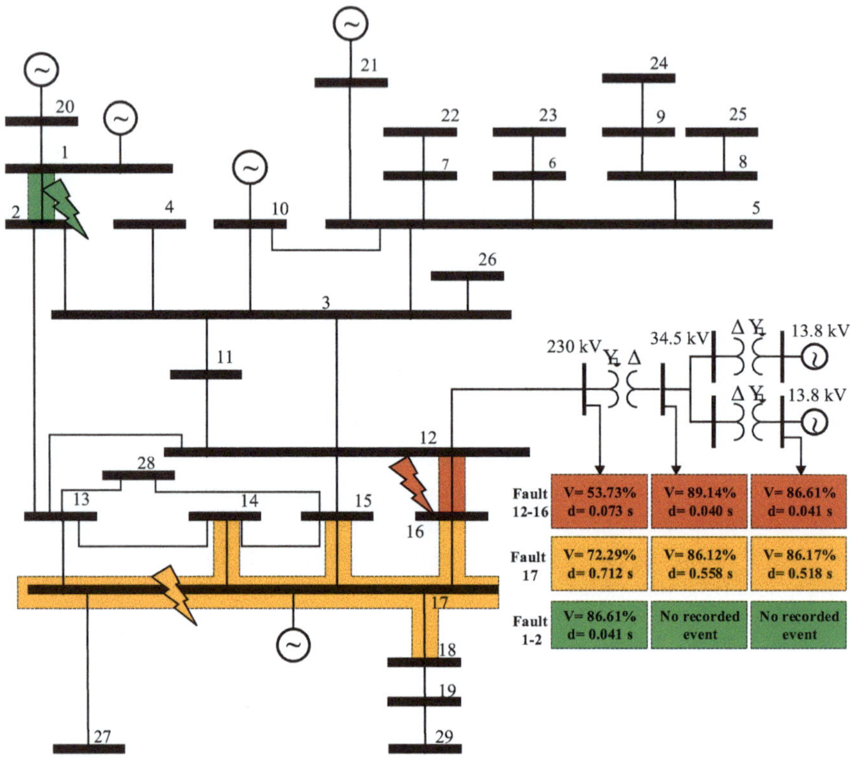

FIGURE 8.9 Examples of voltage dips related to faults in the power system: fault in lines 12–16 (red (dark gray in print version)), fault in bus 17 (yellow (light gray in print version)), and fault in lines 1–2 (red (dark gray in print version)).

voltage dip propagated to the lower voltage levels: 34.5 kV presented a voltage dip with a magnitude of 88.14% and 13.8 kV a voltage dip with a magnitude of 86.61%. The magnitude is higher in the levels of 34.5 kV and 13.8 kV due to the changes in the voltage dip type caused by the transformer windings connection Y–Δ.

For a fault in bus 17 (yellow (light gray in print version)), the propagation to the lower levels was also observed. In this case, the voltage dips presented a magnitude of 72.29% in 230 kV, 86.12% in 34.5 kV, and 86.17% in 13.8 kV. For a fault in the transmission line 1–2 (green (gray in print version)), the magnitude of the resulting voltage dip in 230 kV was 86.61%. No events were recorded for the levels of 34.5 kV and 13.8 kV. In this case, the magnitude is higher than the two previous faults. Probably, when the voltage dip propagated to the levels of 34.5 kV and 13.8 kV the magnitude was higher than or equal to 90%. Thus no events were recorded for the lower voltage levels.

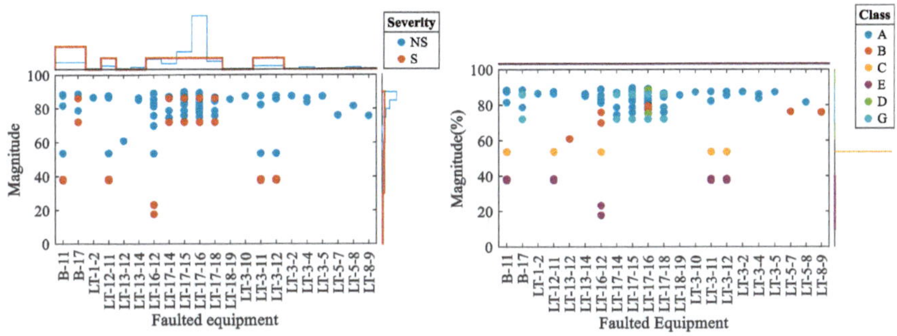

FIGURE 8.10 Scatter histograms relating faulted equipment, dip magnitude, and indexes stratification: severity by IEEE 1564 (*left*) and sensitivity classes by the Brazilian grid code (*right*).

To perform the grouping between the fault causes and the related 140 dips, scatter histograms were generated to visualize the simultaneous events. Fig. 8.10 shows the scatter histograms relating faulted equipment, dip magnitude, severity index, and sensitivity classes. Similarly, Fig. 8.11 shows the scatter histograms relating to faulted equipment, dip duration, and indexes stratification. In Figs. 8.10 and 8.11, the acronym LT means transmission line and B means bus. By analyzing Fig. 8.10 (left), it is possible to see that two ranges of magnitude are related to severe dips. For faults in equipment B-11, B12-11, LT-16-12, LT-3-11, and LT-3-12 the severe dips are related to a magnitude lower than 40%. However, the voltage dips whose magnitude is higher than 70% were also classified as severe for faults in equipment B-17, LT-17-14, LT 17-15, LT 17-16, and LT 17−18. For the voltage dips originated in this equipment, the reason to classify them as severe is related to the duration of the dips and not to their magnitude. This can be analyzed in Fig. 8.11 (left), where the severe events present a duration higher than 0.4 s for equipment B-17, LT-17-14, LT-17-15, LT-17-16, and LT-17-18.

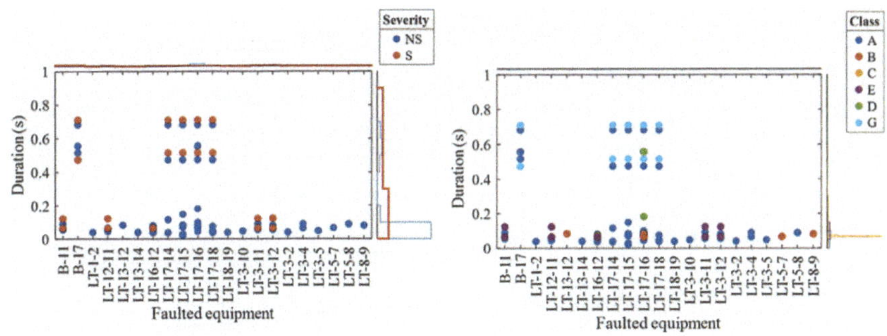

FIGURE 8.11 Scatter histograms relating faulted equipment, dip duration, and indexes stratification: severity by IEEE 1564 (*left*) and sensitivity classes by the Brazilian grid code (*right*).

Regarding the sensitivity classes from the Brazilian grid code, the faults in equipment B-11, B12-11, LT-16-12, LT-3-11, and LT-3-12, whose magnitude was lower than 40%, were classified in class E. Therefore both IEEE 1564 and the Brazilian grid code weighted in a similar way the events with lower magnitude. However, to classify the events with higher duration also in a high weighting class such as E or F, the Brazilian grid code classified as class G the dips from the faults caused by B-17, LT-17-14, LT-17-15, LT-17-16, and LT-17-18.

Fig. 8.12 shows the scatter histogram considering faulted equipment, fault cause, and both indexes stratification. The acronym FE means equipment failures in protection, sensors, and isolators, among others; DI means undetermined; CN stands for natural causes such as rain, wind, and storms; and AVO means animals, vegetation, or objects in contact with the equipment.

In Fig. 8.12 (left) it is possible to see that the severe events are caused by equipment failure in B-11, B-17, LT-12-11, LT-16-12, LT-17-14, LT-17-15, LT-17-16, LT-17-18, LT-3-11, and LT-3-12; by undetermined causes in LT-16-12; and by animals, vegetation, or objects in contact with the equipment in LT-16-12. Analyzing Fig. 8.12 (right), the voltage dips classified in class E are caused by equipment failures in B11, LT-12-11, LT-3-11, and LT-3-12; by undetermined causes in LT-16-12, and by animals, vegetation, or objects in contact with the equipment in LT-16-12.

Fig. 8.13 summarizes the analysis made through the scatter histograms, showing the grouping results for the most critical cases of voltage dips: magnitude lower than 40% (top left), duration higher than 0.4 s (top right), classification in class E of the Brazilian grid code (bottom left), and classification as a severe dip by IEEE 1564 (bottom right). The numbers in the arrows indicate the number of events that satisfy the previous condition of the balloon.

By observing Fig. 8.13 it is possible to conclude that the analysis of the effect of voltage dips by only one characteristic (magnitude or duration) is

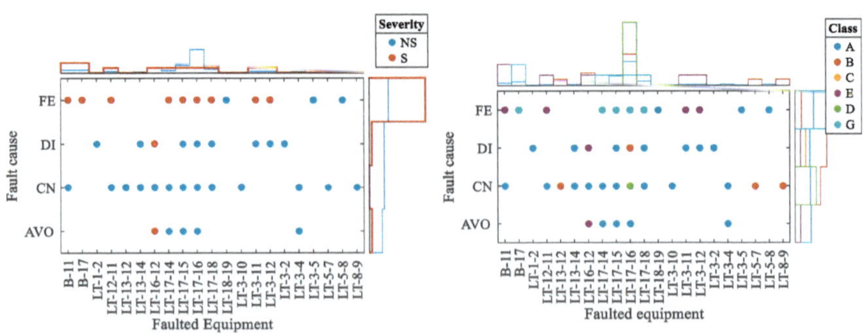

FIGURE 8.12 Scatter histograms relating faulted equipment, fault cause, and indexes' stratification: severity by IEEE 1564 (*left*) and sensitivity classes by the Brazilian grid code (*right*).

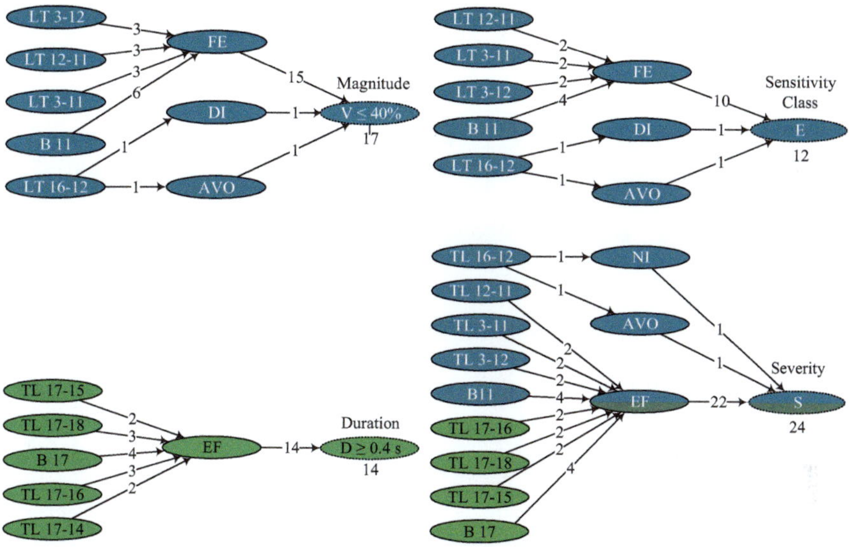

FIGURE 8.13 Grouping results for the critical cases: dip magnitude lower than 40% (*top left*), dip duration higher than or equal to 0.4 s (*top right*), dip in sensitivity class E (*bottom left*), and severe dip (*bottom right*).

very imprecise since the conditions for the magnitude to be small and the duration to be long differ. Thus the use of standards is justified because they allow one to obtain a single indicator that encompasses the conditions of magnitude and duration. For this dataset, only the severity indicator included the critical characteristics of low magnitude and high duration.

8.3.3 Assessment of the voltage dip vulnerability area

From the groupings assessed in the previous subsection, it would be possible to generate a vulnerability area according to the magnitude, duration, severity, and sensitivity class. However, as concluded in the previous analysis, the severity index can incorporate both critical conditions of magnitude and duration into a unique index. Thus the vulnerability area was mapped according to the severity index. Fig. 8.14 shows the assessed vulnerability area: the green (shaded light gray in print version) area indicates faulted equipment that generated dips with severity lower than 1 and the red (shaded dark gray in print version) area indicates faulted equipment that generated dips with severity higher than 1.

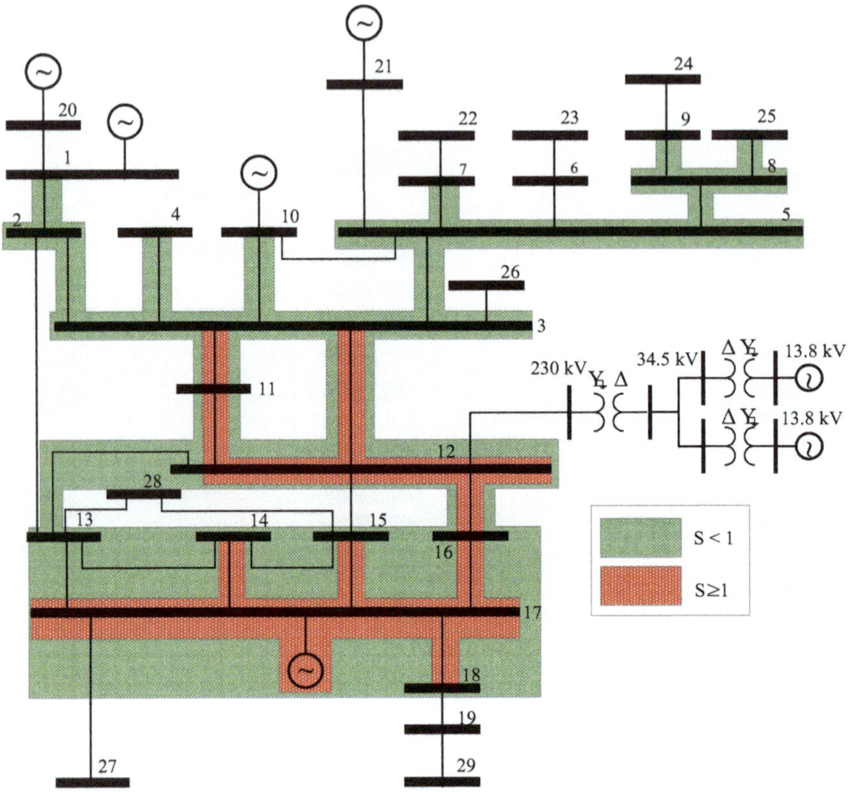

FIGURE 8.14 Vulnerability area for voltage dips according to the severity index: nonsevere dips (green (shaded light gray in print version)) and severe dips (red (shaded dark gray in print version)).

8.4 Conclusions

This chapter dealt with a comparison of the IEEE 1564 standard and the Brazilian grid code. For such comparison, voltage dips were monitored in a paper and pulp plant for 24 months. By the simple quantification of events, a difference was noted in the number of occurrences computed according to each standard. This divergence arose from the definition of a voltage dip according to each standard, since IEEE 1564 considers events lasting up to 1 min and the Brazilian grid code up to 3 min.

Similarly, the time interval for aggregation is different between the procedures, since that of IEEE is carried out for 1 min and that of the Brazilian grid code is carried out for 3 min. Even with divergent computation, in both quantifications the level with the largest number of dips was 230 kV. By applying the FI indicator, it was noted that in the three voltage levels, all months of monitoring showed $FI < 1$. On the other hand, the severity indicator indicated months with $S \geq 1$. This difference was due to the Brazilian grid code FI_{base}.

In this chapter, the grouping between variables of electrical faults and voltage dips was also established. The results of the groupings showed that with the use of the severity index it was possible to incorporate the critical cases of magnitude and duration of the dips. On the contrary, using the sensitivity classes from the Brazilian grid code, the critical conditions established were bias by the voltage dip magnitude. Furthermore, mitigation solutions were recommended to reduce the number of faults in the power systems, consequently reducing the effect of voltage dips on the industrial consumer. Although the measures may already be known to the energy concessionaire, these improvements would be made according to the interests of the industrial consumer.

References

[1] IEEE, Standard 1346: IEEE Recommended Practice for Evaluating Electric Power System Compatibility with Electronic Process Equipment, 1998.

[2] IEEE, Standard 493: IEEE Recommended Practice for the Design of Reliable Industrial and Commercial Power Systems, 2007.

[3] IEC, "Testing and Measurement Techniques - General Guide on Harmonics and Interharmonics Measurements and Instrumentation, for Power Supply Systems and Equipment Connected Thereto." 202AD.

[4] IEC, 61000-4-11 - Electromagnetic Compatibility. Testing and Measurement Techniques - Voltage Dips, Short Interruptions and Voltage Variations Immunity Tests, 2005.

[5] IEC, NRS 048-2: Electricity Supply - Quality of Supply Part 2: Minimum Standards for Application by the National Electricity Regulator, 1996.

[6] IEEE, 1564 Task Force Group, 2017.

[7] ANEEL, Consulta Pública n.05, Brasília, 2011.

[8] ANEEL, PRODIST Módulo 8 - Revisão 7, Brasília, 2016.

[9] ANEEL, Consulta Pública n.018, Brasília, 2014.

[10] D.D. Sabin, M.H.J. Bollen, Overview of IEEE Std 1564-2014 guide for voltage sag indices, in: Harmonics and Quality of Power (ICHQP), 2014 IEEE 16th International Conference on, 2014, pp. 497–501.

[11] S. Arias-Guzmán, et al., Analysis of voltage sag severity case study in an industrial circuit, IEEE Trans. Ind. Appl. 53 (1) (2017) 15–21.

[12] E. de Oliveira Monteiro, A. de L. Ferreira Filho, Comparative analysis of regulatory procedures for classifiying short-duration voltage variations, in: Harmonics and Quality of Power (ICHQP), 2016 17th International Conference on, 2016, pp. 569–574.

[13] T.R. Kempner, Análise da robustez e da sensibilidade de sistemas de distribuição para a alocação otimizada de medidores frente às variações de tensão de curta duração, 2017.

[14] N. Essl, H. Renner, Assessing the severity of voltage sags due to short circuits in transmission and distribution grids, in: Harmonics and Quality of Power, ICHQP), 2016, pp. 278–282.

[15] J. Miranda Filho, J.M. de Carvalho Filho, A.P. Paiva, P.V.G. de Souza, S. Tomasin, A PCA-based approach for substation clustering for voltage sag studies in the Brazilian new energy context, Elec. Power Syst. Res. 136 (2016) 31–42.

[16] T.C. de Oliveira, J.M. de Carvalho Filho, R.C. Leborgne, G.T. Watanabe, J.P.G. de Abreu, J.F. de Oliveira, Analysis of the relation between voltage sags severity and fault characteristics, in: Harmonics and Quality of Power (ICHQP), 2010 14th International Conference on, 2010, pp. 1–8.

[17] M. De Santis, C. Noce, P. Varilone, P. Verde, Analysis of the origin of measured voltage sags in interconnected networks, Elec. Power Syst. Res. 154 (2018) 391–400.

[18] IEEE, Standard 1564: IEEE Guide for Voltage Sag Indices, 2014.

[19] T.C. Farqui, N. Kagan, M.R. gouvêa, Voltage Sag on Pulp and Paper Industry, Esc. Politécnica da Univ. São Paulo, São Paulo, 2005.

[20] B.H. Chowdhury, Power quality, IEEE potentials 20 (2) (2001) 5–11.

[21] ITI, ITI(Cbmea Curve) Application Note, 2000.

[22] P.S. Lab, SEMI F47-0706 Application Note, 2006.

[23] M.H.J. Bollen, Understanding power quality problems, Piscataway, NJ, USA IEEE 1 (2000) 1–35.

[24] Schneider, "Medidor de energia e qualidade de energia PowerLogic ION8650.", 2018.

[25] 61000-4-34 IEC, - Electromagnetic Compatibility. Voltage Dips, Short Interruptions and Voltage Variations Immunity Tests for Equipment with Input Current More than 16 A Per Phase, 2005.

[26] D.C. Montgomery, G.C. Runger, Applied Statistics and Probability for Engineers, John Wiley & Sons, 2010.

Further reading

[1] I. CGTI, Desenvolvimento de Dispositivo Repelente Automático de Pássaros para Linhas de Transmissão, 2016.

Chapter 9

Power quality improvement by a double-source multilevel inverter with reduced device and standing voltage on switches

Mahdi Karimi[1], Paria Kargar[1], Kazem Varesi[1], P. Sanjeevikumar[2]
[1]*Power Electronics Research Lab. (PERL), Electrical Engineering, Sahand University of Technology, Tabriz, East Azerbaijan, Iran;* [2]*Esbjerg Energy Section, Department of Energy Technology, Aalborg University, Esbjerg, Denmark*

Chapter outline

9.1 Introduction **245**
9.2 Proposed structure **248**
9.2.1 Converter configuration 248
9.2.2 Operational scheme 249
9.2.3 Standing voltage on switches 252
9.2.4 Cascaded version of the proposed topology 253
9.3 Modulation technique **254**
9.4 Power loss analysis **260**
9.4.1 Switching losses 260
9.4.2 Conduction losses 261
9.4.3 Capacitor ripple losses 265
9.5 Capacitor design **265**
9.6 Comparative analysis **265**
9.7 Simulation results **270**
9.8 Power quality analysis **272**
9.8.1 Form factor 272
9.8.2 Ripple factor 272
9.8.3 Harmonic factor 274
9.8.4 Total harmonic distortion 276
9.9 Conclusions **279**
References **279**

9.1 Introduction

Multilevel inverters (MLIs) utilize power electronic devices to convert DC to AC voltage. This conversion is done with low distortion, high (power) quality, low power loss, and high efficiency. MLIs provide less blocking voltage on

Power Quality in Modern Power Systems. https://doi.org/10.1016/B978-0-12-823346-7.00004-9

switches and require smaller filters [1–4]. These converters have recently attracted a great deal of attention from industries, such as: renewable energy systems (photovoltaic or wind turbine applications), high-voltage DC transmission systems, uninterruptible power supplies, high- or medium-voltage motor drives, automotive applications, and flexible AC transmission systems [5–8]. Some examples of the applications of MLIs are illustrated in Fig. 9.1.

These converters are divided into three categories: floating capacitors, neutral point clamped, and cascaded H-bridge. Each of these structures has disadvantages, the most important being the need for a large number of devices to achieve higher voltage levels at the output [9–12]. Therefore researchers have been interested in providing structures that produce the highest number of steps with the least number of devices and the simplest control strategy. One of the important devices is DC voltage sources. DC voltage sources are more expensive, heavier, and bulkier than other devices. Therefore to reduce cost, volume, etc., an element that plays a role similar to DC voltage sources should be used. Capacitors are the best alternatives to DC voltage sources [13]. So, recently, structures based on switched capacitors have been introduced and are called switched-capacitor multilevel inverters

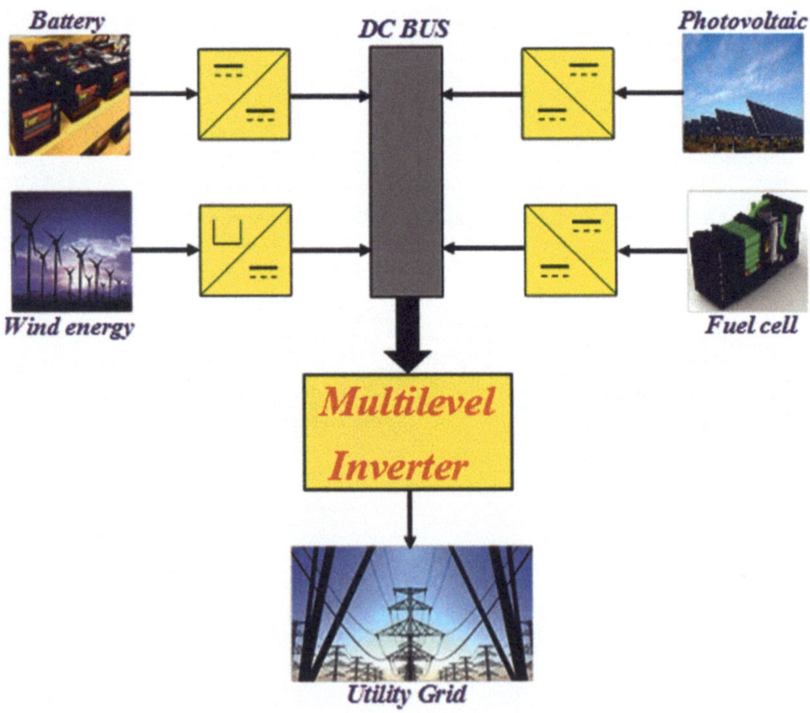

FIGURE 9.1 Typical applications of multilevel inverters.

(SCMLIs) [14,15]. Switched capacitor-based MLIs are divided into two major categories: high-gain SCMLIs [16–18] and unity-gain switched capacitor MLIs (UG-SCMLIs) [19–21]. The first group is dedicated to applications such as solar cells because they have low magnitude of voltage and by using these structures the number of converters used to increase the input voltage range is reduced. However, the second group focuses on not increasing but reducing the number of sources and is dedicated to applications that only require input voltage conversion. In [22] a novel switched capacitor-based module for asymmetric MLIs has been proposed, in which the inherent production of negative voltage levels, step-up capability, and self-voltage balancing of the capacitors are its advantages. However, the high ratio of the number of metal–oxide–semiconductor field-effect transistors (MOSFETs) per number of levels is one of the most important disadvantages. Two unity-gain switched-capacitor-based MLIs have been proposed in [23,24]. These structures use the six-switch H-bridge to produce negative voltage levels. The merit of these two structures is the natural voltage balancing of capacitors without any additional circuits. In contrast, the high ratio of the number of DC sources per number of levels is their important drawback. A new switched-source module for symmetric and asymmetric MLIs has been proposed in [25], where producing negative voltage levels without an H-bridge and low total standing voltage (TSV) are the main benefits, and a high number of DC sources leading to high weight, volume, and size of the converter are significant shortcomings. A unity-gain T-type cascaded H-bridge switched capacitor-based MLI has been proposed in [26], in which the H-bridge needed for producing negative voltage levels and the voltage of capacitors have been naturally balanced. A generalized switched capacitor-based MLI for high-frequency AC power distribution has been presented in [27], which benefits from self-voltage balancing of capacitors and high step-up capability. On the other hand, to achieve a higher number of output voltage levels, a DC source string is required. Also, high TSV and the inability to generate negative voltage levels at the output of the converter without using an H-bridge are the main disadvantages of [27]. A unity-gain modified T-type switched capacitor-based MLI has been presented in [28]. Producing negative voltage levels and a low ratio of the number of MOSFETs per number of levels are important benefits and high TSV is the main drawback. A high step-up switched capacitor-based MLI has been presented in [29]. Self-voltage balancing of capacitors and boosting capability are the significant merits. However, high ratio of number of MOSFETs per number of levels and producing negative voltage levels with an H-bridge, leading to high standing voltage (SV) on switches, are the main disadvantages. A hybrid switched source-based MLI has been presented in [30]. Low number of MOSFETs per number of levels is the important advantage. On the other hand, the high number of DC sources and the H-bridge needed at the end of the converter are the main disadvantages.

One of the topics that has recently attracted the attention of researchers is improving power quality using multilevel converters [31]. So, this study presents a two-source 15-level UG-SCMLI with reduced number of MOSFETs and DC voltage sources per steps, low voltage ripple on capacitors, minimum on-state switches, and improved power quality of the output. One of the main advantages of the proposed structure is the natural voltage balancing of capacitors. The voltage balancing of capacitors is done without using external control circuits. It should also be noted that this structure produces negative voltage levels without the need for an H-bridge.

9.2 Proposed structure

9.2.1 Converter configuration

The proposed 15-level SCMLI shown in Fig. 9.2 consists of eight unidirectional switches, one bidirectional switch, two capacitors (C_1, C_2), and two DC voltage sources (V_L, V_R). Since the bidirectional switch consists of two MOSFETs, the total number of MOSFETs is 10. The unidirectional and common source bidirectional switches demand a single driver circuit. So, the driver count is the same as the switch count. The key parameters of the proposed structure are:

$$N_{Level} = 15, N_{Source} = 2, N_{Capacitor} = 2, N_{Switch} = N_{Driver} = 9, N_{MOSFET} = 10 \tag{9.1}$$

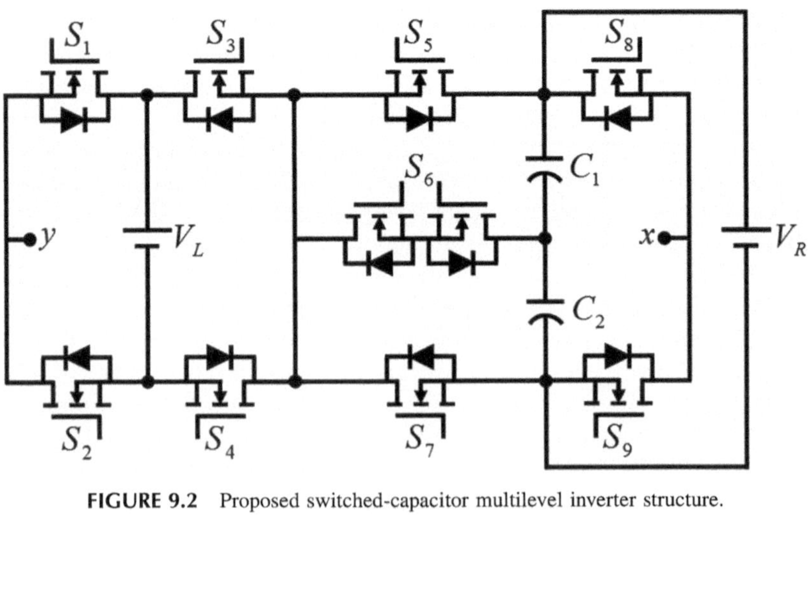

FIGURE 9.2 Proposed switched-capacitor multilevel inverter structure.

It should be noted that the MOSFETs have a reversed-parallel diode in their structure called the body diode. Therefore no diodes are used in this structure.

In the proposed structure, the output voltage is $V_o = V_x - V_y$. Left and right DC voltage sources are considered as $V_L = V_{dc}$, $V_R = 6V_{dc}$ to achieve the highest number of levels at the output port.

9.2.2 Operational scheme

The switching states and charging or discharging modes of capacitors are shown in Table 9.1. It should be noted that "0" and "1" are the OFF and ON modes for the proposed structure switches, respectively. Also, "▲" and "▼" represent the charging and discharging states of capacitors, respectively.

According to Table 9.1, the C_1 and C_2 capacitors are discharged in six successive operational modes, but it should be noted that the source is always

TABLE 9.1 Switching states, output voltage, and charge/discharge status of capacitors in the proposed structure.

No.	S_1	S_2	S_3	S_4	S_5	S_6	S_7	S_8	S_9	V_o	C_1	C_2
A	0	1	1	0	0	0	1	1	0	$7V_{dc}$	▼	▼
B	0	1	0	1	0	0	1	1	0	$6V_{dc}$	▼	▼
C	1	0	0	1	0	0	1	1	0	$5V_{dc}$	▼	▼
D	0	1	1	0	0	1	0	1	0	$4V_{dc}$	▼	▲
E	0	1	0	1	0	1	0	1	0	$3V_{dc}$	▼	▲
F	1	0	0	1	0	1	0	1	0	$2V_{dc}$	▼	▲
G	0	1	1	0	1	0	0	1	0	V_{dc}	▲	▲
H	1	0	1	0	1	0	0	1	0	0^+	▲	▲
I	0	1	0	1	0	0	1	0	1	0^-	▲	▲
J	1	0	0	1	0	0	1	0	1	$-V_{dc}$	▲	▲
K	0	1	1	0	0	1	0	0	1	$-2V_{dc}$	▲	▼
L	1	0	1	0	0	1	0	0	1	$-3V_{dc}$	▲	▼
M	1	0	0	1	0	1	0	0	1	$-4V_{dc}$	▲	▼
N	0	1	1	0	1	0	0	0	1	$-5V_{dc}$	▼	▼
O	1	0	1	0	1	0	0	0	1	$-6V_{dc}$	▼	▼
P	1	0	0	1	1	0	0	0	1	$-7V_{dc}$	▼	▼

connected in parallel with the capacitors and prevents a deep voltage decrement in the capacitors. This reduces the voltage ripple of the capacitors. Based on Table 9.1, the maximum output voltage of the proposed topology is:

$$V_{o,max} = V_L + V_{C1} + V_{C2} = 7V_{dc} \tag{9.2}$$

Generation of $V_o = 7\ V_{dc}$: In this state, by turning on the S_2, S_3, S_7, and S_8 switches the required energy of the load is supplied through the addition of the V_1 source and stored energy in the C_1, C_2 capacitors.

Generation of $V_o = -7\ V_{dc}$: In this state, by turning on the S_1, S_4, S_5, and S_9 switches the required energy of the load is supplied through the addition of the V_1 source and stored energy in the C_1, C_2 capacitors.

Generation of $V_o = 6\ V_{dc}$: In this state, by turning on the S_2, S_4, S_7, and S_8 switches the required energy of the load is supplied through the addition of stored energy in the C_1, C_2 capacitors.

Generation of $V_o = -6\ V_{dc}$: In this state, by turning on the S_1, S_3, S_5, and S_9 switches the required energy of the load is supplied through the addition of stored energy in C_1, C_2 capacitors.

Generation of $V_o = 5\ V_{dc}$: In this state, by turning on the S_1, S_4, S_7, and S_8 switches the required energy of the load is supplied through the subtraction of the V_1 source and stored energy in the C_1, C_2 capacitors.

Generation of $V_o = -5\ V_{dc}$: In this state, by turning on the S_2, S_3, S_5, and S_9 switches the required energy of the load is supplied through the subtraction of the V_1 source and stored energy in the C_1, C_2 capacitors.

Generation of $V_o = 4\ V_{dc}$: In this state, by turning on the S_2, S_3, S_6, and S_8 switches the required energy of the load is supplied through the addition of the V_1 source and stored energy in the C_1 capacitor.

Generation of $V_o = -4\ V_{dc}$: In this state, by turning on the S_1, S_4, S_6, and S_9 switches the required energy of the load is supplied through the addition of the V_1 source and stored energy in the C_2 capacitor.

Generation of $V_o = 3\ V_{dc}$: In this state, by turning on the S_2, S_4, S_6, and S_8 switches the required energy of the load is supplied through the stored energy in the C_1 capacitor.

Generation of $V_o = -3\ V_{dc}$: In this state, by turning on the S_1, S_3, S_6, and S_9 switches the required energy of the load is supplied through the stored energy in the C_2 capacitor.

Generation of $V_o = 2\ V_{dc}$: In this case, by turning on the S_1, S_4, S_6, and S_8 switches the required energy of the load is supplied through the subtraction of the V_1 source and stored energy in the C_1 capacitor.

Generation of $V_o = -2\ V_{dc}$: In this case, by turning on the S_2, S_3, S_6, and S_9 switches the required energy of the load is supplied through the subtraction of the V_1 source and stored energy in the C_2 capacitor.

Generation of $V_o = V_{dc}$: In this state, by turning on the S_2, S_3, S_5, and S_8 switches the required energy of the load is supplied through the V_1 source.

Generation of $V_o = -V_{dc}$: In this state, by turning on the S_1, S_4, S_7, and S_9 switches the required energy of the load is supplied through the V_1 source.

Generation of $V_o = 0^+$ ($I_o \geq 0$): In this state, by turning on the S_1, S_3, S_5, and S_8 switches the voltage at both ends of the load is zero.

Generation of $V_o = 0^-$ ($I_o < 0$): In this state, by turning on the S_2, S_4, S_7, and S_9 switches the voltage at both ends of the load is zero.

As shown in Figs. 9.3–9.10, capacitors are always charged at 3 V_{dc} and no switching is required to charge the capacitors. Also, the maximum number of ON switches to produce a step is four. Therefore the proposed structure has little switching losses.

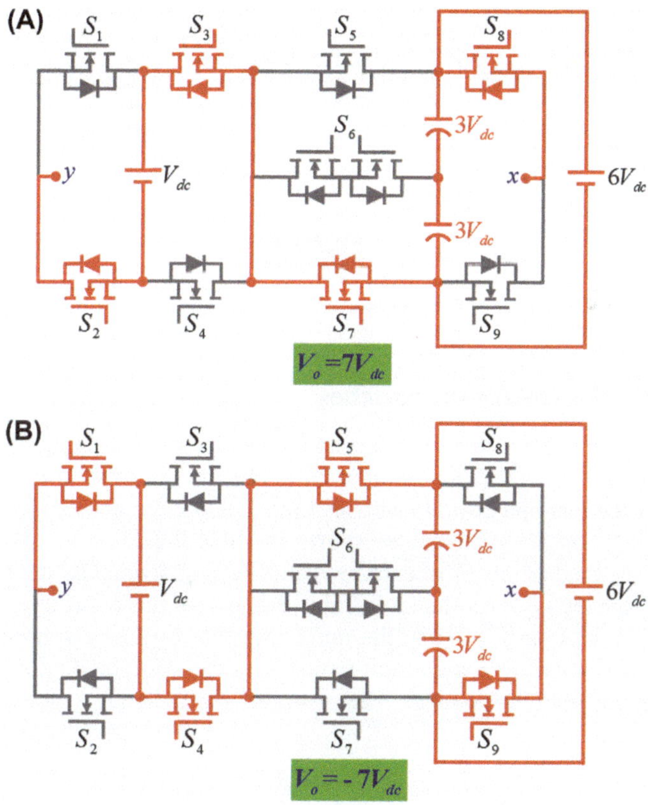

FIGURE 9.3 State of switches during generation of $\pm 7\ V_{dc}$ output voltage levels.

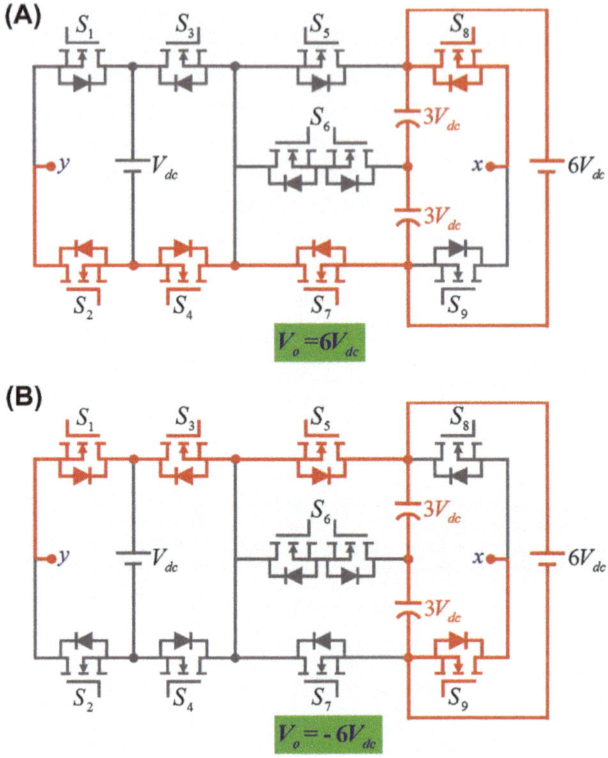

FIGURE 9.4 State of switches during generation of $\pm 6\ V_{dc}$ output voltage levels.

9.2.3 Standing voltage on switches

The highest voltage on the switch during its inactive state is called SV. The SV of switches is one of the most important parameters in multilevel converters that affects the loss and cost of switches [32]. The SV and normalized standing voltage (NSV) on the switches are shown in Table 9.2.

NSV is defined as the SV of each switch (SV_{Switch}) divided by the maximum output voltage ($V_{o,max}$). As shown in Table 9.2, five of the switches (S_1-S_4, S_6) have an NSV less than 50% and the remaining four (S_5, S_7-S_9) have an NSV more than the others. It is important to note that the switches do not withstand the maximum output voltage. The average normalized standing voltage (ANSV) is:

$$ANSV = \frac{\sum NSV}{N_{Switch}} = \frac{442.5\%}{9} = 49.2\% \tag{9.3}$$

The ANSV of switches is less than 50%, which is acceptable.

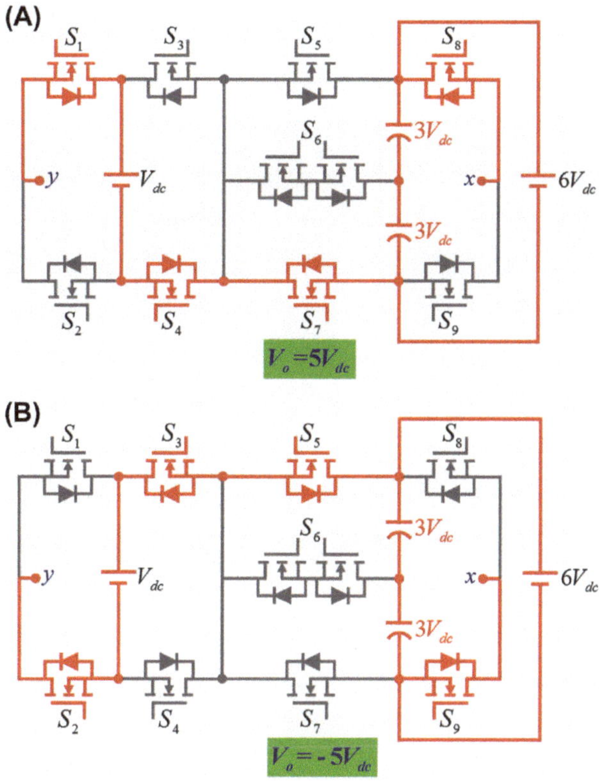

FIGURE 9.5 State of switches during generation of $\pm5\ V_{dc}$ output voltage levels.

9.2.4 Cascaded version of the proposed topology

Modular configuration of the proposed structure can be used to achieve higher levels. Fig. 9.11 shows the modular configuration of the proposed SCMLI. In the cascade version, the number of levels is significantly high, which leads to a reduction in the total harmonic distortion (THD) of the proposed structure. Thus the dimensions of the output filter can be reduced [33]. In contrast, more switches and sources are needed, which increases the cost, size, and volume of the converter. Important parameters of the proposed converter in cascade form are:

$$N_{Level} = (3 \times 5)^{\mathrm{m}}, N_{Source} = 2\mathrm{m},\ N_{Capacitor} = 2\mathrm{m}, N_{Switch} = N_{Driver} = 9\mathrm{m}, N_{MOSFET} = 10\mathrm{m} \tag{9.4}$$

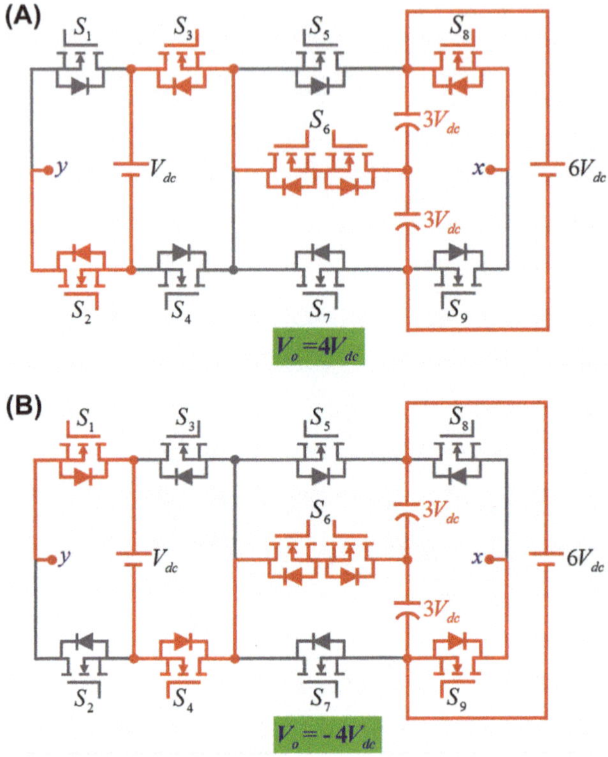

FIGURE 9.6 State of switches during generation of ±4 V_{dc} output voltage levels.

9.3 Modulation technique

In the literature, various modulation methods have been presented for generating the switching pulses of multilevel converter switches [34–36]. In this study, the phase disposition pulse width modulation (PD-PWM) method has been used to produce the gate pulses of the proposed structure. This method leads to a reduced THD output voltage waveform, compared with other methods [37]. In this method, the sinusoidal reference waveform with a fundamental frequency of $f_{ref} = 50$ Hz is compared with triangular carrier waveforms with an amplitude of A_{car} and a frequency of f_{car}. As shown in Fig. 9.12, in the proposed 15-level topology, a sinusoidal reference waveform with $A_{ref} = 6.9$ and $f_{ref} = 50$ Hz and 14 carrier waves ($C_1 - C_{14}$)

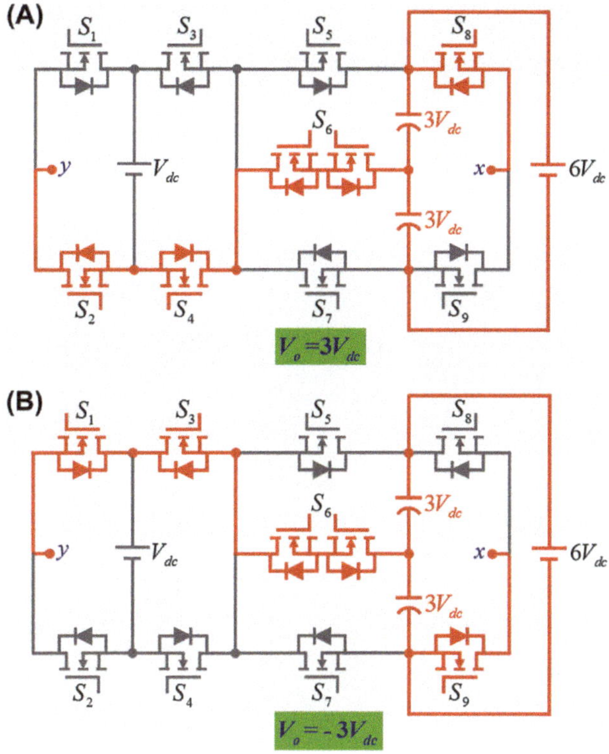

FIGURE 9.7 State of switches during generation of $\pm 3\ V_{dc}$ output voltage levels.

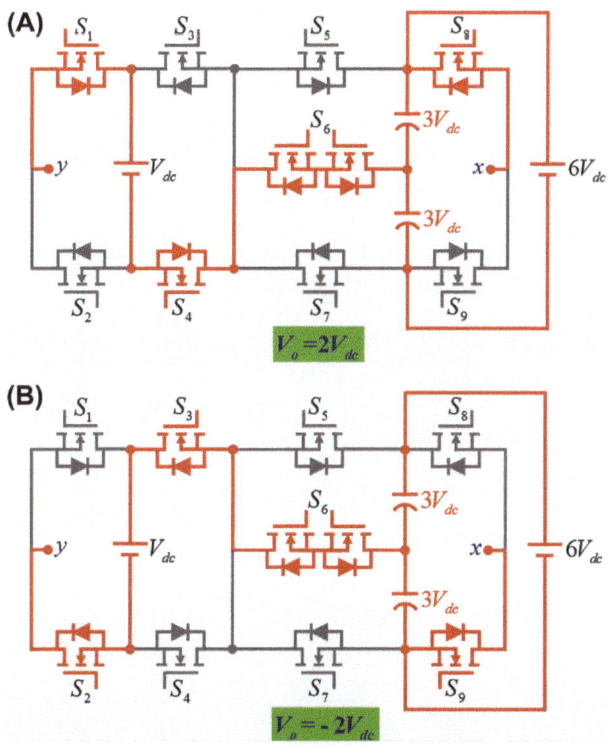

FIGURE 9.8 State of switches during generation of $\pm 2\ V_{dc}$ output voltage levels.

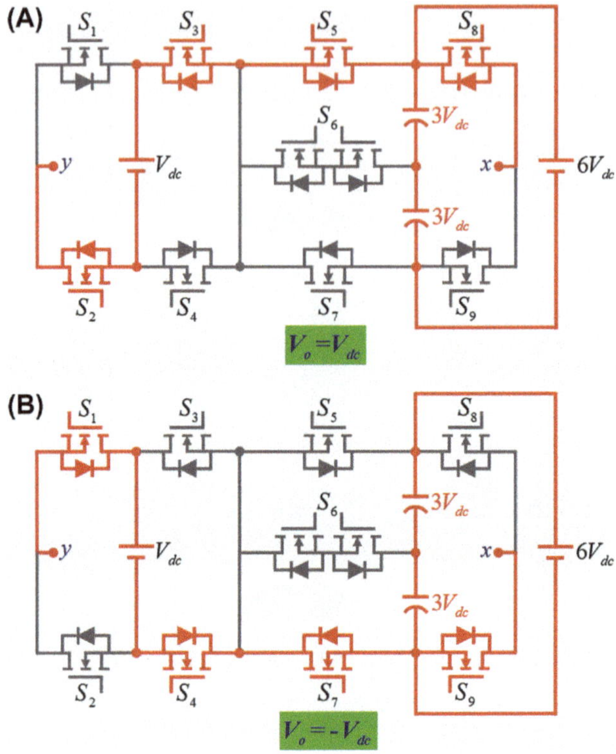

FIGURE 9.9 State of switches during generation of $\pm V_{dc}$ output voltage levels.

with $A_{car} = 1$ and $f_{car} = 2.5$ kHz are required to produce proper switching pulses. The amplitude of reference waveform is determined by a factor called modulation index (M), which is always equal or smaller than one, as defined in:

$$M = \frac{A_{ref}}{V_{o,max}}, \quad (0 \leq M \leq 1) \tag{9.5}$$

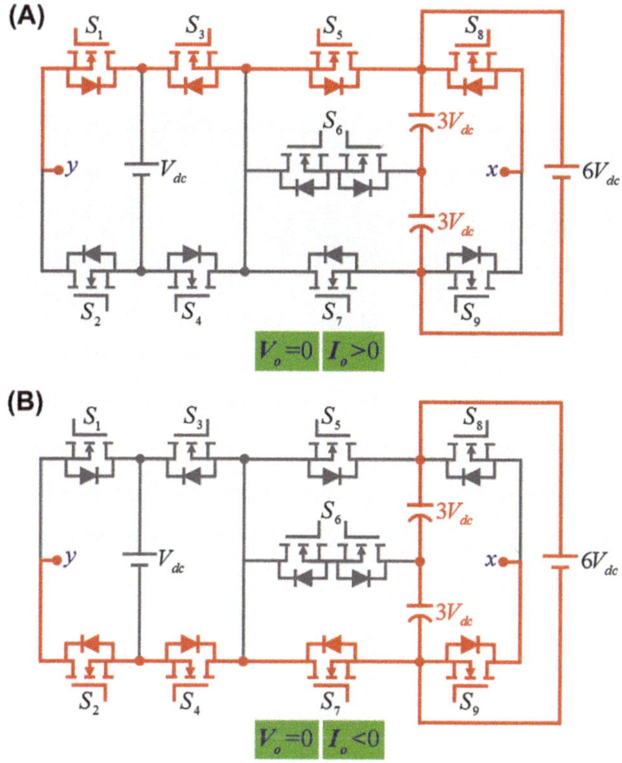

FIGURE 9.10 State of switches during generation of 0^+ ($I_o \geq 0$) and 0^- ($I_o < 0$) output voltage levels.

TABLE 9.2 Standing voltage of switches.

Switches	SV	NSV (%)
S_1	V_{dc}	14.2
S_2	V_{dc}	14.2
S_3	V_{dc}	14.2
S_4	V_{dc}	14.2
S_5	6 V_{dc}	85.7
S_6	3 V_{dc}	42.9
S_7	6 V_{dc}	85.7
S_8	6 V_{dc}	85.7
S_9	6 V_{dc}	85.7
Total standing voltage	31 V_{dc}	

NSV, Normalized standing voltage; *SV*, standing voltage.

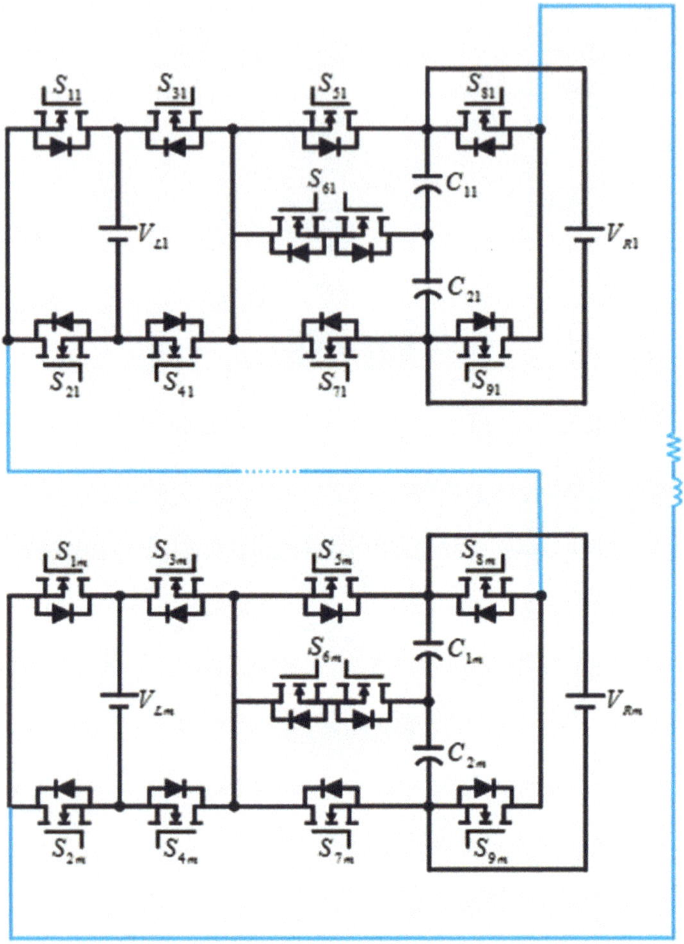

FIGURE 9.11 Cascaded configuration of the proposed topology.

The relationships between the reference waveform and the carriers for generating each of the output voltage levels are shown in Table 9.3. Switching pulses are generated based on Table 9.3. For example, the S_1 switch conducts during $C_3 < Ref < C_2$, $C_6 < Ref < C_5$, $C_7 < Ref < C_6$, $C_8 < Ref < C_7$, $C_9 < Ref < C_8$, $C_{11} < Ref < C_{10}$, or $C_{12} < Ref < C_{11}$ cases.

The switching pulses of the proposed 15-level structure are shown in Fig. 9.13. As seen, the S_5–S_9 switches work at low frequencies, which leads to lower switching losses.

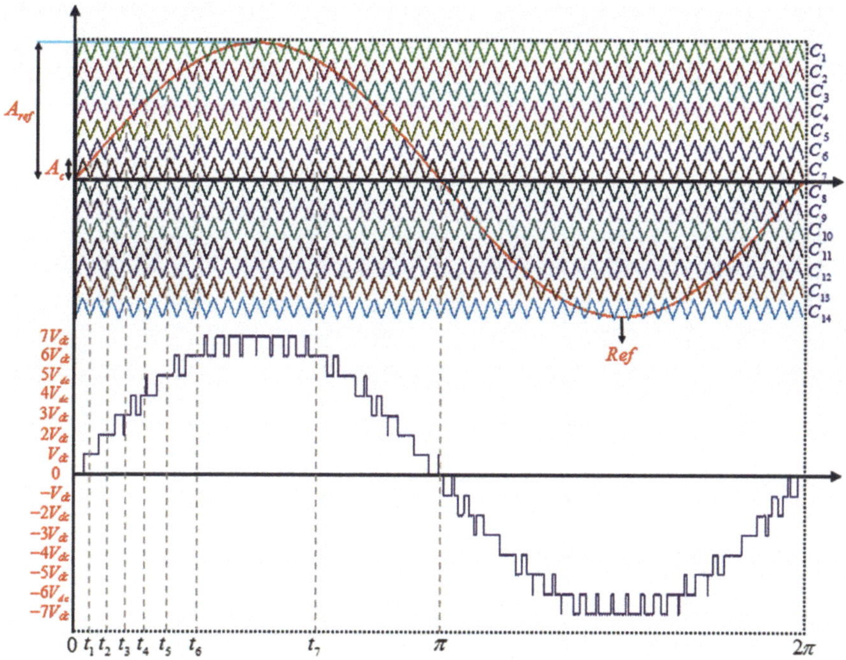

FIGURE 9.12 Reference, carrier, and output voltage waveforms.

TABLE 9.3 Relationships for producing switching pulses.

Relations between reference and carriers	ON switches	V_o
$C_1 < Ref$	S_2, S_3, S_7, S_8	$7\ V_{dc}$
$C_2 < Ref < C_1$	S_2, S_4, S_7, S_8	$6\ V_{dc}$
$C_3 < Ref < C_2$	S_1, S_4, S_7, S_8	$5\ V_{dc}$
$C_4 < Ref < C_3$	S_2, S_3, S_6, S_8	$4\ V_{dc}$
$C_5 < Ref < C_4$	S_2, S_4, S_6, S_8	$3\ V_{dc}$
$C_6 < Ref < C_5$	S_1, S_4, S_6, S_8	$2\ V_{dc}$
$C_7 < Ref < C_6$	S_1, S_3, S_7, S_8	V_{dc}
$C_8 < Ref < C_7$	S_1, S_3, S_5, S_8	0^+
	S_2, S_4, S_7, S_9	0^-
$C_9 < Ref < C_8$	S_1, S_4, S_7, S_9	$-V_{dc}$
$C_{10} < Ref < C_9$	S_2, S_3, S_6, S_9	$-2\ V_{dc}$
$C_{11} < Ref < C_{10}$	S_1, S_3, S_6, S_9	$-3\ V_{dc}$
$C_{12} < Ref < C_{11}$	S_1, S_4, S_6, S_9	$-4\ V_{dc}$
$C_{13} < Ref < C_{12}$	S_2, S_3, S_5, S_9	$-5\ V_{dc}$
$C_{14} < Ref < C_{13}$	S_1, S_3, S_5, S_9	$-6\ V_{dc}$
$Ref < C_{14}$	S_1, S_4, S_5, S_9	$-7\ V_{dc}$

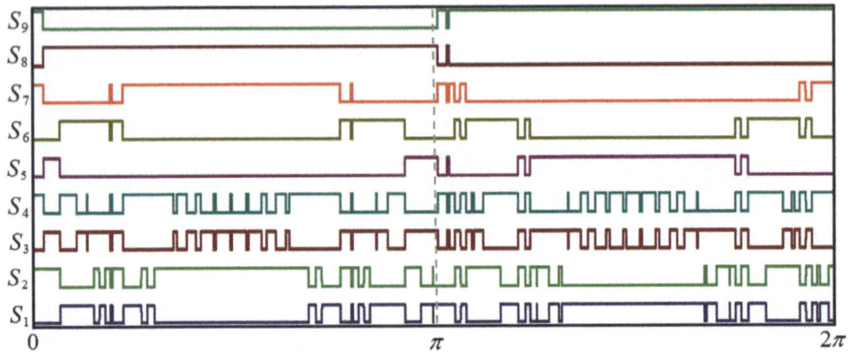

FIGURE 9.13 Gate pulses of the proposed 15-level switched-capacitor multilevel inverter.

9.4 Power loss analysis

The SCMLIs have three types of losses, including switching losses, conduction losses, and capacitor voltage ripple losses. These three types form the total loss of the converter.

9.4.1 Switching losses

The switching losses (symbolized by $P_{Switching}$) create a significant part of total loss, especially at high-frequency PWM-based converters. The switching losses happen during the turning on and off process of semiconductors. This type of loss is wasted in the form of heat on the switch. So, it is necessary to keep this loss as low as possible. Switching loss of each switch can be derived from [38]:

$$\begin{cases} P_{Switching,on} = f_S \int_0^{t_{on}} v_S(t) i_S(t) dt = f_S \int_0^{t_{on}} \left(\frac{v_S}{t_{on}} t\right)\left(-\frac{I_S^{on}}{t_{on}}(t - t_{on})\right) dt = \frac{1}{6} f_S V_S^{off} I_S^{on} t_{on} \\ P_{Switching,off} = f_S \int_0^{t_{off}} v_S(t) i_S(t) dt = f_S \int_0^{t_{off}} \left(\frac{v_S}{t_{off}} t\right)\left(-\frac{I_S^{off}}{t_{off}}(t - t_{off})\right) dt = \frac{1}{6} f_S V_S^{off} I_S^{off} t_{off} \end{cases} \tag{9.6}$$

According to Eq. (9.6), the switching losses are divided into two parts related to the turning-on and turning-off mode of switches. As is well known, both modes depend on the switching frequency, current, and voltage associated with each mode. t_{on} and t_{off} represent the turning-on and turning-off transient period of the switch, respectively. V_S^{off} shows standing voltage of the switch. Also, I_S^{on} is the switch current at the end of the t_{on} mode of the switch and I_S^{off}

is the switch current at the beginning of the t_{off}. switching frequency or f_s is obtained as [39]:

$$f_{s,on} = N_{s,on} \times f_{ref} \tag{9.7}$$

$$f_{s,off} = N_{s,off} \times f_{ref} \tag{9.8}$$

where $N_{s,on}$ and $N_{s,off}$ denote the number of turnings ON and OFF of each switch in a cycle. f_{ref} is the frequency of the reference waveform. $N_{s,on}$ and $N_{s,off}$ are calculated based on:

$$N_{s,on} = \frac{t_{on,switch}}{2\pi} \times N_t \times \frac{f_{car}}{f_{ref}} \tag{9.9}$$

$$N_{s,off} = \frac{t_{off,switch}}{2\pi} \times N_t \times \frac{f_{car}}{f_{ref}} \tag{9.10}$$

where $t_{on,switch}$ is the ON time of the switch. $t_{off,\ switch}$ is the time the switch is OFF, N_t is the total number of turns ON or turns OFF in each switching period, and f_{car} is the switching frequency of the carrier waveforms.

Finally, switching losses are derived by:

$$P_{Switching,total} = \sum_{i=1}^{N_{Switch}} \left(P_{Switching,on(i)} + P_{Switching,off(i)}\right) \tag{9.11}$$

9.4.2 Conduction losses

Conduction losses ($P_{Conducting}$) are highly dependent on the on-state resistance of semiconductors ($R_{on,MOS}$) and the equivalent series resistance (ESR) of capacitors. According to Figs. 9.3–9.10, the equivalent resistance of a converter at each state can be achieved according to the number of ON switches. Therefore the equivalent resistances at each output voltage level are listed in Table 9.4, in which $R_{on,MOS}$ and R_{ESR}, respectively, are about 0.05 and 0.03 Ω.

Based on Fig. 9.12, for $|Ref| < A_c$, the output voltage changes between the 0 and 1 levels. In this case, the energy loss between the 0 and t_1 times is obtained through Eq. (9.12) [40]. The values of A_c, A_{ref}, and f_{ref} are considered to be 1, 6.9, and 50 Hz, respectively.

$$E_{0,1} = \int_0^{t_1} \left[I_{Load}\sin(2\pi f_{ref}t)\right]^2 \times \left[\left(4R_{on,MOS}\right)\left(\frac{A_{ref}\sin(2\pi f_{ref}t)}{A_{car}}\right) + \left(4R_{on,MOS}\right)\left(1 - \frac{A_{ref}\sin(2\pi f_{ref}t)}{A_{car}}\right)\right] dt \tag{9.12}$$

TABLE 9.4 Equivalent resistance in each level.

Output voltage level	Equivalent resistance	Value (Ω)
± 0	$4R_{on,MOS}$	0.2
$\pm V_{dc}$	$4R_{on,MOS}$	0.2
$\pm 2\ V_{dc}$	$5R_{on,MOS} + R_{ESR}$	0.28
$\pm 3\ V_{dc}$	$5R_{on,MOS} + R_{ESR}$	0.28
$\pm 4\ V_{dc}$	$5R_{on,MOS} + R_{ESR}$	0.28
$\pm 5\ V_{dc}$	$4R_{on,MOS}+2R_{ESR}$	0.26
$\pm 6\ V_{dc}$	$4R_{on,MOS}+2R_{ESR}$	0.26
$\pm 7\ V_{dc}$	$4R_{on,MOS}+2R_{ESR}$	0.26

t_1 is obtained from:

$$t_1 = \frac{\sin^{-1}\left(\frac{A_{car}}{A_{ref}}\right)}{2\pi f_{ref}} = 4.63 \times 10^{-4}\ s \tag{9.13}$$

Thus:

$$E_{0,1} = 6.5 \times 10^{-7}\left(\frac{P_{out}}{V_{in}}\right)^2 \tag{9.14}$$

Similarly, the energy loss in other time periods can be calculated from:

$$E_{1,2} = \int_{t_1}^{t_2} [I_{Load}\sin(2\pi f_{ref}t)]^2 \times \left[(5R_{on,MOS} + R_{ESR})\left(\frac{A_{ref}\sin(2\pi f_{ref}t) - A_{car}}{A_{car}}\right) + (4R_{on,MOS})\left(1 - \frac{A_{ref}\sin(2\pi f_{ref}t) - A_{car}}{A_{car}}\right)\right]dt \tag{9.15}$$

$$E_{2,3} = \int_{t_2}^{t_3} [I_{Load}\sin(2\pi f_{ref}t)]^2 \times \left[(5R_{on,MOS} + R_{ESR})\left(\frac{A_{ref}\sin(2\pi f_{ref}t) - 2A_{car}}{A_{car}}\right) + (5R_{on,MOS} + R_{ESR})\left(1 - \frac{A_{ref}\sin(2\pi f_{ref}t) - 2A_{car}}{A_{car}}\right)\right]dt \tag{9.16}$$

$$E_{3,4} = \int_{t_3}^{t_4} [I_{Load} \sin(2\pi f_{ref} t)]^2 \times \left[(5R_{on,MOS} + R_{ESR}) \left(\frac{A_{ref} \sin(2\pi f_{ref} t) - 3A_{car}}{A_{car}} \right) + (5R_{on,MOS} + R_{ESR}) \left(1 - \frac{A_{ref} \sin(2\pi f_{ref} t) - 3A_{car}}{A_{car}} \right) \right] dt \tag{9.17}$$

$$E_{4,5} = \int_{t_4}^{t_5} [I_{Load} \sin(2\pi f_{ref} t)]^2 \times \left[(4R_{on,MOS} + R_{ESR}) \left(\frac{A_{ref} \sin(2\pi f_{ref} t) - 4A_{car}}{A_{car}} \right) + (5R_{on,MOS} + R_{ESR}) \left(1 - \frac{A_{ref} \sin(2\pi f_{ref} t) - 4A_{car}}{A_{car}} \right) \right] dt \tag{9.18}$$

$$E_{5,6} = \int_{t_5}^{t_6} [I_{Load} \sin(2\pi f_{ref} t)]^2 \times \left[(4R_{on,MOS} + R_{ESR}) \left(\frac{A_{ref} \sin(2\pi f_{ref} t) - 5A_{car}}{A_{car}} \right) + (4R_{on,MOS} + R_{ESR}) \left(1 - \frac{A_{ref} \sin(2\pi f_{ref} t) - 5A_{car}}{A_{car}} \right) \right] dt \tag{9.19}$$

$$E_{6,7} = \int_{t_6}^{t_7} [I_{Load} \sin(2\pi f_{ref} t)]^2 \times \left[(4R_{on,MOS} + R_{ESR}) \left(\frac{A_{ref} \sin(2\pi f_{ref} t) - 6A_{car}}{A_{car}} \right) + (4R_{on,MOS} + R_{ESR}) \left(1 - \frac{A_{ref} \sin(2\pi f_{ref} t) - 6A_{car}}{A_{car}} \right) \right] dt \tag{9.20}$$

Also, t_2–t_7 are derived from:

$$t_2 = \frac{\sin^{-1}\left(\frac{2A_{car}}{A_{ref}} \right)}{2\pi f_{ref}} = 9.36 \times 10^{-4}\ s \tag{9.21}$$

$$t_3 = \frac{\sin^{-1}\left(\frac{3A_{car}}{A_{ref}} \right)}{2\pi f_{ref}} = 1.432 \times 10^{-3}\ \text{sec} \tag{9.22}$$

$$t_4 = \frac{\sin^{-1}\left(\frac{4A_{car}}{A_{ref}}\right)}{2\pi f_{ref}} = 1.968 \times 10^{-3}\ \text{sec} \tag{9.23}$$

$$t_5 = \frac{\sin^{-1}\left(\frac{5A_{car}}{A_{ref}}\right)}{2\pi f_{ref}} = 2.58 \times 10^{-3}\ \text{sec} \tag{9.24}$$

$$t_6 = \frac{\sin^{-1}\left(\frac{6A_{car}}{A_{ref}}\right)}{2\pi f_{ref}} = 3.356 \times 10^{-3}\ \text{sec} \tag{9.25}$$

$$t_7 = \frac{\pi - \sin^{-1}\left(\frac{6A_{car}}{A_{ref}}\right)}{2\pi f_{ref}} = 6.64 \times 10^{-3}\ \text{sec} \tag{9.26}$$

Therefore the energy losses during the production of consecutive levels are:

$$E_{1,2} = 6.24 \times 10^{-6}\left(\frac{P_{out}}{V_{in}}\right)^2 \tag{9.27}$$

$$E_{2,3} = 1.84 \times 10^{-5}\left(\frac{P_{out}}{V_{in}}\right)^2 \tag{9.28}$$

$$E_{3,4} = 3.94 \times 10^{-5}\left(\frac{P_{out}}{V_{in}}\right)^2 \tag{9.29}$$

$$E_{4,5} = 5.97 \times 10^{-5}\left(\frac{P_{out}}{V_{in}}\right)^2 \tag{9.30}$$

$$E_{5,6} = 1.31 \times 10^{-9}\left(\frac{P_{out}}{V_{in}}\right)^2 \tag{9.31}$$

$$E_{6,7} = 7.81 \times 10^{-9}\left(\frac{P_{out}}{V_{in}}\right)^2 \tag{9.32}$$

Finally, due to the symmetry in the staircase output voltage waveform, the conduction losses for a cycle are calculated according to:

$$P_{conduction} = \left[4E_{0,1} + 4E_{1,2} + 4E_{2,3} + 4E_{3,4} + 4E_{4,5} + 4E_{5,6} + 2E_{6,7}\right]f_{ref} = 0.12 \tag{9.33}$$

9.4.3 Capacitor ripple losses

Ripple losses (P_{Ripple}) are caused by the voltage difference between the input DC source and the capacitor during the capacitor discharging period. By having the values related to the capacitance of the capacitors and determining the allowable size of the ripple voltage of the capacitors, the ripple losses can be calculated from:

$$P_{Ripple} = \frac{f_{ref}}{2}\left(\sum_{i=1}^{N_{Capacitor}} C_i(\Delta V_{Ripple} \times V_C)^2\right) \tag{9.34}$$

It should be noted that in Eq. (9.34), V_C is the magnitude of the voltage at which the capacitor is charged. In this converter, $V_C = 3\ V_{dc}$.

Thus the total loss of the proposed converter is calculated by:

$$P_{Loss} = P_{Conduction} + P_{Switching} + P_{Ripple} \tag{9.35}$$

Finally, the efficiency is obtained from:

$$\eta = \frac{P_{out}}{P_{in}} = \frac{P_{out}}{P_{out} + P_{Loss}} \tag{9.36}$$

9.5 Capacitor design

The maximum discharge of a capacitor in a cycle is denoted by ΔQ_C, which can be obtained from:

$$\Delta Q_C = \int_{t_x}^{t_y} I_{Load}\sin(2\pi f_{ref} t)dt \tag{9.37}$$

$[t_x - t_y]$ is the discharging interval of the capacitor. In the design of capacitance, the deepest discharge process is considered, which happens during the longest discharging period (LDP). According to Fig. 9.12, the LDP of the C_1 or C_2 capacitor is $[t_6 - t_7]$. So, the capacitances by which the voltage ripple is restricted in the desired range can be computed through Eq. (9.38). Note that 3 V_{dc} is the nominal voltage of the capacitors:

$$C \geq \frac{\Delta Q_C}{\Delta V_{ripple} \times 3V_{dc}} \tag{9.38}$$

9.6 Comparative analysis

To examine the suitability of the proposed structure, it is compared with a traditional cascaded H-bridge (CHB) and similar structures presented in [22–30]. The total description of the proposed topology and other structures is presented in Table 9.5. The comparisons have been conducted from the

TABLE 9.5 Comparison parameters of the proposed topology, cascaded H-bridge (CHB), and [22–30].

Topology	N_{Level}	N_{Source}	$N_{Capacitor}$	N_{MOSFET}	N_{Diode}	N_{Driver}	N_{Device}	$TSV(\times V_{dc})$	$V_{o,max}(\times V_{dc})$	N_V	BF	Production of negative levels
Proposed	15	2	2	10	–	9	23	31	7	2	1	Inherit
[22]	13	2	2	14	–	11	29	32	6	2	2	Inherit
[23]	11	2	2	7	4	7	22	25	5	2	1	Switch H-bridge
[24]	9	2	1	7	4	7	21	17	2	1	1	Switch H-bridge
[25]	15	5	–	10	–	10	25	28	7	3	1	Inherit
[26]	9	2	4	12	–	10	28	10	2	1	1	H-bridge
[27]	19	3	2	12	2	12	31	66	9	1	3	H-bridge
[28]	17	3	5	10	8	10	36	47	8	3	1	Inherit
[29]	13	2	4	18	2	18	44	34	6	1	3	H-bridge
[30]	15	7	–	10	6	10	33	40	7	1	1	H-bridge
CHB	15	3	–	12	–	12	27	28	7	3	1	H-bridge

following aspects: number of levels (N_{Level}), number of sources (N_{Source}), capacitors ($N_{Capacitor}$), MOSFETs (N_{MOSFET}), diodes (N_{Diode}), gate driver circuits (N_{Driver}), total components (N_{Device}), total standing voltage (TSV), maximum output voltage ($V_{o,max}$), variety of DC sources (N_V), boosting factor (BF), and dependency on an H-bridge to produce a bipolar voltage waveform.

For a better understanding of comparisons, the number of sources, semiconductors (MOSFETs and diodes), devices, and TSV_{pu} per number of levels is presented in Table 9.6. TSV_{pu} is defined as:

$$TSV_{pu} = \frac{TSV}{V_{o,\max}} \tag{9.39}$$

According to Table 9.5, the topology presented in [22] generates 13 voltage levels by means of two DC sources, two capacitors, 14 MOSFETs, 11 gate drivers, and total devices of 29. Its TSV on switches is 32 V_{dc}. Due to application of two unequal sources, the variety of sources is equal to 2. The maximum output voltage and boosting factor (of [22]) are 6 V_{dc} and 2, respectively. The double gain of [22] is an important advantage over the proposed structure. Both the proposed topology and [22] generate negative voltage levels without an H-bridge. According to Table 9.6, [22] has a higher number of sources, semiconductors, devices, and TSV_{pu} per levels than the proposed structure, which leads to higher cost, size, and volume.

TABLE 9.6 Comparison between proposed topology, cascaded H-bridge (CHB), and the presented topologies in [22–30].

Topology	$\frac{N_{Source}}{N_{Level}}$	$\frac{N_{Semiconductor}}{N_{Level}}$	$\frac{N_{Device}}{N_{Level}}$	$\frac{TSV_{pu}}{N_{Level}}$
Proposed	0.13	0.67	1.53	0.3
[22]	0.15	1.08	2.23	0.36
[23]	0.18	1	2	0.45
[24]	0.22	1.22	2.33	0.94
[25]	0.33	0.67	1.67	0.27
[26]	0.22	1.33	3.11	0.56
[27]	0.16	0.74	1.63	0.39
[28]	0.18	1.06	2.12	0.35
[29]	0.15	1.54	3.38	0.44
[30]	0.47	1.07	2.2	0.38
CHB	0.2	0.8	1.8	0.27

Table 9.5 shows that the structure presented in [23] can produce 11 levels with two unequal DC sources, two capacitors, seven MOSFETs, four diodes, seven gate driver circuits, and total components of 22. The TSV on switches is 25 V_{dc}. Also, the maximum output voltage is equal to 5 V_{dc}. This structure ([23]) acts as a unity-gain topology (boosting factor of 1). [23] applies a developed six-switch H-bridge to generate negative voltage levels. As evident from Table 9.6, the number of sources, semiconductors, devices, and TSV_{pu} per levels of [23] is larger than that of the proposed structure. This means that for achieving the same levels, [23] utilizes more devices than the proposed topology. This results in increased size and cost.

As seen from Table 9.5, the structure presented in [24] requires two identical DC sources, one capacitor, seven MOSFETs, four diodes, seven gate drivers, and total devices of 21 to produce nine output voltage levels. The TSV on its switches is 17 V_{dc}. The maximum output voltage is 2 V_{dc}. Since two equal DC sources have been applied, the variety of DC sources is one. This topology has unity gain, the same as the proposed topology. Similar to [23], [24] produces the negative voltage levels by means of a developed six-switch H-bridge. Table 9.6 confirms that the number of sources, semiconductors, devices, and TSV_{pu} per levels in [24] is larger than that of the suggested structure. So, with the same components, the suggested topology can produce more levels than [24].

The topology presented in [25] generates 15 levels with five DC sources, 10 MOSFETs, 10 gate driver circuits, and total devices of 25. Three of the sources are identical but the other two sources have uneven magnitudes. So, the variety of sources is three. The TSV on the switches of [25] is equal to 28 V_{dc}. The peak output voltage and boosting factor of [25] are respectively about 7 V_{dc} and 1. The generation of negative voltage levels takes place inherently, without an H-bridge. Table 9.6 certifies that the number of semiconductors per levels in [25] is the same as the proposed topology, but it has a larger number of devices per levels than the proposed topology. This implies that at equal-level versions, [25] requires more devices than the proposed topology. The TSV_{pu} per number of levels in [25] is smaller than that of the proposed topology. Also, the number of sources per levels of [25] is nearly three times larger than that of the proposed topology. So, with the same number of sources, the number of levels of the proposed topology will be triple that of [25].

The symmetric structure presented in [26] produces nine levels by means of two identical DC sources, four capacitors, 12 MOSFETs, 10 gate drivers, and a total component count of 28. The TSV on switches as well as the maximum output voltage of [26] are respectively 10 V_{dc} and 2 V_{dc}. Since the topology is symmetric, the variety of DC sources is one. This topology has a unity boosting factor and also requires an H-bridge to produce the negative voltage levels. It is seen from Table 9.6 that the number of sources, semiconductors, devices, and TSV_{pu} per levels in [26] is higher than that of the suggested structure. So, [26] is expected to have higher cost, size, and losses than the suggested topology.

Table 9.5 indicates that the structure presented in [27] demands three DC sources, two capacitors, 12 MOSFETs, two diodes, 12 gate drivers, and total components of 31 to generate 19 output voltage levels. All the DC sources are identical, so the variety of sources is equal to one. The triple boosting factor is the main superiority of [27] over the suggested topology. This topology requires an H-bridge to create negative voltage steps. The TSV on switches as well as the maximum output voltage of [27] are 66 V_{dc} and 9 V_{dc}, respectively. Table 9.6 proves that [27] has a higher ratio of number of sources, semiconductors, devices, and TSV_{pu} to the number of levels than the proposed structure. This can result in increased cost, size, and volume.

It is seen from Table 9.5 that the topology presented in [28] creates 17 voltage levels using three DC sources, five capacitors, 10 MOSFETs, eight diodes, 10 gate driver circuits, and a total component count of 36. This topology is asymmetric and utilizes three different DC voltage sources, so $N_V = 3$. The TSV on semiconductors and peak output voltage are respectively equal to 47 V_{dc} and 8 V_{dc}. The boosting factor of [28] is 1. This structure can inherently produce a bipolar voltage waveform without any H-bridge. Table 9.6 shows that the number of sources, semiconductors, devices, and TSV_{pu} per levels in [28] is higher than that of the proposed structure, leading to larger size, volume, and cost.

Based on Table 9.5, the structure presented in [29] can produce 13 voltage levels with two DC sources, four capacitors, 18 MOSFETs, two diodes, 18 gate drivers, and total devices of 44. The TSV on semiconductors is 36 V_{dc}. Also, the maximum output voltage and variety of number of input sources are 6 V_{dc} and 1, respectively. Table 9.6 validates that the number of sources, semiconductors, devices, and TSV_{pu} per levels in [29] is greater than that of the proposed structure. So, with the same number of devices, the suggested topology can create more steps than [29].

Based on Table 9.5, the structure presented in [30] provides 15 voltage levels by means of seven DC sources, 10 MOSFETs, six diodes, 10 gate driver circuits, and total components of 33. The TSV on switches is equal to 40 V_{dc}. This structure has a maximum output voltage of 7 V_{dc} with a unity boosting factor. The magnitude of DC voltage sources is the same, so the variety of sources is one. This structure employs an H-bridge for generating negative voltage steps. According to Table 9.6, the number of sources, semiconductors, devices, and TSV_{pu} per number of levels in [30] is larger than that of the proposed topology. So, at equal-level versions, the suggested topology utilizes fewer devices and has less TSV on its switches than [30].

As seen from Table 9.5, the asymmetric three-input version of a traditional CHB employs 12 MOSFETs, 12 gate driver circuits, and 27 total devices to generate 15 voltage levels at the output. In such condition, the variety of DC sources is three. The TSV on switches is equal to 28 V_{dc}. The maximum output voltage and boosting capability of CHB are 7 V_{dc} and 1, respectively. CHB generates negative voltage levels with an H-bridge. According to Table 9.6, the number of semiconductors, sources, and devices per number of levels in CHB is higher than that of the proposed topology but the TSV_{pu} per number of levels in CHB is smaller than the proposed topology, which shows the superiority of CHB in this case over the proposed structure.

9.7 Simulation results

The proposed 15-level structure is simulated in PSCAD/EMTDC software to evaluate its correct function. The PD-PWM method is used for generating the switching pulses of the proposed structure. The simulation parameters are listed in Table 9.7.

The output voltage and current waveforms of the proposed structure are shown in Fig. 9.14. As seen, the proposed structure has successfully produced 15 output voltage levels, including 0, ±30, ±60, ±90, ±120, ±150, ±180, and

TABLE 9.7 Simulation parameters.

Parameter	Value
Input DC sources	30 V, 180 V
Switching frequency (f_{car})	5 kHz
Modulation index (m)	0.986
C_1 and C_2 capacitances	2200 μF
Load (R-L)	100 Ω and 300 mH

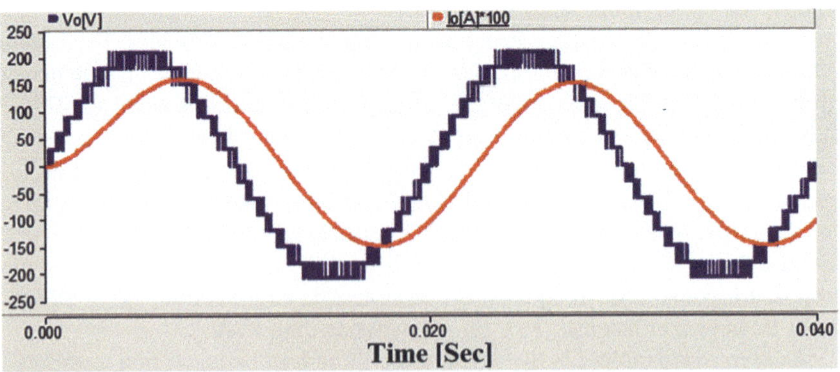

FIGURE 9.14 Output voltage and load current waveforms of the proposed switched-capacitor multilevel inverter.

±210 V. The maximum output voltage and load current are respectively about 210 V and 1.51 A. Fig. 9.14 shows a phase difference of about 43.3° degrees between output voltage and load current waveforms. This phase difference is confirmed by $\Delta\varphi = \arctan(L\omega/R)$. Based on Fig. 9.14, the proposed structure can sufficiently feed the low power factor (*R-L*) loads. It should be noted that for better visibility, the output current has been shown with a scaling factor of 100. The measured efficiency in this case is about 96.4%, which validates the suitability of the proposed structure.

The voltage waveforms of capacitors are shown in Fig. 9.15. It is seen that the voltage of both capacitors is regulated on 90 V, which is certified by theoretical concepts. The voltage balancing process of the capacitor is done naturally. Also, the voltage ripple on both capacitors is small, leading to high output-voltage quality. The low capacitances used in the proposed SCMLI reduce the total cost and dimensions of the proposed topology.

Fig. 9.16 shows the harmonic spectrum of output voltage of the proposed topology. Due to employment of PD-PWM, the magnitude of most of the harmonics is less than 0.5%. This results in a very low THD of about 1.3051%. The low THD and consequently the high-quality output voltage is one of the major benefits of the proposed structure, which results in a significant reduction in filter dimensions at the output.

One of the most important parameters in SCMLIs is SV on switches. As shown in Fig. 9.17 the SVs of switches are about: $SV_{S1} = 30$ V, $SV_{S2} = 30$ V, $SV_{S3} = 30$ V, $SV_{S4} = 30$ V, $SV_{S5} = 180$ V, $SV_{S6} = 90$ V, $SV_{S7} = 180$ V, $SV_{S8} = 180$ V, and $SV_{S9} = 180$ V. The obtained results are confirmed by Table 9.2.

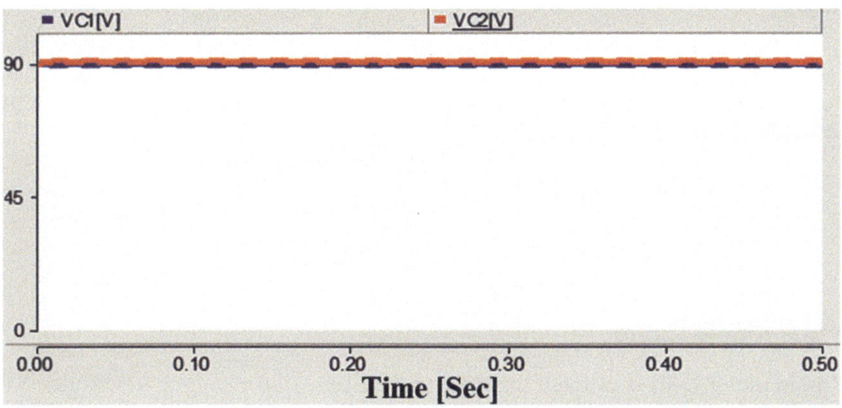

FIGURE 9.15 Voltage waveform of capacitors in the proposed switched-capacitor multilevel inverter.

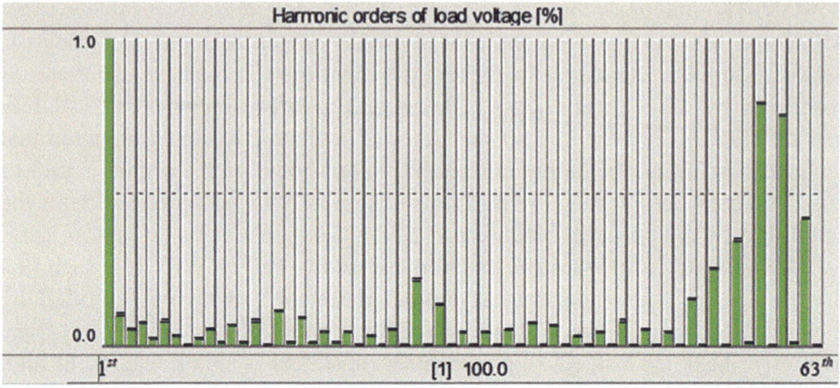

FIGURE 9.16 Harmonic spectrum of output voltage of the proposed switched-capacitor multi-level inverter.

Fig. 9.18 shows the current waveform of switches in the proposed structure. It is seen from Fig. 9.18 that the current stress of all switches is equal to the magnitude of load current, which is about 1.5 A.

9.8 Power quality analysis

Different definitions can be found for "power quality" in books and standards that can be considered for the voltage or current waveforms. "Power quality" is generally defined as the measuring, analyzing, and improving of the quality of bus voltage to achieve a sinusoidal waveform with nominal frequency and amplitude. Some popular power quality parameters as well as their formulas are presented next [41,42].

9.8.1 Form factor

Form factor (FF) measures the shape of the waveform and is defined as [41,42]:

$$FF = \frac{V_{rms}}{V_{dc}} \tag{9.40}$$

9.8.2 Ripple factor

Ripple factor (RF) is utilized to measure the ripple of the output waveform. RF is computed from [41,42]:

$$RF = \frac{V_{ac}}{V_{dc}} = \frac{\sqrt{(V_{rms})^2 - (V_{dc})^2}}{V_{dc}} = \sqrt{\frac{(V_{rms})^2}{(V_{dc})^2} - \frac{(V_{dc})^2}{(V_{dc})^2}} = \sqrt{FF^2 - 1} \tag{9.41}$$

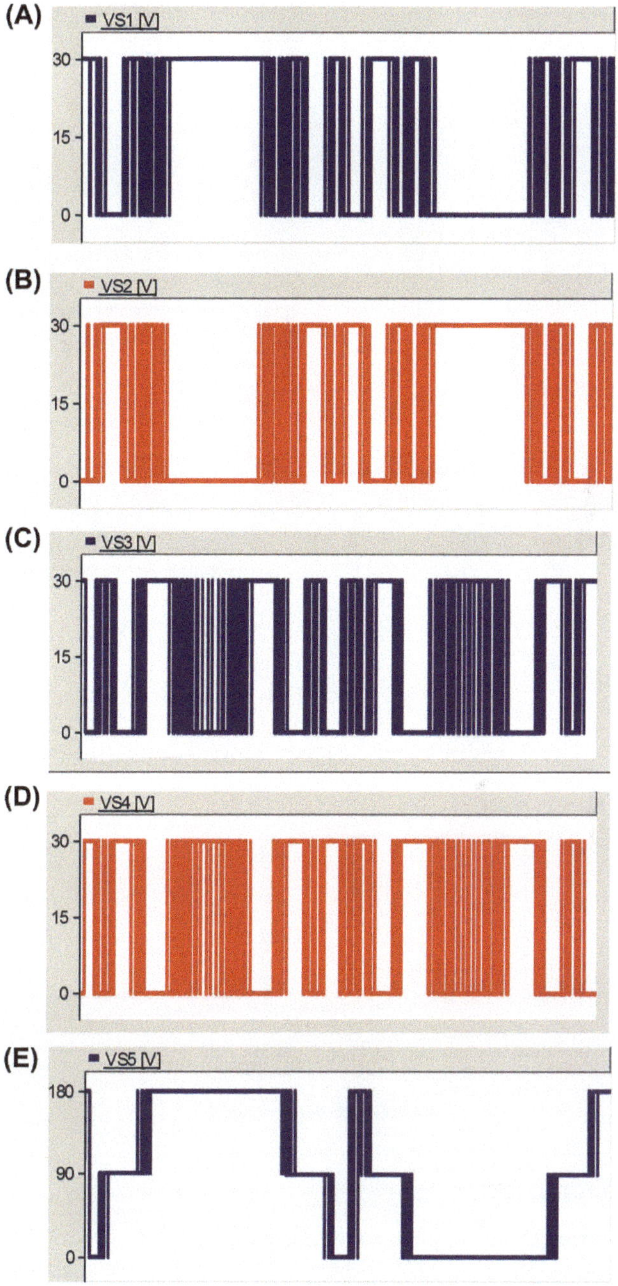

FIGURE 9.17 Voltage waveform of switches in the proposed switched-capacitor multilevel inverter.

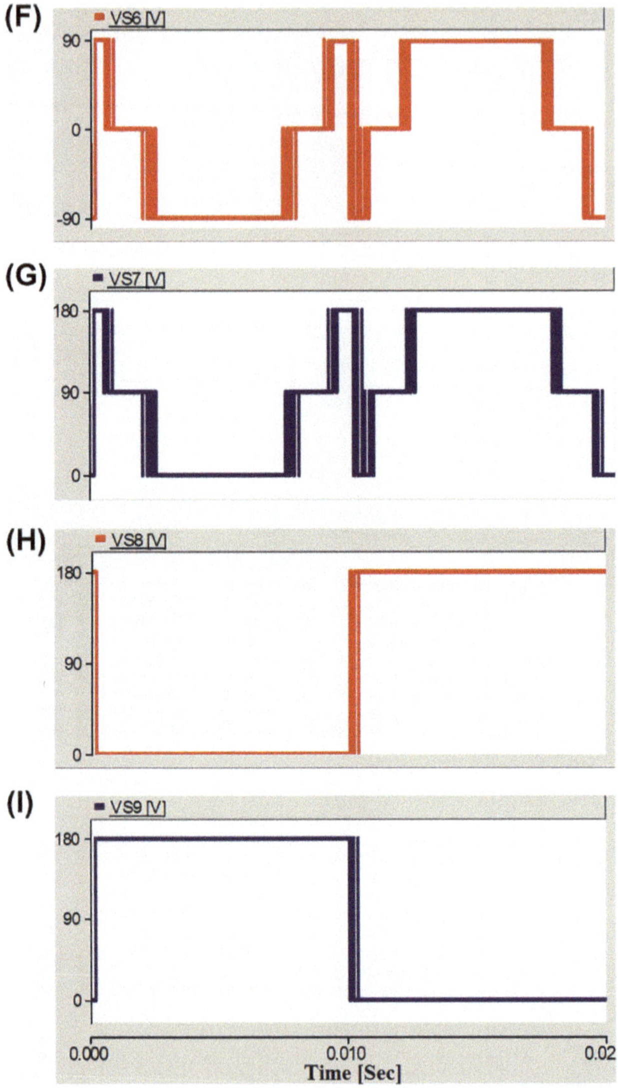

FIGURE 9.17 cont'd

9.8.3 Harmonic factor

Harmonic factor (HF) is applied to measure the content of individual harmonic orders, as shown in Eq. (9.42), where V_{rms}^1 and V_{rms}^h denote the root mean square voltage of first (fundamental frequency) and hth harmonic orders, respectively [41,42]:

$$HF_h = \frac{V_{rms}^h}{V_{rms}^1} \tag{9.42}$$

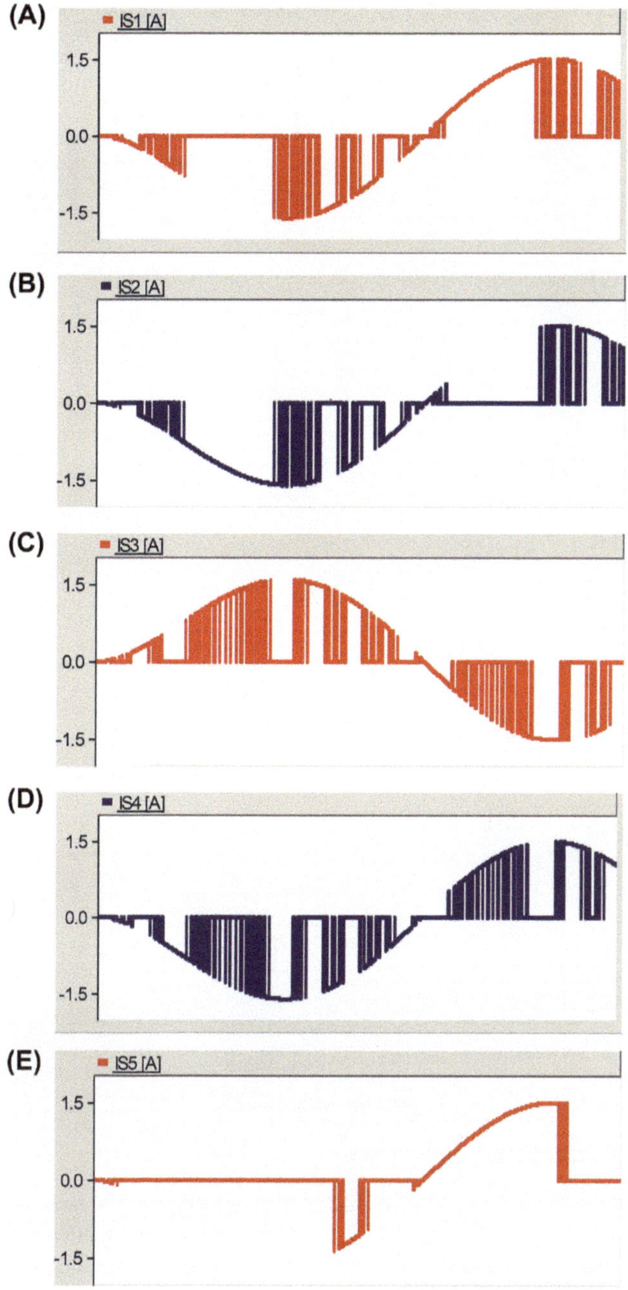

FIGURE 9.18 Current waveform of switches in the proposed switched-capacitor multilevel inverter.

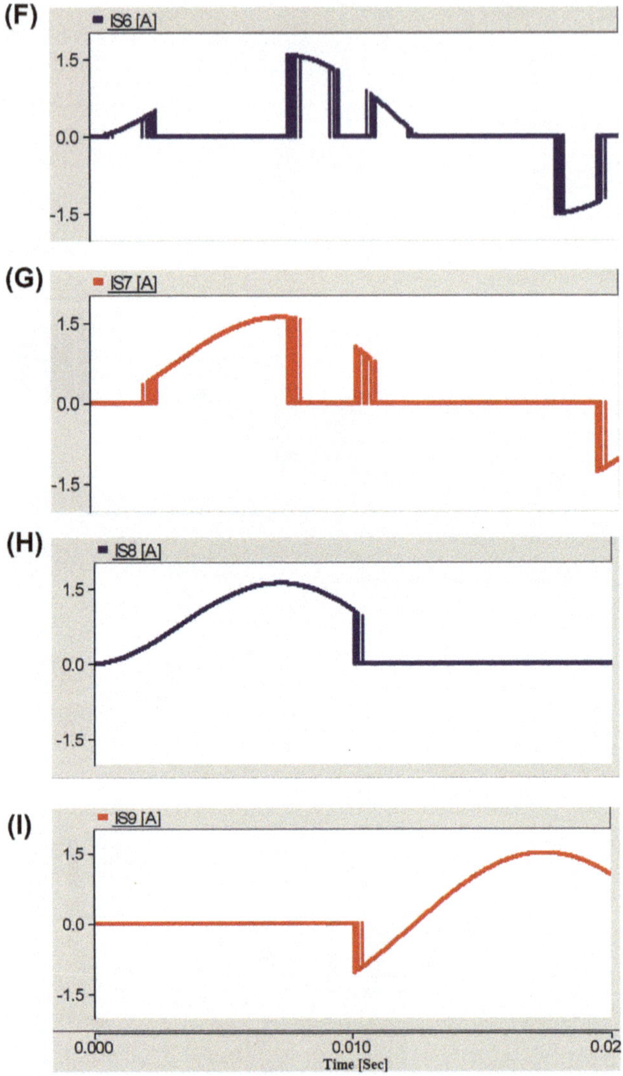

FIGURE 9.18 cont'd

9.8.4 Total harmonic distortion

The most popular and frequent indicator used to display the distortion content of a waveform is called THD. THD is calculated from Eq. (9.43) [41,42]. In power electronic converters, it is intended to reduce the THD as much as possible.

$$THD = \frac{\sqrt{\sum_{h=2}^{\infty} \left(V_{rms}^{h}\right)^{2}}}{V_{rms}^{1}} \tag{9.43}$$

TABLE 9.8 Comparison between simulation results and standard values.

Harmonic order (*h*)	Simulation value (%)	Standard limit (%)	Harmonic order (*h*)	Simulation value (%)	Standard limit (%)	Harmonic order (*h*)	Simulation value (%)	Standard limit (%)
1	100	–	22	0.0429	0.2	43	0.0044	0.49
2	0.0995	2	23	0.0047	1	44	0.0448	0.2
3	0.0505	5	24	0.0275	0.2	45	0.0042	0.2
4	0.0686	1	25	0.0039	1.5	46	0.0752	0.2
5	0.0173	6	26	0.0471	0.2	47	0.0012	0.47
6	0.0804	0.5	27	0.0029	0.2	48	0.0546	0.2
7	0.0287	5	28	0.2080	0.2	49	0.0016	0.46
8	0.0025	0.5	29	0.0034	0.63	50	0.0446	0.2
9	0.0249	1.5	30	0.1278	0.2	51	0.0019	0.2
10	0.0533	0.5	31	0.0022	0.6	52	0.1542	0.2
11	0.0072	3.5	32	0.0367	0.2	53	0.0038	0.44
12	0.0580	0.2	33	0.0040	0.2	54	0.2506	0.2
13	0.0082	3	34	0.0350	0.2	55	0.0023	0.43
14	0.0801	0.2	35	0.0018	0.56	56	0.3365	0.2
15	0.0019	0.3	36	0.0460	0.2	57	0.0061	0.2

Continued

TABLE 9.8 Comparison between simulation results and standard values.—cont'd

Harmonic order (*h*)	Simulation value (%)	Standard limit (%)	Harmonic order (*h*)	Simulation value (%)	Standard limit (%)	Harmonic order (*h*)	Simulation value (%)	Standard limit (%)
16	0.1108	0.2	37	0.0019	0.54	58	0.7865	0.2
17	0.0082	2	38	0.0673	0.2	59	0.0009	0.41
18	0.0897	0.2	39	0.0023	0.2	60	0.7515	0.2
19	0.0050	1.5	40	0.0017	0.2	61	0.0069	0.4
20	0.0358	0.2	41	0.0255	0.5	62	0.4114	0.2
21	0.0063	0.2	42	0.0044	0.2	63	0.0031	0.2

As seen, the power quality parameters are dependent on the amplitude of harmonic orders. To guarantee a desired power quality, the magnitude of harmonic orders (h) should be restricted. Based on the application and voltage/power level of the suggested topology, the harmonic values presented in the IEC 61000-2-2 standard have been considered [43].

Table 9.8 presents the magnitude of 63 harmonic orders of output voltage of the suggested topology as well as the standard harmonic values stated in IEC 61000-2-2 [43]. It is observed from Table 9.8 that the amplitude of all harmonic orders ($h > 1$) of generated voltage waveform are less than standard values. Also, the THD of output voltage of suggested topology is only about 1.3051%.

The results presented in Table 9.8 approve the high output voltage (power) quality of the suggested topology.

9.9 Conclusions

This study proposed a 15-level switched capacitor-based inverter, which benefits from simple control, self-voltage balancing of capacitors, increased steps per devices, low voltage ripple on capacitors, reduced on-state switches, and high-quality output voltage waveform. The extended version was also proposed by cascading the basic 15-level units, in which the ratio of number of levels to devices increased considerably. Also, the standing voltage and current stress of semiconductors remained in an acceptable range. The proposed structure was capable of supplying low power factor loads. The increased levels of proposed topology reduced the THD of the output voltage waveform, which accordingly resulted in downsized end-side filters. This also reduced the cost, size, and weight of the suggested converter. The comparative analysis certified the superiority of the proposed topology over previously presented similar structures. Also, the correct operation of the proposed 15-level inverter was validated by simulation results.

References

[1] T. Roy, P.K. Sadhu, A. Dasgupta, Cross-switched multilevel inverter using novel switched capacitor converters, IEEE Trans. Ind. Electron. 66 (11) (2019) 8521–8532.

[2] M. Vijeh, M. Rezanejad, E. Samadaei, K. Bertilsson, A general review of multilevel inverters based on main submodules: structural point of view, IEEE Trans. Power Electron. 34 (10) (2019) 9479–9502.

[3] E. Babaei, S.H. Hosseini, New cascaded multilevel inverter topology with minimum number of switches, Energy Convers. Manag. 50 (11) (2009) 2761–2767.

[4] T. Roy, P.K. Sadhu, A step-up multilevel inverter topology using novel switched capacitor converters with reduced components, IEEE Trans. Ind. Electron. (2020).

[5] L.G. Franquelo, J. Rodriguez, J.I. Leon, S. Kouro, R. Portillo, M.A. Prats, The age of multilevel converters arrives, IEEE Indus. Electron. Magazine 2 (2) (2008) 28–39.

[6] J.-S. Lai, F.Z. Peng, Multilevel converters-a new breed of power converters, IEEE Trans. Ind. Appl. 32 (3) (1996) 509–517.

[7] K. Varesi, M. Karimi, P. Kargar, A new cascaded 35-level inverter with reduced switch count, in: Iranian Conference on Renewable Energy & Distributed Generation, ICREDG), 2019.

[8] F. Esmaeili, K. Varesi, A modified single-phase multi-level inverter with increased number of steps, in: Iranian Conference on Renewable Energy & Distributed Generation, ICREDG), 2019.

[9] E. Babaei, S.S. Gowgani, Hybrid multilevel inverter using switched capacitor units, IEEE Trans. Ind. Electron. 61 (9) (2013) 4614–4621.

[10] R.S. Alishah, S.H. Hosseini, E. Babaei, M. Sabahi, G.B. Gharehpetian, New high step-up multilevel converter topology with self-voltage balancing ability and its optimization analysis, IEEE Trans. Ind. Electron. 64 (9) (2017) 7060–7070.

[11] R. Barzegarkhoo, H.M. Kojabadi, E. Zamiry, N. Vosoughi, L. Chang, Generalized structure for a single phase switched-capacitor multilevel inverter using a new multiple DC link producer with reduced number of switches, IEEE Trans. Power Electron. 31 (8) (2015) 5604–5617.

[12] M. Karimi, P. Kargar, K. Varesi, A novel sub-multilevel cell (SMC) with increased ratio of number of levels to number of sources and switches, in: Iranian Conference on Renewable Energy & Distributed Generation, ICREDG), 2019.

[13] K. Varesi, M. Karimi, P. Kargar, A new basic step-up cascaded 35-level topology extendable to higher number of levels, in: 10th International Power Electronics, Drive Systems and Technologies Conference (PEDSTC), IEEE, 2019, pp. 291–296, 2019.

[14] M. Karimi, P. Kargar, K. Varesi, A novel high-gain switched-capacitor based 11-level inverter topology, in: International Power System Conference, PSC), Tehran, Iran, 2019, pp. 404–409, 2019.

[15] M. Karimi, P. Kargar, K. Varesi, Two novel switched-capacitor based multi-level inverter topologies, in: International Power System Conference, PSC), Tehran, Iran, 2019, pp. 391–396, 2019.

[16] P. Kargar, M. Karimi, K. Varesi, A novel boost switched-capacitor based multi-level inverter structure, in: 2020 11th International Power Electronics, Drive Systems and Technologies Conference (PEDSTC), IEEE, Tehran, Iran, 2020, pp. 1–6.

[17] S. Deliri Khatoonabad, K. Varesi, An extended high step-up switched-capacitor based multi-level inverter topology, in: International Power System Conference (PSC), IEEE, Tehran, Iran, 2019, pp. 459–464, 2019.

[18] M. Saeedian, M.E. Adabi, S.M. Hosseini, J. Adabi, E. Pouresmaeil, A novel step-up single source multilevel inverter: topology, operating principle, and modulation, IEEE Trans. Power Electron. 34 (4) (2019) 3269–3282.

[19] S. Majumdar, B. Mahato, K.C. Jana, Implementation of an optimum reduced components multicell multilevel inverter (MC-MLI) for lower standing voltage, IEEE Trans. Ind. Electron. 67 (4) (2019) 2765–2775.

[20] M.M. Zaid, J.-S. Ro, Switch Ladder modified H-bridge multilevel inverter with novel pulse width modulation technique, IEEE Access 7 (2019) 102073–102086.

[21] A. Taheri, A. Rasulkhani, H.-P. Ren, An asymmetric switched capacitor multilevel inverter with component reduction, IEEE Access 7 (2019) 127166–127176.

[22] E. Samadaei, M. Kaviani, K. Bertilsson, A 13-levels module (K-type) with two DC sources for multilevel inverters, IEEE Trans. Ind. Electron. 66 (7) (2018) 5186−5196.

[23] S.P. Gautam, L. Kumar, S. Gupta, Hybrid topology of symmetrical multilevel inverter using less number of devices, IET Power Electron. 8 (11) (2015) 2125−2135.

[24] S.P. Gautam, L. Kumar, S. Gupta, Single-phase multilevel inverter topologies with self-voltage balancing capabilities, IET Power Electron. 11 (5) (2018) 844−855.

[25] B. Mahato, S. Majumdar, S. Vatsyayan, K. Jana, A new and generalized structure of MLI topology with half-bridge cell with minimum number of power electronic devices, IETE Tech. Rev. (2020) 1−12.

[26] A. Amir, A. Amir, J. Selvaraj, N. Abd Rahim, Grid-connected photovoltaic system employing a single-phase T-type cascaded H-bridge inverter, Sol. Energy 199 (2020) 645−656.

[27] Y.C. Fong, S.R. Raman, Y. Ye, K.W.E. Cheng, Generalized topology of a hybrid switched-capacitor multilevel inverter for high-frequency AC power distribution, IEEE J. Emerg. Select. Top. Power Electron. 8 (3) (2019) 2886−2897.

[28] B. Mahato, S. Majumdar, K.C. Jana, Single-phase Modified T-type−based multilevel inverter with reduced number of power electronic devices, Int. Trans. Electric. Energy Syst. 29 (11) (2019) e12097.

[29] A. Ahmad, et al., Realization of a generalized switched-capacitor multilevel inverter topology with less switch requirement, Energies 13 (7) (2020) 1556.

[30] F. Sedaghati, S.H.L. Majareh, A multilevel inverter based on cascade connection of sub-multilevel units with reduced switch count, Int. J. Circ. Theor. Appl. 47 (7) (2019) 1152−1172.

[31] P.R. Bana, K.P. Panda, G. Panda, Power quality performance evaluation of multilevel inverter with reduced switching devices and minimum standing voltage, IEEE Trans. Indus. Inform. 16 (8) (2019) 5009−5022.

[32] S. Deliri Khatoonabad, K. Varesi, A novel dual-input switched-capacitor based 27-level boost inverter topology, in: 28th Iranian Conference on Electrical Engineering (ICEE2020), IEEE, Tabriz, Iran, 2020.

[33] F. Esmaeili, K. Varesi, A novel single-phase multi-level inverter topology based on bridge-type connected sources with enhanced number of levels per number of devices, J. Energy Manage. Technol. 4 (3) (2020) 37−47.

[34] J. Rodriguez, J.-S. Lai, F.Z. Peng, Multilevel inverters: a survey of topologies, controls, and applications, IEEE Trans. Ind. Electron. 49 (4) (2002) 724−738.

[35] I. Colak, E. Kabalci, R. Bayindir, Review of multilevel voltage source inverter topologies and control schemes, Energy Conver. Manage. 52 (2) (2011) 1114−1128.

[36] F. Esmaeili, K. Varesi, An asymmetric multi-level inverter structure with increased steps per devices, in: 2020 11th International Power Electronics, Drive Systems and Technologies Conference (PEDSTC), IEEE, Tehran, Iran, 2020, pp. 1−5.

[37] C.-H. Hsieh, T.-J. Liang, S.-M. Chen, S.-W. Tsai, Design and implementation of a novel multilevel DC−AC inverter, IEEE Trans. Ind. Appl. 52 (3) (2016) 2436−2443.

[38] R. Barzegarkhoo, M. Moradzadeh, E. Zamiri, H.M. Kojabadi, F. Blaabjerg, A new boost switched-capacitor multilevel converter with reduced circuit devices, IEEE Trans. Power Electron. 33 (8) (2017) 6738−6754.

[39] A. Taghvaie, J. Adabi, M. Rezanejad, A self-balanced step-up multilevel inverter based on switched-capacitor structure, IEEE Trans. Power Electron. 33 (1) (2017) 199−209.

[40] M. Saeedian, E. Pouresmaeil, E. Samadaei, E. Manuel Godinho Rodrigues, R. Godina, M. Marzband, An innovative dual-boost nine-level inverter with low-voltage rating switches, Energies 12 (2) (2019) 207.

[41] E. Fuchs, M.A. Masoum, Power Quality in Power Systems and Electrical Machines, Academic press, 2011.

[42] S. Bhattacharyya, Power Quality Requirements and Responsibilities at the Point of Connection, 2011.

[43] B. Singh, A. Chandra, K. Al-Haddad, Power Quality: Problems and Mitigation Techniques, John Wiley & Sons, 2014.

Chapter 10

E-STATCOM (energy storage + STATCOM): a solution to integrate large-scale wind farms into the grid at medium and high power levels

Anil Bharadwaj, Suman Maiti, Nirmalya Dhal, S. Chakraborty, Sreekesh K. Pillai
Department of Electrical Engineering, Indian Institute of Technology Kharagpur, Kharagpur, West Bengal, India

Chapter outline

10.1 Introduction 284
10.2 Challenges of renewable power-dominated grids 285
10.3 Existing methods to provide grid codes 285
10.4 Concept of an E-STATCOM 286
10.4.1 Energy storage system in an E-STATCOM 287
10.4.2 Two-/three-level converter-based E-STATCOM 287
10.4.3 Modular multilevel converter-based E-STATCOM 288
10.5 System for performance study 289
10.6 Control methodology of an E-STATCOM 291
10.6.1 Operation of an MMC-based E-STATCOM 292
10.6.2 Operation of hybrid storage system 297
10.7 Discussion of results 302
10.8 Conclusion 304
References 309

Power Quality in Modern Power Systems. https://doi.org/10.1016/B978-0-12-823346-7.00007-4

10.1 Introduction

Power generation from renewable energy resources (RERs) like solar parks and wind farms has witnessed enormous growth during the last decade. Figs. 10.1 and 10.2 show the layout of a large-scale power generation system using solar and wind energy, respectively. In the case of photovoltaic (PV) integration, the PV panels are connected to the grid through a two-/three-level dc—ac converter. Similarly, for wind power integration, a doubly fed induction generator is generally used where stator terminals are connected to the grid directly, whereas the rotor side is connected to the same grid through back-to-back converters. Power electronic converters capture maximum available power from RERs at any instant. A common bus is formed to collect the total power generated by RERs, which is fed to the grid through a transmission network [1,2].

The power generated by these plants varies in nature due to fluctuations in solar irradiance and wind velocity. To extract maximum power at any instant, the power electronic converters are generally operated with maximum power point tracking (MPPT) control. In [3,4], various types of MPPT methods have been proposed to execute MPPT operation, which results in the injection of fluctuating power into the grid.

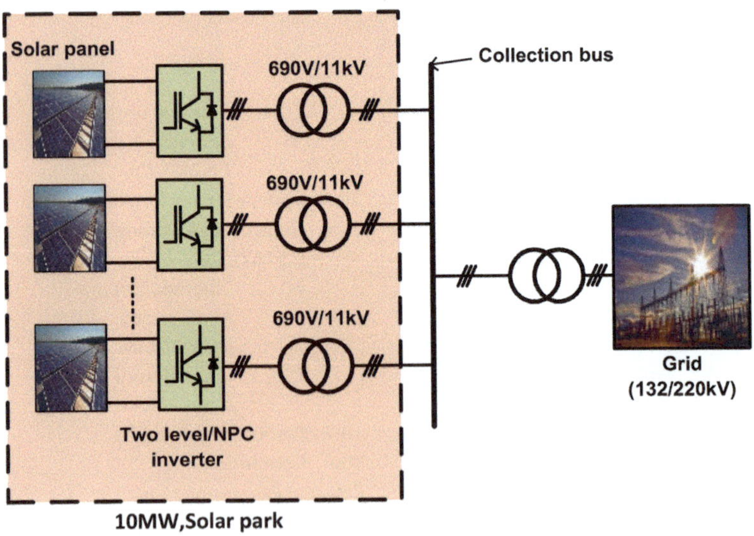

FIGURE 10.1 Layout of a solar energy-based power generation system. *NPC*, neutral point clamp; *WT*, wind turbine.

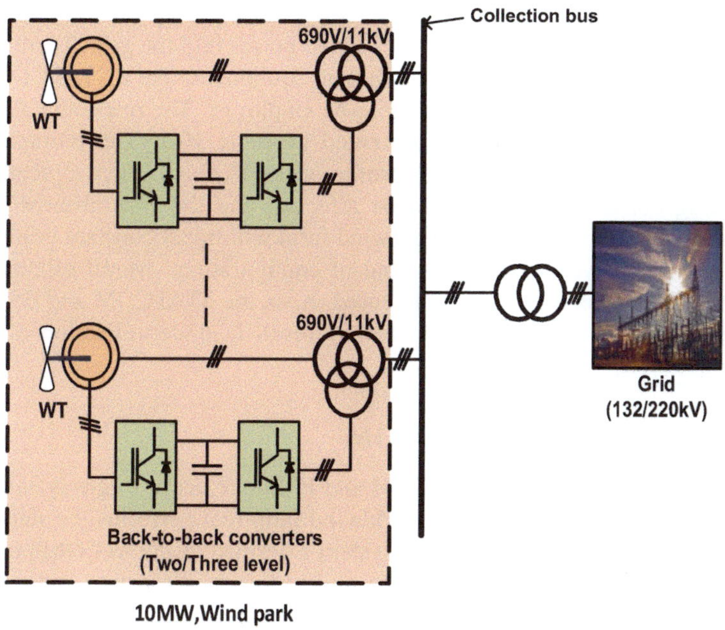

FIGURE 10.2 Layout of a wind energy-based power generation system.

10.2 Challenges of renewable power-dominated grids

With the injection of variable power into the grid, operation of the existing electrical network becomes more challenging and scheduling of conventional power plants becomes more complex. Also, this affects the voltage and frequency stability of the grid. Thus to improve the voltage and frequency stability of the renewable power-dominated grids, various grid codes have evolved over time. The following are some of the important grid codes that need to be satisfied during large-scale integration of RERs with the grid [5,6]:

1. Voltage regulation at the point of common coupling (PCC)
2. Frequency support by reducing the fluctuation caused by RERs
3. Unbalance current/voltage mitigation at the PCC
4. Harmonic compensation
5. Low-voltage ride-through capability during a fault at the PCC

10.3 Existing methods to provide grid codes

To satisfy the aforesaid requirements at the PCC, multifunctional power electronic converters have been suggested, which can be employed in solar parks/wind farms [7,8]. Since the converters have limited dc-link voltage and power delivering capability, the interfacing converters in the solar park and wind farm are employed only to inject active power into the grid. Therefore to satisfy the grid codes at the PCC, additional equipment, like a STATCOM and an energy storage system (ESS) need to be employed. The respective features of these individual systems are presented in Table 10.1. The purpose of a STATCOM is to maintain voltage profile at the PCC such that operation of interfacing converters [9] will not be affected, and the ESS absorbs power fluctuations, such that the solar park/wind farm will inject constant power into the grid [10]. However, with this kind of configuration, overall efficiency is reduced and component count is increased. Also, the STATCOM and ESS may not be operated continuously, which may result in underutilization of these systems.

10.4 Concept of an E-STATCOM

To obtain the features of a STATCOM and ESS together, a single system may be installed at the PCC of a solar park/wind farm that performs the same jobs as listed in Table 10.1. This kind of system is termed "E-STATCOM" (energy storage + STATCOM), which has the ability to provide active and reactive power support for a specified duration [11]. An E-STATCOM is formed by integrating an ESS into a power electronic converter. The topology of the power electronic converter plays an important role in the formation of an E-STATCOM because of scalability, modularity, resiliency, efficiency, etc. Two different kinds of power converter topologies are considered in this work: one with a conventional two-level converter and the other based on a modular multilevel converter (MMC), as discussed in the following sections. Before looking at converter topologies in detail, the ESS is discussed in brief in the following subsection.

TABLE 10.1 Grid support features of a STATCOM and energy storage system (ESS).

STATCOM	Reactive power support
	Harmonic mitigation
	Unbalanced compensation
	Low-voltage ride-through capability
ESS	Active power support
	Frequency regulation
	Power oscillation damping

10.4.1 Energy storage system in an E-STATCOM

In [12], the authors presented different types of storage system, such as battery, ultracapacitor (UC), fuel cell, supermagnetic energy storage, and flywheel-based storage system. Each of these has its own advantages and limitations. Among them, the battery-based storage system (BSS) is mostly used in power system applications due to low investment cost, less maintenance, and ease of control [13]. In recent times, the UC-based storage system (BSS) has attracted the attention of researchers due to its features, such as high charging/discharging cycles, longer lifespan, and less maintenance [14]. Due to the variety of application areas, it is expected that the ESS of an E-STATCOM should have a longer lifespan, low investment, less maintenance, and ability to support high-power density as well as high-energy density demands. However, the storage system is not capable of providing these features alone. Hence, it is efficient to combine two or more kinds of storage systems together, which are known as hybrid storage systems (HSS). Among the different types of HSSs, battery energy storage system and UC-based HSS are mostly employed to supply the active power needed [15]. In subsequent sections, the integration of HSS with power electronic converters will be discussed in detail along with relevant control methods.

10.4.2 Two-/three-level converter-based E-STATCOM

The configuration of an E-STATCOM using a two-level voltage source converter (VSC) is presented in Fig. 10.3. The HSS is connected at the dc-link of the converter using a bidirectional boost dc–dc converter [15], which helps to regulate power flow between the dc-link of the VSC and the storage modules. This configuration can be used up to an ac voltage of 690 V and power level of 1 MVA [16] considering switching devices rated at 1.2 kV. The following techniques may be adopted to enhance the power rating of a two-level VSC:

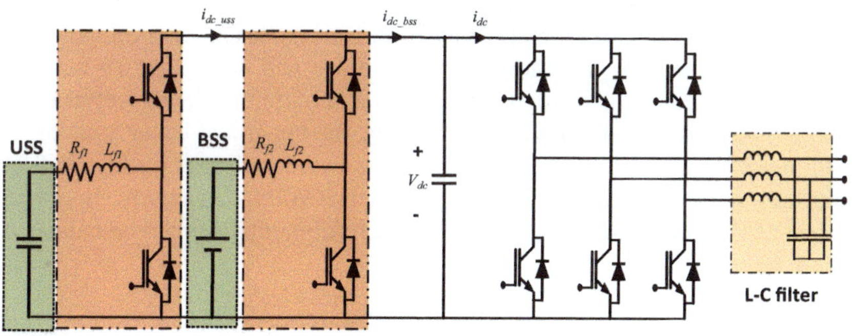

FIGURE 10.3 Configuration of a two-level converter-based E-STATCOM with hybrid storage system. *BSS*, battery storage system; *USS*, ultracapacitor storage system.

- By increasing voltage rating of the dc-link, which can be possible through series connection of switching devices.
- By increasing current rating, which requires paralleling of switching devices.
- Several VSCs connected in parallel to accomplish the desired power rating.

Common problems that a two-level VSC may encounter at higher power level are low efficiency, higher losses, need of large passive filters, higher electromagnetic interference, etc. Considering these limitations, use of a two-level VSC for high-power applications is not recommended [17].

10.4.3 Modular multilevel converter-based E-STATCOM

In [18], the use of multilevel converters for ESS integration at high voltage/power levels has been analyzed. With the invention of the MMC by Prof. Marquardt in 2003, the paradigm of high-power converters has completely changed due to potential advantages of the MMC like modularity, scalability, redundancy, and better efficiency [19]. Due to these advantages, the MMC can be considered as a potential candidate for energy storage integration at the grid scale [20–24]. In [25], an MMC-based configuration is proposed to integrate battery and UC with the grid, which is shown in Fig. 10.4. In this configuration, the battery storage system (BSS) is distributed at the dc-link, while the UC modules are used in place of submodule capacitors. Shortcomings of the topology are that current associated with both active and reactive power is flowing through the UCs, which significantly increases the size of the ESS. To overcome this limitation, a topology using MMC which is embeded with HSS has been presented.

A phase leg of the MMC has two identical arms that are connected through inductors to limit the circulating currents. An arm can be formed by cascading half-bridge cells or full-bridge cells. The full-bridge cell-based structures are mostly preferred for high-voltage direct current and drive applications. On the other hand, for STATCOM and ESS integration, half-bridge cell-based submodules are mostly employed. During steady-state operation of an MMC, each leg has to generate an ac voltage and dc voltage so that the desired voltage can be generated at the input and output of the converter [20]. To obtain these ac and dc equivalent voltages, the upper and lower arms of an MMC need to produce a voltage that can be expressed through Eqs. (10.1) and (10.2), respectively. The sum of the capacitor voltage in an arm should be charged at V_{dc} and hence the capacitor of each submodule should be balanced around $\frac{V_{DC}}{N}$.

$$v_{armu} = \left(\sum_{i=1}^{N} V_{ci}\right)\left(\frac{1 - m\sin(\omega t)}{2}\right) \tag{10.1}$$

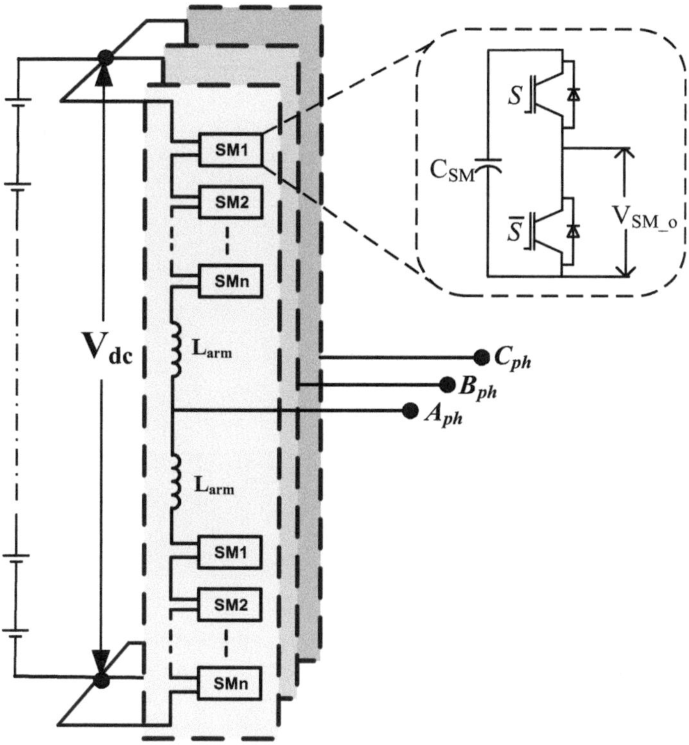

FIGURE 10.4 Configuration of modular multilevel converter with lumped battery storage system at the dc-link.

$$v_{arml} = \sum_{i=1}^{N} V_{ci}\left(\frac{1 + m\sin(\omega t)}{2}\right) \tag{10.2}$$

where m is the modulation index, N is the number of submodules per arm, V_{ci} is the capacitor voltage of ith submodule, and V_{DC} is the voltage across the dc-link.

10.5 System for performance study

Fig. 10.6 shows the system that has been employed to investigate the performance of the presented configurations in this chapter. The wind farm is formed by connecting smaller units of a wind power generation system, which consists of a wind turbine connected to either a doubly fed induction generator or squirrel cage induction generator (SCIG). In this work, an SCIG is considered that is connected to a common collection bus through an ac/dc/ac conversion stage. The SCIG is operated with MPPT control to extract maximum power at any instant [26]. Out of the two converters, the machine-side converter

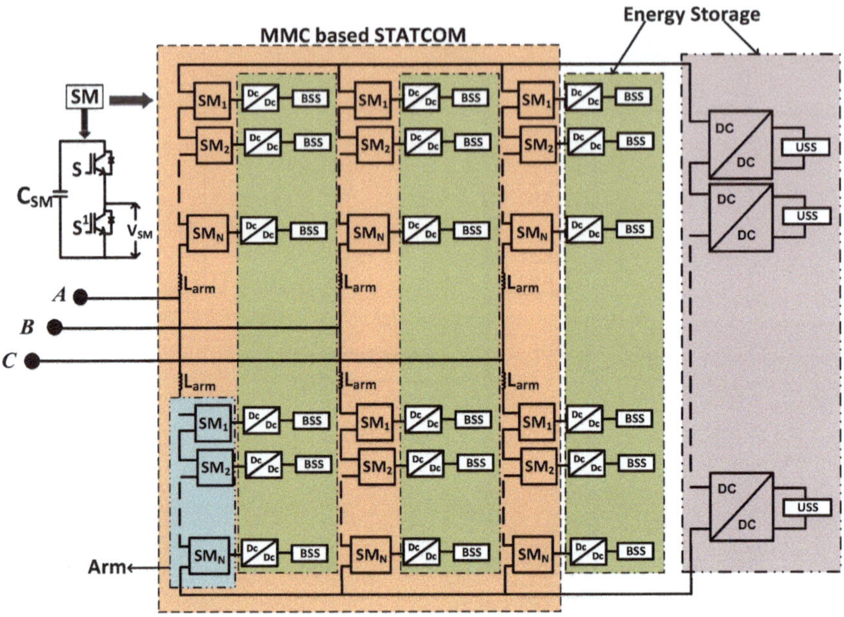

FIGURE 10.5 MMC-based E-STATCOM with distributed BSS and ultracapacitor modules at the dc-link and submodule. *BSS*, battery storage system; *MMC*, modular multilevel converter; *USS*, ultracapacitor storage system.

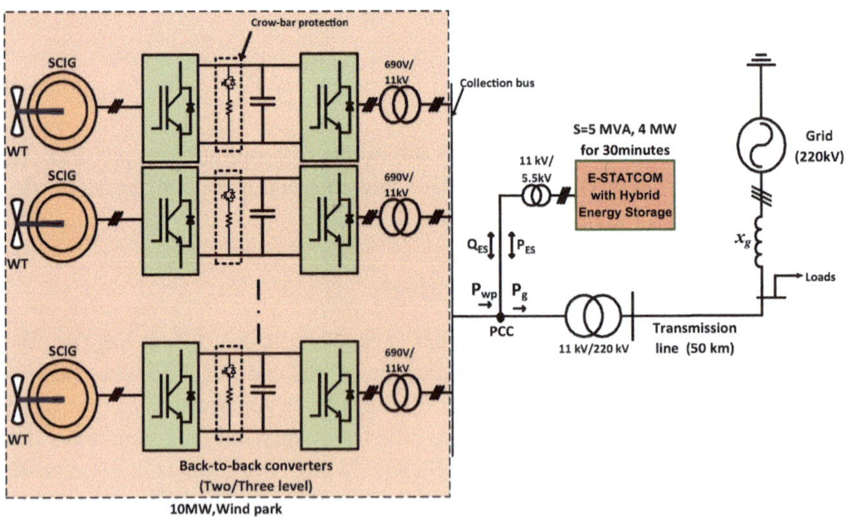

FIGURE 10.6 Grid-connected wind park with an E-STATCOM. *SCIG*, squirrel cage induction generator; *WT*, wind turbine.

helps to regulate the speed of the machine, while the grid-side converter is used to regulate the intermediate dc-link voltage. The wind park is integrated with the grid through a transmission network. Each SCIG is rated at 1.5 MW, and the total capacity of the wind farm is 10 MW. Here, the objectives of an E-STATCOM are (1) power smoothening through active power support, (2) voltage regulation at the PCC through reactive power support, and (3) negative sequence current compensation to balance the grid current.

10.6 Control methodology of an E-STATCOM

To meet the objectives set for the E-STATCOM (e.g., power smoothening, voltage regulation), precise active and reactive power control are the main requirements, which are put in the top level of the control hierarchy. Along with this, control methods are also incorporated to provide other ancillary services, like negative sequence current compensation and low-voltage ride-through capability. The various layers of control for operation of the E-STATCOM are shown in Fig. 10.7. Complete control is broadly divided into higher-level control and lower-level control. In the higher-level control, the power generated by the wind farm is estimated for the next 30 min [27]. Depending on the amount of available energy in the HSS, using Eq. (10.3) the active power needed to inject constant active power into the grid can be estimated:

$$P_{es} = P_g - P_{wp} \tag{10.3}$$

where P_g, P_{wp}, and P_{es} are grid power, power generated by wind park, and active power supplied by the E-STATCOM, respectively.

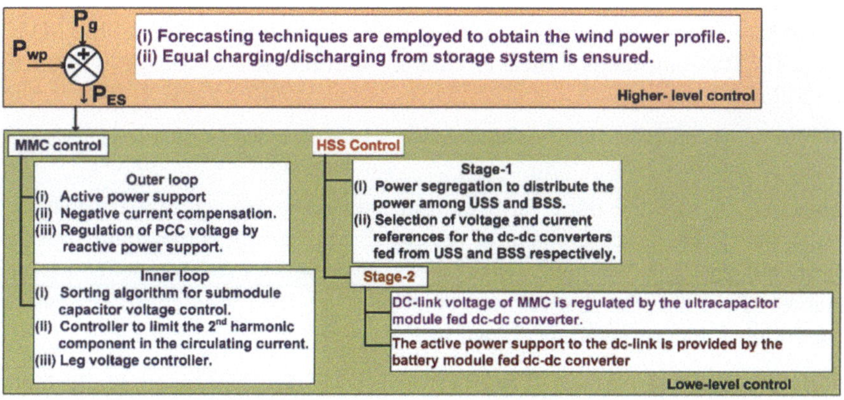

FIGURE 10.7 Overview of control methodology to operate an E-STATCOM. *BSS*, battery storage system; *MMC*, modular multilevel converter; *PCC*, point of common coupling; *USS*, ultracapacitor storage system.

10.6.1 Operation of an MMC-based E-STATCOM

The control block diagram to operate the MMC-based E-STATCOM is shown in Fig. 10.8. The controller is implemented in a synchronously rotating $d-q$ frame, where the active and reactive power are controlled by regulating i_d and i_q, respectively. The generation of references for positive and negative sequence current components is the key of this control method, which is covered in subsequent sections of this chapter.where P_{es} is the active power demand, and V_{dg}^{+} is the positive sequence component of the PCC voltage.

1. Generation of positive sequence current reference: The magnitude of $i_{d_{ref}}^{+}$ and $i_{q_{ref}}^{+}$ can be determined using the following expressions:
2. Extraction of sequence components: To provide the required positive and negative sequence powers, it is necessary to extract both positive and negative sequence components. A synchronous reference frame-based phase-locked loop is used here to extract the positive and negative sequence components in the PCC voltage and line currents. The respective control block diagram is shown in Fig. 10.8.
3. PCC voltage controller: To regulate the voltages at the PCC within a specified limit, the voltage controller decides the reactive power reference, Q_{es}^{*}, which needs to be injected into the grid. The $i_{q_{ref}}^{+}$ is then generated from Q_{es}^{*} according to Eq. (10.5) as shown in Fig. 10.8.

$$i_{d_{ref}}^{+} = \frac{P_{es}}{1.5 \times V_{dg}^{+}} \tag{10.4}$$

$$i_{q_{ref}}^{+} = \frac{Q_{es}}{1.5 \times V_{dg}^{+}} \tag{10.5}$$

4. Circulating current control: A differential current is flowing in between the dc source and phase leg of an MMC due to a difference in instantaneous voltages, which are generated by upper and lower arms of the MMC [28]. This current is called circulating current, which can be written as follows:

$$i_{cira} = \frac{i_{ua} + i_{la}}{2} \tag{10.6}$$

where i_{ua}, i_{la} are the upper arm current and lower arm current of phase a, respectively. The dc component of the circulating current is inevitable, but the harmonic part is unwanted, which needs to be minimized for better performance. To limit the harmonic component of the circulating current, a controller is incorporated as shown in Fig. 10.9.

5. Leg-voltage controller: The BSS and USS are distributed at the submodules and dc-link of the MMC (as shown in Fig. 10.5) to supply required active power support. Under steady-state conditions, the total power drawn from the dc-link is equally shared among the distributed

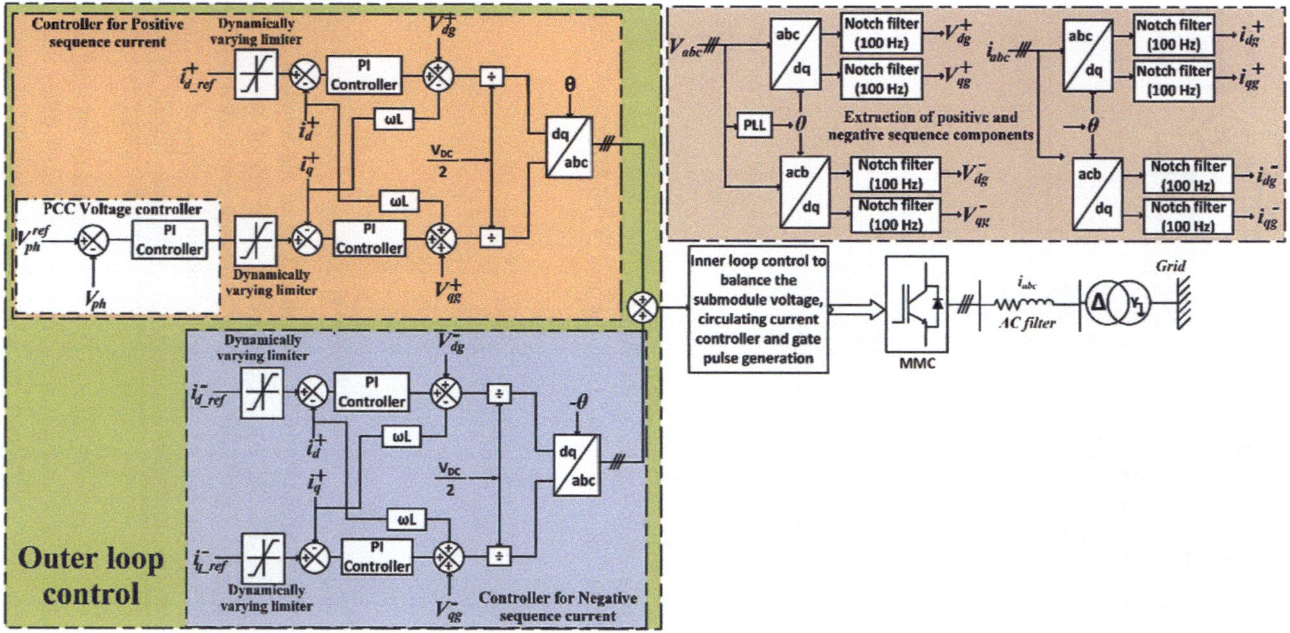

FIGURE 10.8 Control block diagram of a modular multilevel converter-based E-STATCOM. *MMC*, modular multilevel converter; *PCC*, point of common coupling; *PI*, proportional integral; *PLL*, phase-locked loop.

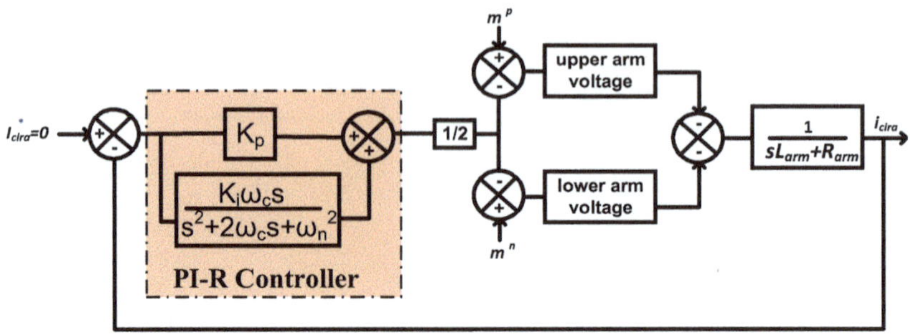

FIGURE 10.9 Block diagram for a circulating current control. *PI-R*, proportinal integral-resonant controller.

storage modules. However, the following nonideal situation may appear, which may have a detrimental effect on the performance of the MMC.

- Fault in any one of the storage modules
- Fault in any one of the dc–dc converters

During the aforesaid situation, the corresponding dc–dc converter has to be bypassed, which leads to a reduction in dc-link voltage of the MMC and the amount of active power supplied by the USS. According to Eq. (10.1), there will be a reduction in average submodule capacitor voltage, which causes reduction in the arm voltages of the MMC. The voltage at the output of a submodule (*i*th) in the positive and negative arm of the MMC can be expressed as follows:

$$v_{sm_{pi}} = V_{C_{nom}}\left(\frac{d}{2} - \frac{m \sin \omega t}{2}\right) \tag{10.7}$$

$$v_{sm_{pi}} = V_{C_{nom}}\left(\frac{d}{2} + \frac{m \sin \omega t}{2}\right) \tag{10.8}$$

$$V_{dc} = N(d \times V_{C_{nom}}) \tag{10.9}$$

where d is the dc modulation index and m is the ac modulation index at an operating condition. The dc equivalent voltage in a leg of the MMC can be expressed through Eq. (10.9) from which it may be concluded that any variations in dc-link voltage can be adjusted by varying d without altering V_{c_nom}. Fig. 10.10 shows the control block diagram to regulate the dc component of the modulation index.

6. Balancing of submodule capacitors and generation of gating signals: Submodule capacitor voltages may diverge from each other due to some nonidealities, like difference in equivalent series resistance, unequal semiconductor switching characteristics, etc. To perform capacitor voltage

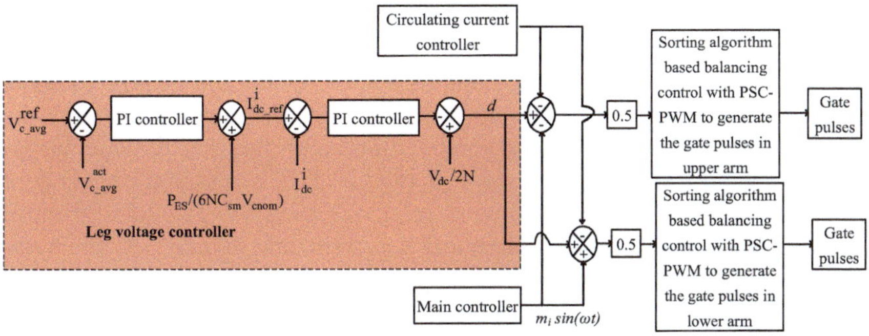

FIGURE 10.10 Inner-loop control of modular multilevel converter with adjustment of the dc modulation index. *PI*, proportional integral; *PSC-PWM*, phase shift carrier-pulse width modulation.

balancing, a sorting algorithm [21] is used here that inserts or bypasses cells based on capacitor voltage and direction of arm current, as shown in Fig. 10.11. As a result, the submodule capacitor voltages stay at their nominal value with the ripple component due to switching of the arm current.

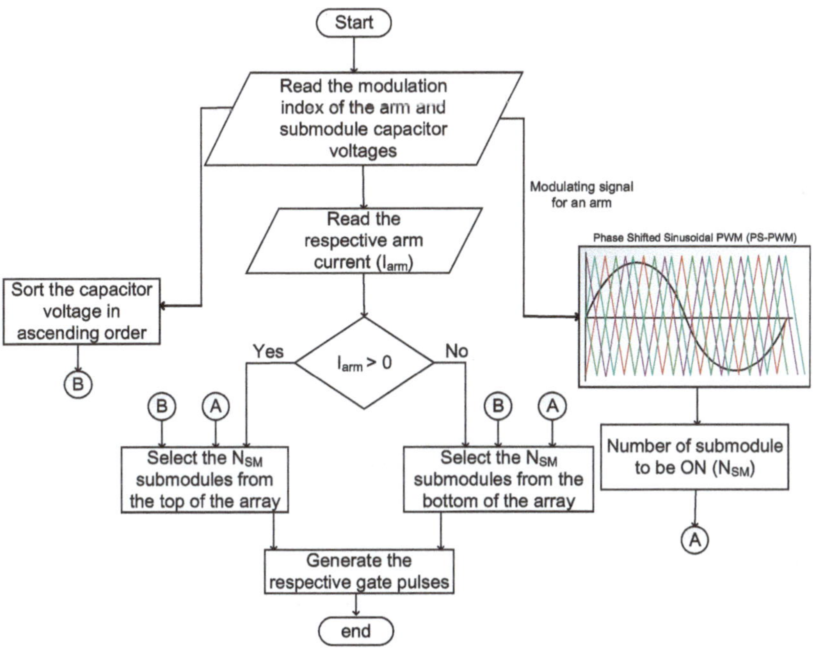

FIGURE 10.11 Flowchart of the sorting algorithm to balance submodule capacitor voltages in an arm of the modular multilevel converter. *PWM*, pulse width modulation.

7. Negative sequence current compensation: With the presence of an unbalance load at the collection bus, unbalance current is drawn from the grid, which leads to voltage unbalance at the PCC. The E-STATCOM at the PCC provides a negative sequence current component and makes the grid current balanced. The corresponding controller that provides negative sequence currents is shown in Fig. 10.8. The per-phase equivalent circuit of the grid-connected wind farm with an E-STATCOM is given in Fig. 10.12. By applying Kirchhoff's current law at node A, the relationship between grid current, load current, and converter current can be expressed as:

$$i_{load} = i_g + i_{conv} \tag{10.10}$$

Decomposing Eq. (10.10) into positive and negative sequence components, it may be written as follows:

$$\begin{bmatrix} i^+_{load_d} \\ i^+_{load_q} \end{bmatrix} = \begin{bmatrix} i^+_{g_d} \\ i^+_{g_q} \end{bmatrix} + \begin{bmatrix} i^+_{conv_d} \\ i^+_{conv_q} \end{bmatrix} \tag{10.11}$$

$$\begin{bmatrix} i^-_{load_d} \\ i^-_{load_q} \end{bmatrix} = \begin{bmatrix} i^-_{g_d} \\ i^-_{g_q} \end{bmatrix} + \begin{bmatrix} i^-_{conv_d} \\ i^-_{conv_q} \end{bmatrix} \tag{10.12}$$

There would be no negative sequence component in the grid current under balanced operating conditions. At that time, the negative sequence component of the load current should be entirely supplied by the converter, which yields the following expression:

$$\begin{bmatrix} i^*_{d_{ref}} \\ i^*_{q_{ref}} \end{bmatrix} = \begin{bmatrix} i^-_{conv_d} \\ i^-_{conv_q} \end{bmatrix} = \begin{bmatrix} i^-_{load_d} \\ i^-_{load_q} \end{bmatrix} \tag{10.13}$$

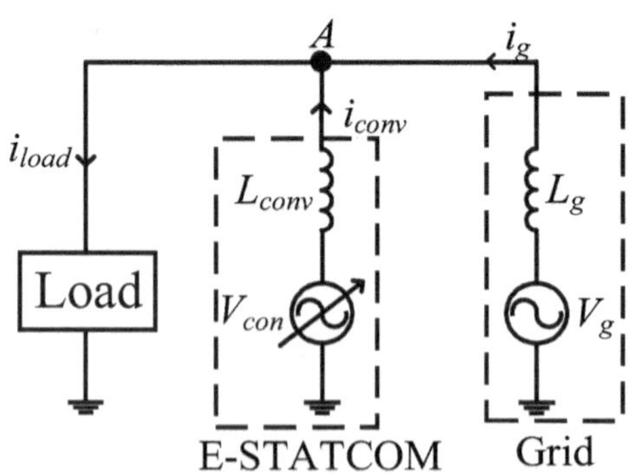

FIGURE 10.12 Per-phase equivalent circuit of wind farm along with an E-STATCOM.

8. Low-voltage ride-through capability: PCC voltage may fall to a very low value during faults in the network. According to a literature review, reactive power support can play an important role in the restoration of PCC voltage. So, complete capacity of the E-STATCOM is employed to supply reactive power only. To perform this task, the limits of *d*- and *q*-axis currents are adjusted so that the converter is able to deliver more reactive power compared to active power. The limiting values of *d*- and *q*-axis currents can be determined from PCC voltage, and the respective values are given in the flowchart as shown in Fig. 10.13.

10.6.2 Operation of hybrid storage system

The HSS is distributed at the dc-link of the MMC to provide required active power support. Since the charging/discharging characteristics of battery and UC modules are different, this requires appropriate distribution of active power among these energy storage modules.

1. Power segregation: The P_{es} is supplied by battery and UC modules as given in Eq. (10.14). The active power should be distributed properly among battery and UC modules considering their respective charge/discharge characteristics so that a fast dynamic response can be achieved.

$$P_{es} = P_{bss} + P_{uss} \tag{10.14}$$

The methodology employed to divide power among battery and UC modules is shown in Fig. 10.14. A low-pass filter is used that segregates P_{es} into low- and high-frequency components. The low-frequency component is supplied by the battery, and the remaining part, i.e., P_{uss} (where $P_{uss} = P_{es} - P_{bss}$), is given by UCs. The low-frequency component (i.e., P_{bss}) is then passed

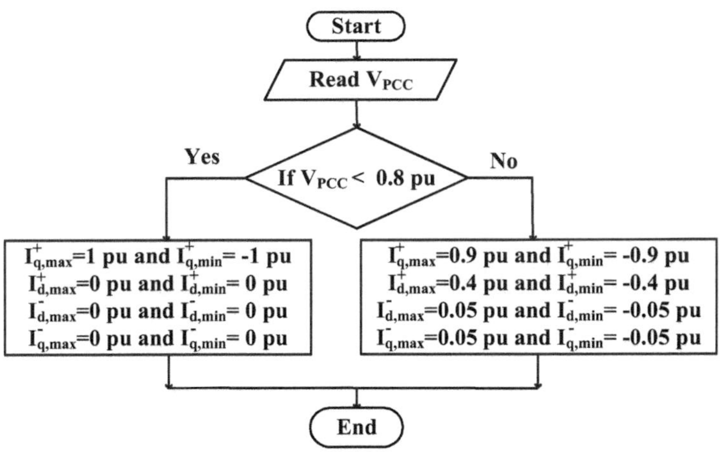

FIGURE 10.13 Procedure to determine limits for positive and negative sequence controller.

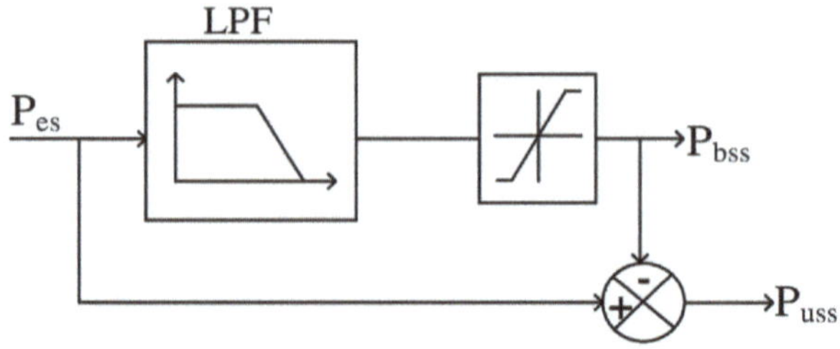

FIGURE 10.14 Block diagram of power segregation among the battery storage system and *USS*, ultracapacitor storage system; *LPF*, low-pass filter.

through a hard limiter that clamps the power according to the size of the BSS. The cut-off frequency of the low-pass filter is chosen as 1.25 Hz, and the limiting values of P_{bss} will be derived in the subsequent sections. The power segregation method is pictorially shown in Fig. 10.15.

2. Control of UC-based storage system: The dc–dc converters fed from UC modules are operated in voltage mode control (VMC) as shown in Fig. 10.16.

According to the total dc bus voltage can be expressed as:

$$V_{DC} = \sum_{k=1}^{M} V_{o_k} \tag{10.15}$$

$$V_{DC} = M \times V \tag{10.16}$$

where V_{o_k} is the output voltage of the kth dc–dc converter at the dc-link of the MMC, and M is the total number of dc–dc converters. It is assumed that all of the dc–dc converters are identical, and hence, it can be envisaged that total dc bus voltage is shared equally by the dc–dc converters. Therefore:

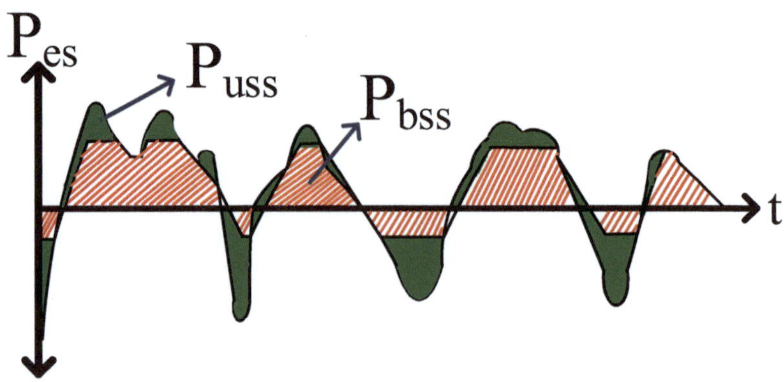

FIGURE 10.15 Distribution of power among the battery storage system and *USS*, ultracapacitor storage system.

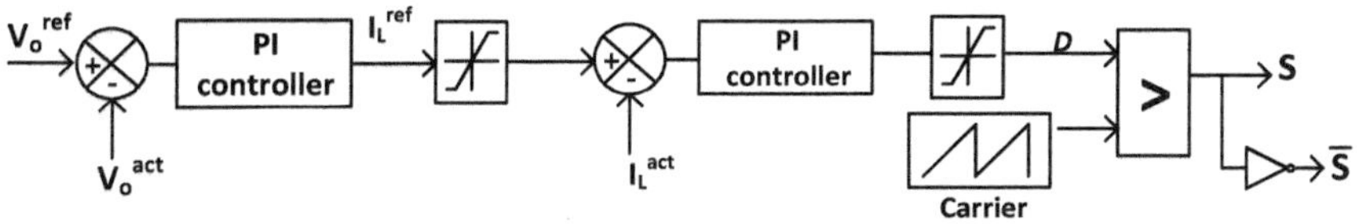

FIGURE 10.16 Control block diagram of a voltage mode control ultracapacitor module-fed dc–dc converter. *PI*, proportional integral.

$$V = \frac{V_{DC}}{M} \tag{10.17}$$

where V is the output of one dc–dc converter. The reference output voltage of each converter can be set according to Eq. (10.17) such that the dc-link voltage is maintained at the desired value.

3. Controller for the BSS: The distributed BSS modules at the dc bus of the MMC are controlled using current mode control (CMC). This helps to control the amount of current drawn from the respective battery module, and the low-frequency component of the active power comes from the battery. The control block diagram of the converter operated in CMC is shown in Fig. 10.17, where a term "α" is used to initiate the controller. Since the HSS is distributed at the dc-link of the MMC, P_{es} in Eq. (10.18) can be expressed as:

$$P_{es} = V_{DC} \times \left(I_{dc}^{bat} + I_{dc}^{uc}\right) \tag{10.18}$$

where I_{dc}^{bat} and I_{dc}^{uc} are the components of dc-link current supplied by battery and UC modules, respectively. Here, P_{bss} is supplied by M numbers of dc–dc converters, and hence, the power (P_o^{bat}) given by each dc–dc converter can be expressed in Eq. (10.19):

$$P_o^{bat} = \frac{P_{bss}}{M} \tag{10.19}$$

Under ideal operating conditions, input and output powers of each dc–dc converter are identical, which may yield the following expression:

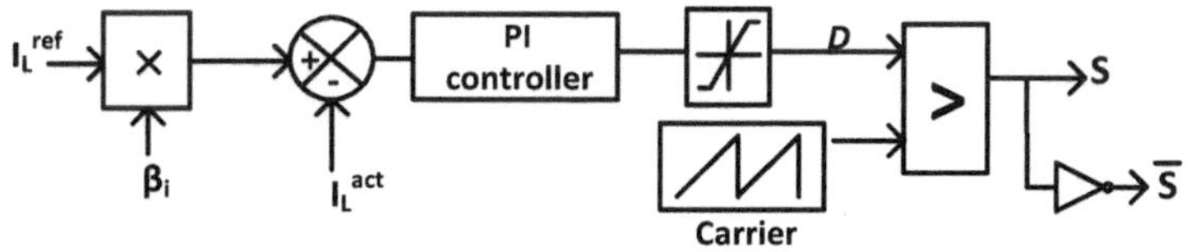

FIGURE 10.17 Control block diagram for battery-fed dc–dc converters. *PI*, proportional integral.

$$\frac{P_{bss}}{M} = V_{bat} \times I_L \tag{10.20}$$

where V_{bat} is the battery voltage. The value of inductor current reference (I_L^{ref}) can be calculated from Eq. (10.21):

$$I_L^{ref} = \frac{P_{bat}}{V_{bat} \times M} \tag{10.21}$$

4. State-of-charge (SoC) balancing among distributed energy storage modules: The SoC of the distributed storage modules may diverge from each other due to differences in their operating conditions and manufacturing tolerances. To avoid this problem, an algorithm has been developed to converge the SoCs of UC and battery modules, which is presented in this subsection. Initially, the algorithm is implemented to converge the SoCs of UC modules; later it is extended to the battery modules. The reference value of SoC can be determined using the following expression:

$$SoC_{ref} = \frac{\sum_{i=1}^{M} SoC_i}{M} \tag{10.22}$$

The actual SoC of a UC can be estimated using the following expression:

$$SoC_i = \frac{V_{uci}^2}{V_{uc_{max}}^2} \tag{10.23}$$

where V_{uci} is the voltage of the ith UC module and $V_{uc_{max}}$ is the voltage of the UC module corresponding to 100% SoC. In this work, voltage references of dc–dc converters, which control charge/discharge of UC modules, are adjusted to update the SoC of UC modules. The voltage reference for the ith dc–dc converter can be computed through Eq. (10.23) where determination of "α_i" plays a key role. A flowchart is provided in Fig. 10.18, which elaborates the calculation of "α_i" for the ith UC module. The SoCs of battery modules, which have been distributed at the dc-link of the MMC, may also diverge from each other due to various nonidealities as stated earlier. Hence, an algorithm has to be used to overcome this issue, which is covered in brief in the subsequent section.

$$V_{oi}^{ref} = \alpha_i \left(\frac{V_{dc}}{M} \right) \tag{10.24}$$

The SoC of a battery module can be determined as:

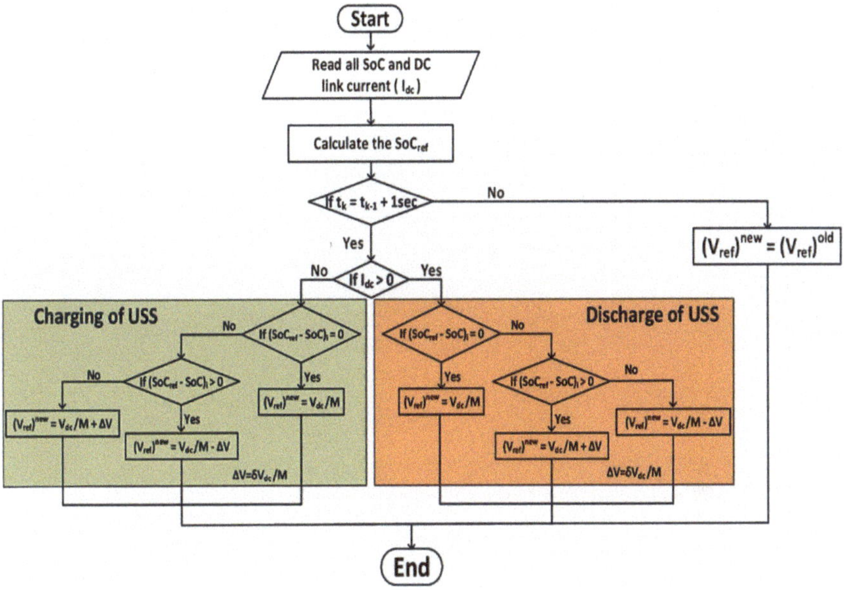

FIGURE 10.18 Algorithm to converge the SoC of ultracapacitor modules. *SoC*, state of charge; *USS*, ultracapacitor storage system.

$$SoC_i = \frac{1}{C_{bat_{nom}}} \int_0^t i_{bat}(t)dt \tag{10.25}$$

where C_{bat} is the nominal capacity of the respective battery module and i_{bat} is the current associated with the battery module.

Using Laplace transformation, the SoC of the ith battery module at any instant can be rewritten as:

$$SoC_i = SoC_i(0) + \alpha_i i_{bat} e^{-t/\tau} \tag{10.26}$$

where $SoC_i(0)$ is the SoC at $t = 0+$ and τ is the time constant of the filter inductor. By varying α_i, the inductor current can be controlled, which in turn adjusts the charge/discharge rates of battery modules, so that SoC converges to the desired reference value. The corresponding flowchart to determine the magnitude of α_i is presented in Fig. 10.19.

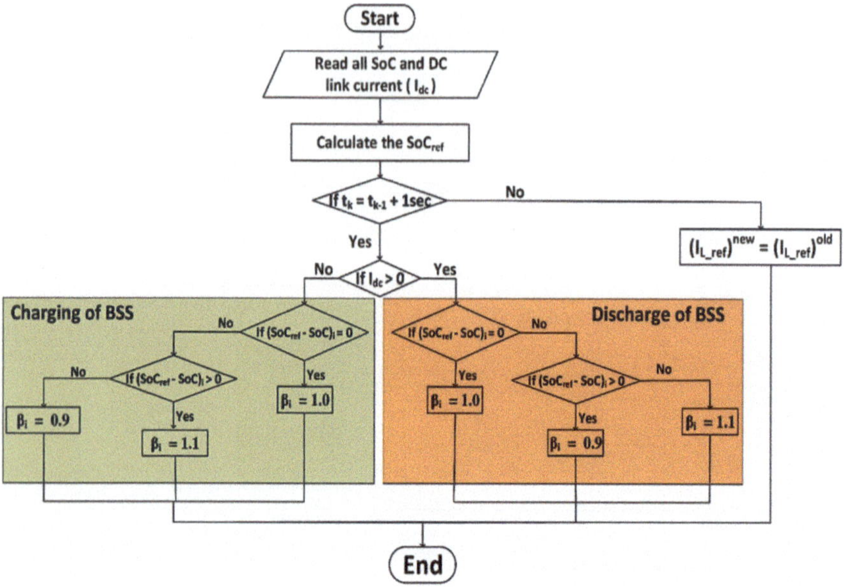

FIGURE 10.19 Algorithm to converge the SoC of battery modules. *BSS*, battery storage system; *SoC*, state of charge.

10.7 Discussion of results

A simulation model of the system is developed in PSCAD/EMTDC to check the performance of the proposed configurations along with developed control algorithms. The rating and parameters used for simulation are provided in Tables 10.2–10.4. The data related to the wind farm is taken from [29]. The power generated by the wind farm is presented in Fig. 10.20A, from where it can be observed that the power generated by the wind farm is variable in nature. To inject constant active power into the grid, the active power support from an E-STATCOM is given in Fig. 10.20B.

An energy management system (EMS) is presented in Fig. 10.14, and active power supplied by the BSS is given in Fig. 10.21. From the figure, it can be noticed that the low-frequency component of the E-STATCOM is supplied by the BSS and the remaining power is given by UC modules. Note that the ultracapacitor-fed dc–dc converters hold the dc-link voltage of the MMC at reference value (i.e., 12 kV) as shown in Fig. 10.22.

Performance of a leg-voltage controller of an MMC is presented in Fig. 10.23. During $t = 4$ s to $t = 6$ s, voltage of the dc-link is reduced to 9 kV from its rated value of 12 kV due to a fault in one dc–dc converter. Though the

TABLE 10.2 Parameters of a modular multilevel converter.

Parameter	Value
Converter capacity (S)	5 MVA
Active power support for 30 min	4 MW
DC-link voltage	12 kV
Line voltage	5.5 kV
Number of submodules	4
Arm inductance (0.1 pu)	2.5 mH
Nominal voltage of a submodule	3 kV
Capacitance of a submodule capacitor	5 mF
Switching frequency of a submodule	500 Hz
Voltage rating of the transformer	5.5 kV/11 kV

TABLE 10.3 Rating of battery energy storage system.

Parameter	Value
Capacity of each battery module	0.625 MWh
No. of battery modules	4
Active power support from battery modules	2.5 MW
Nominal voltage of each battery module	1.75 kV
Rating of each dc–dc converter	0.5 MW
Output voltage of dc–dc converter	3 kV
Filter inductance of a dc–dc converter	6 mH
Switching frequency	2.5 kHz

dc-link voltage falls, the submodule capacitor voltages are regulated at their nominal value of 3.75 kV (as shown in Fig. 10.23) with the help of the leg-voltage controller. To demonstrate the performance of a dynamically varying limiter, PCC voltage and the corresponding active and reactive power generated by an E-STATCOM are plotted in Figs. 10.24 and 10.25. Due to

TABLE 10.4 Rating of an ultracapacitor (UC)-based energy storage system (ESS).

Parameter	Value
Active power support from UC-ESS	1.5 MW
No. of UC modules	4
Maximum voltage of each UC module	2 kV
Minimum voltage of each UC module	1.75 kV
Capacity of each dc–dc converter	0.375 MW
Output voltage of dc–dc converter	3 kV
Filter inductance of a dc–dc converter	10 mH
Switching frequency	2.5 kHz

severe load disturbance or faults in the network, the PCC voltage falls below the acceptable limit of 0.8 pu between 2 and 2.35 s. This results in a reduction of active power generated by the wind farm from 9.6 to 7.6 MW as shown in Fig. 10.25. During this condition, the main objective of the E-STATCOM is to maintain the PCC voltage at its reference. Thus the complete capacity of the E-STATCOM is utilized to generate reactive power. Fig. 10.26 shows reactive and active power supplied by an E-STATCOM during the aforesaid duration (Fig. 10.27).

Similarly, the performance of the proposed SoC balancing algorithm is presented in Fig. 10.28, which shows that SoCs of UC and BSS modules converge to their respective references after activation of the balancing algorithm.

10.8 Conclusion

This chapter presented various configurations of an E-STATCOM using an MMC and hybrid ESS. The configuration can concurrently cater for the needs of active and reactive powers. The ESS is kept at the dc-link, which ensures equal sharing of active power among the phases. With the presence of both battery and UC modules, the configuration is useful for both high-power and high-energy applications. Also, the configuration has some interesting merits like higher efficiency, fewer components to integrate the HSS, and better transient performance. Along with this, the power level of the configuration can be increased by adding more submodules in the arms of the MMC. P-Q control in the *d-q* reference frame is employed for operation of the MMC, and CMC/VMC is used for control of the dc–dc converters. An EMS is formed by combining a low-pass filter and peak-saving methods to distribute active power among the

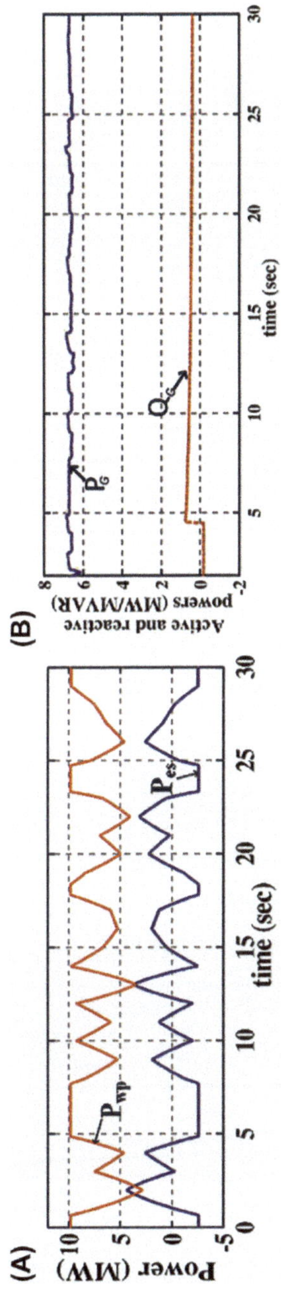

FIGURE 10.20 (A) Power generated by wind farm and an E-STATCOM and (B) P and Q injected into the grid.

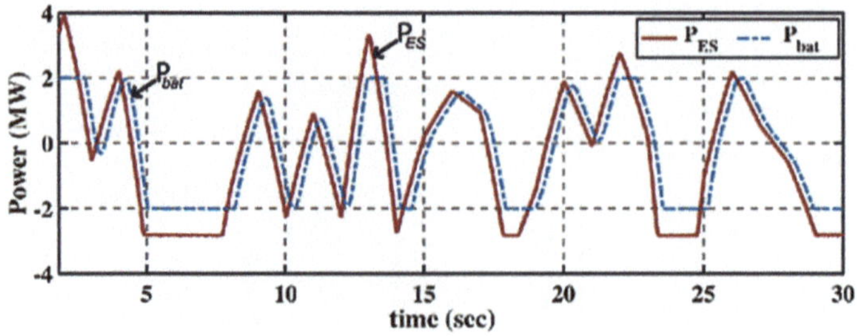

FIGURE 10.21 Active power supplied by battery storage system.

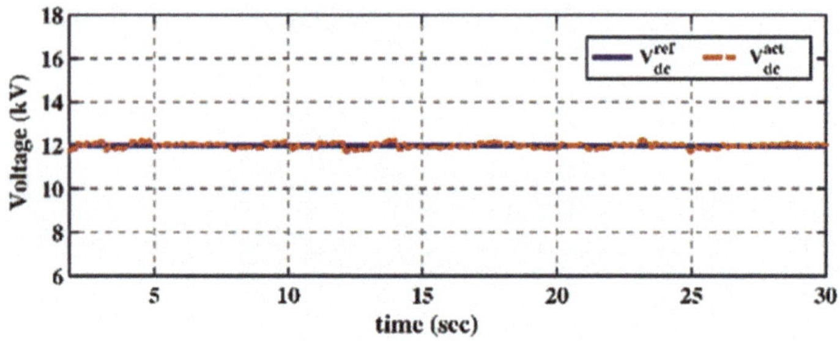

FIGURE 10.22 DC-link voltage of modular multilevel converter with actual and reference values.

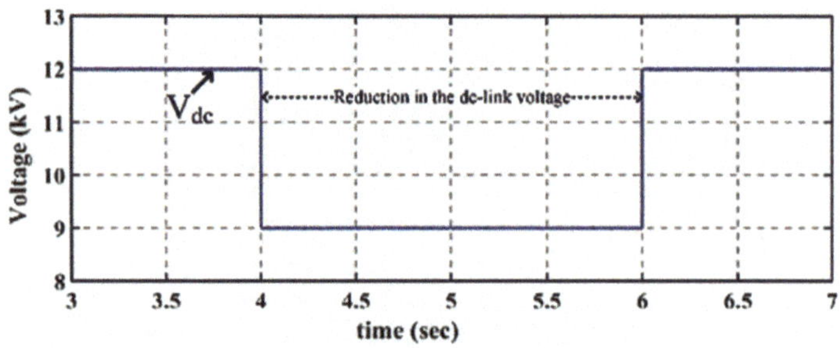

FIGURE 10.23 DC-link voltage of modular multilevel converter during outage of one dc–dc converter.

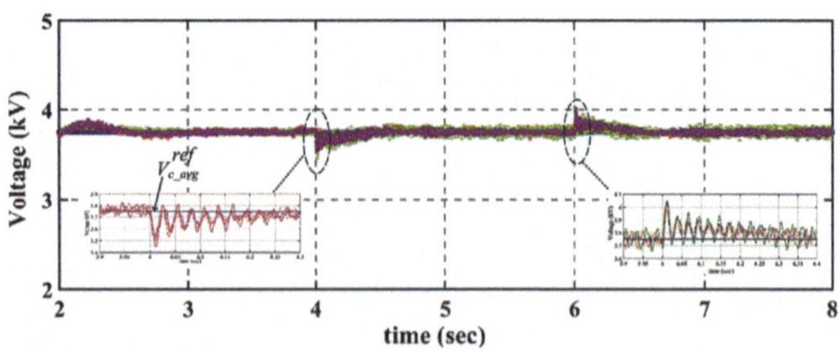

FIGURE 10.24 Average submodule capacitor voltages in modular multilevel converter under the variation of dc-link voltage.

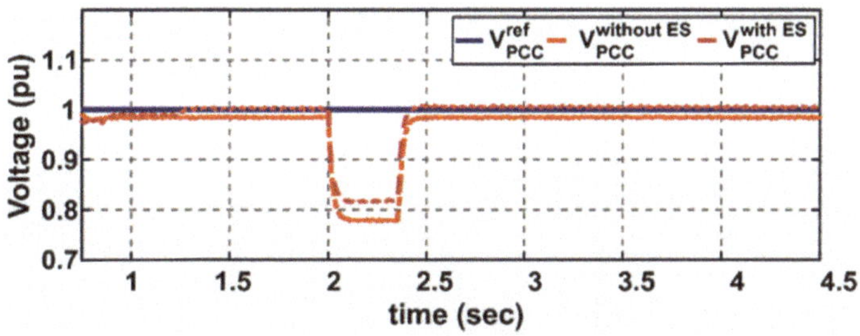

FIGURE 10.25 Point of common coupling voltage during a fault in the electrical network or severe load disturbance.

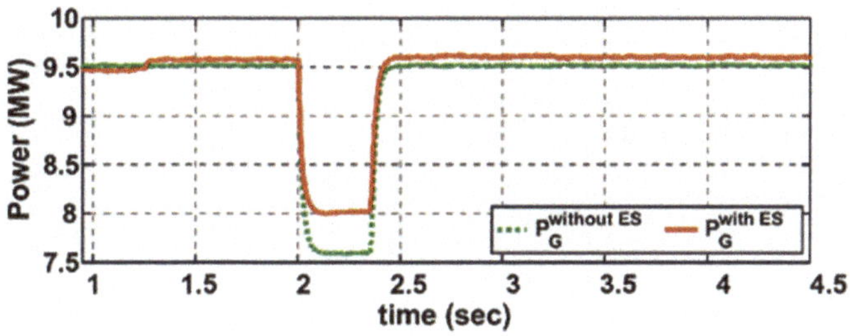

FIGURE 10.26 Active power generated by a wind farm during severe load disturbance (or fault) with and without an E-STATCOM.

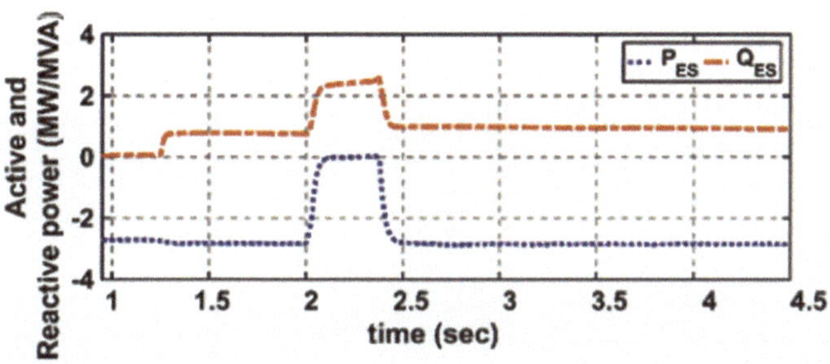

FIGURE 10.27 *P* and *Q* supplied by an E-STATCOM during severe load disturbance (or fault).

BSS and USS. This ensures that battery modules provide only low-frequency components of total active power and UC modules supply high-frequency components along with peak power. Different control techniques for the MMC have been formulated to perform PCC voltage regulation, load unbalance compensation, and low-voltage ride-through operation. To address the load unbalance, a negative sequence current controller is incorporated. Dynamically

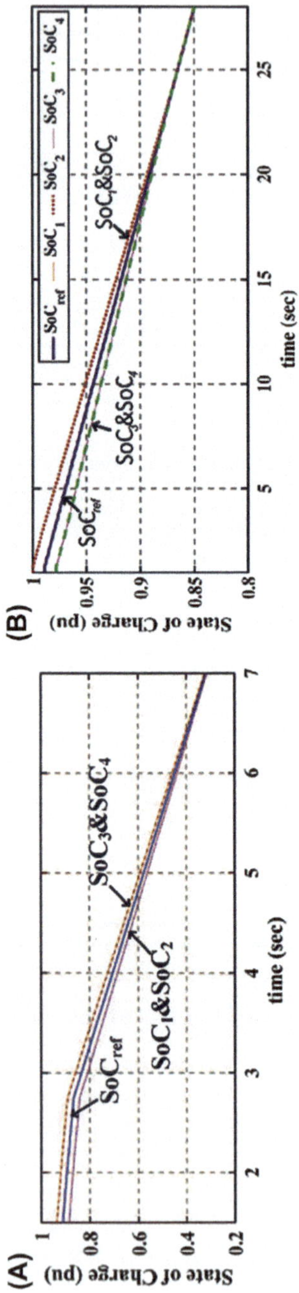

FIGURE 10.28 (A) State of charge (SoC) of distributed battery modules and (B) SoC of distributed ultracapacitor modules.

varying limiters have been used, which enable low-voltage ride-through operation by utilizing complete capacity of the E-STATCOM to provide positive sequence reactive power only. From the performance analysis, it may be concluded that an MMC-based E-STATCOM is efficient for high-voltage/power applications where both active and reactive powers are needed.

References

[1] G. Brinkman, P. Denholm, E. Drury, R. Margolis, M. Mowers, Toward a solar-powered grid, IEEE Power Energy Mag. 9 (3) (May-June 2011) 24–32.

[2] E.A. DeMeo, W. Grant, M.R. Milligan, M.J. Schuerger, Wind plant integration [wind power plants, IEEE Power Energy Mag. 3 (6) (Nov.-Dec. 2005) 38–46.

[3] B. Subudhi, R. Pradhan, A comparative study on maximum power point tracking techniques for photovoltaic power systems, IEEE Trans. Sustain. Energy 4 (1) (Jan. 2013) 89–98.

[4] D. Kumar, K. Chatterjee, A review of conventional and advanced MPPT algorithms for wind energy systems, Renew. Sustain. Energy Rev. 55 (06) (2016) 957–970.

[5] M. Mohseni, S.M. Islam, Review of international grid codes for wind power integration: diversity, technology and a case for global standard, Renew. Sustain. Energy Rev. 16 (06) (2012) 3876–3890.

[6] M.A. Saqib, A.Z. Saleem, Power-quality issues and the need for reactive-power compensation in the grid integration of wind power, Renew. Sustain. Energy Rev. 43 (01) (2015) 51–64.

[7] M.B. Latran, A. Teke, Investigation of multilevel multifunctional grid connected inverter topologies and control strategies used in photovoltaic systems, Renew. Sustain. Energy Rev. 42 (01) (2015) 361–376.

[8] F. Mwasilu, J.J. Justo, K.S. Ro, J.W. Jung, Improvement of dynamic performance of doubly fed induction generator-based wind turbine power system under an unbalanced grid voltage condition, IET Renew. Power Gener. 6 (6) (November 2012) 424–434.

[9] C. Han, et al., STATCOM impact study on the integration of a large wind farm into a weak loop power system, IEEE Trans. Energy Convers. 23 (1) (March 2008) 226–233.

[10] D. Stenclik, P. Denholm, B. Chalamala, Maintaining balance: the increasing role of energy storage for renewable integration, IEEE Power Energy Mag. 15 (6) (Nov.-Dec. 2017) 31–39.

[11] A. Arulampalam, J.B. Ekanayake, N. Jenkins, Application study of a STATCOM with energy storage, IEE Proc. Generat. Transm. Distrib. 150 (3) (May 13, 2003) 373–384.

[12] X. Luo, J. Wang, M. Dooner, J. Clarke, Overview of current development in electrical energy storage technologies and the application potential in power system operation, Appl. Energy 137 (03) (January 2015) 511–536.

[13] K.C. Divya, J. Østergaard, Battery energy storage technology for power systems—an overview, Elec. Power Syst. Res. 79 (04) (2009) 511–520.

[14] M.D. Mufti, S.J. Iqbal, S.A. Lone, Q. Ain, Supervisory adaptive predictive control scheme for supercapacitor energy storage system, IEEE Syst. J. 9 (3) (Sept. 2015) 1020–1030.

[15] Q. Tabart, I. Vechiu, A. Etxeberria, S. Bacha, Hybrid energy storage system microgrids integration for power quality improvement using four-leg three- level NPC inverter and second-order sliding Mode control, IEEE Trans. Ind. Electron. 65 (1) (2018) 424–435.

[16] S. Wang, Y. Tang, J. Shi, K. Gong, Y. Liu, L. Ren, J. Li, Design and advanced control strategies of a hybrid energy storage system for the grid integration of wind power generations, IET Renew. Power Gener. 09 (02) (2015) 89–98.
[17] Solar inverters by ABB [online]. Available from: http://new.abb.com/powerconverters-inverters/solar/central/pvs980.
[18] Y. Cheng, C. Qian, M.L. Crow, S. Pekarek, S. Atcitty, A comparison of diode-clamped and cascaded multilevel converters for a STATCOM with energy storage, IEEE Trans. Ind. Electron. 53 (5) (October 2006) 1512–1521.
[19] M. Priya, P. Ponnambalam, K. Muralikumar, Modular-multilevel converter topologies and applications a review, IET Power Electron. 12 (2) (2019) 170–183.
[20] M. Hagiwara, H. Akagi, Control and experiment of pulsewidth-modulated modular multilevel converters, IEEE Trans. Power Electron. 24 (07) (July 2009) 1734–1746.
[21] M. Saeedifard, R. Iravani, Dynamic performance of a modular multilevel back-to-back HVDC system, IEEE Trans. Power Deliv. 25 (04) (October 2010) 2903–2912.
[22] S. Du, J. Liu, A study on DC voltage control for chopper-cell-based modular multilevel converters in D-STATCOM application, IEEE Trans. Power Deliv. 28 (04) (October 2013) 2030–2038.
[23] M. Vasiladiotis, A. Rufer, Analysis and control of modular multilevel converters with integrated battery energy storage, IEEE Trans. Power Electron. 30 (01) (January 2015) 163–175.
[24] T. Soong, P.W. Lehn, Evaluation of emerging modular multilevel converters for BESS applications, IEEE Trans. Power Deliv. 29 (05) (October 2014) 2086–2094.
[25] F. Guo, R. Sharma, A modular multilevel converter with half-bridge submodules for hybrid energy storage systems integrating battery and UltraCapacitor, in: Proc. IEEE Applied Power Electronics Conference and Exposition (APEC), Bangalore, 2015, pp. 3025–3030.
[26] G.C. Konstantopoulos, A.T. Alexandridis, Full-scale modeling, control, and analysis of grid-connected wind turbine induction generators with back-to-back AC/DC/AC converters, IEEE Trans. Emerg. Sel. Topics Power Electron. 02 (04) (2014) 739–748.
[27] R.J. Bessa, V. Miranda, J. Gama, Entropy and correntropy against minimum square error in offline and online three-day ahead wind power forecasting, IEEE Trans. Power Syst. 24 (4) (2009) 1657–1666.
[28] K. Ilves, A. Antonopoulos, S. Norrga, H. Nee, Steady-state analysis of interaction between harmonic components of arm and line quantities of modular multilevel converters, IEEE Trans. Power Electron. 27 (1) (2012) 57–68.
[29] Per-second wind velocity data. http://www.imd.gov.in/section/nhac/aws/aws.html. (Accessed 30 September 2010).

Chapter 11

PQ disturbance detection and classification combining advanced signal processing and machine learning tools

Md Shafiullah, Meer A.M. Khan, Shakir D. Ahmed
Center of Research Excellence in Renewable Energy, King Fahd University of Petroleum and Minerals, Dhahran, Eastern Province, Saudi Arabia

Chapter outline

11.1 Introduction **311**
11.2 Methods **313**
11.2.1 Advanced signal processing techniques 313
11.2.1.1 Wavelet transformation 314
11.2.1.2 Stockwell transformation 314
11.2.2 Artificial neural network 315
11.3 Proposed PQD detection and classification scheme **317**
11.3.1 Mathematical models of PQD 317
11.3.2 Feature extraction explanation 318
11.4 Results and discussion **322**
11.4.1 PQD detection results 322
11.4.2 PQD classification results 326
11.4.3 Validation of the proposed scheme 327
11.5 Conclusions **331**
11.6 List of abbreviations **332**
Acknowledgments **333**
References **333**

11.1 Introduction

Reliability, security, stability, and power quality (PQ) are the pillars of power system performance. Electrical current, voltage, and frequency characteristics assessed against a set of technical metrics at any point within an electric power network are referred to as PQ [1,2]. In general, PQ can be considered as the compatibility between the electrical outlet and the load connected to it [3]. Sophistication of the grid, coupled with fluctuations in climate, generation,

Power Quality in Modern Power Systems. https://doi.org/10.1016/B978-0-12-823346-7.00012-8

demand, and other factors, to transfer electrical energy from the point of production to consumption, provides many opportunities for compromising supply quality [4,5]. Poor PQ not only increases energy consumption but also may have a detrimental effect on huge industrial facilities and data centers. Power quality disturbance (PQD) also causes a wide range of problems, including instability, failure, malfunction, and short lifetimes of electrical equipment, especially sensitive medical equipment. In a variety of cases, this issue reduces the production pace in industries and small businesses, thereby increasing production costs [6].

Massive development and deployment of distributed generators, renewable energy resources, and microgrid systems in recent years require power electronics interface devices to be integrated with the power system networks. Additionally, the implementation of energy efficiency improvement techniques also requires power electronics devices to dynamically adjust the load requirements. These power electronics device-interfaced loads, generators, and systems are the new addition to the old and traditional PQD producers [7,8]. The traditional sources of PQD are power systems faults, lightning strikes, energization of transformers and large capacitors, switching off large loads, and so on [9]. PQD includes sag, swell, interruption, notch, transients, and harmonics with/without other disturbances. In view of the importance of the matter, international organizations working on PQD set standards for defining acceptable disturbances, including harmonics and voltage distortions such as "IEEE-519 and IEC-61000." They also continually provide updated guidelines to monitor and classify different PQDs in power system networks [10–12]. However, PQD detection and classification have become an evergreen research topic due to the gravity of the topic, the recent addition of PQD producers, and the nonstationary nature of the PQD signals [13–17].

Before the 1990s, researchers used to use manual techniques for monitoring and managing the PQ of the supplied electricity. Over time, the evolution of technologies, including smart signal processing, data mining, and artificial intelligence techniques along with high computing devices, paved the way for automated PQ monitoring and management techniques [15]. In most cases, researchers employed different signal processing techniques for feature extractions from the targeted signals. Then, they trained and tested the machine learning approaches to monitor PQ events. Heydt et al. [18] applied windowed fast Fourier transform for PQD assessment and reported that the narrowness of window widths was suitable for detailed analysis. Fourier and wavelet transforms (FT and WT) were used to extract useful features in the characterization of PQ events in [19]. The discrete short-time Fourier transform (STFT) was employed to analyze voltage disturbance events [20]. It was observed that the proper selection of window size played a vital role in analyzing the events. In [21], the authors used Kalman filters to detect and analyze power system PQ events. With time, many other techniques were also investigated for PQD analysis, including Gabor transform [22], wavelet transform based on neurofuzzy network and extreme machine learning [23,24], wavelet packet transform (WPT)-based on support vector machine [25], Hilbert Huang transform-based balanced neural tree [26], Stockwell transform (ST)-based decision tree [27], and wavelet-based deep learning techniques [28]. Recently, PQ events monitoring and analyzing in renewable energy resources-integrated power systems has received great attention [29,30].

Most authors have reported that appropriate selection of features using advanced signal processing (ASP) and appropriate configuration of machine learning tools (MLTs) could improve the overall accuracy and robustness of the techniques significantly. However, in the authors' best knowledge, there is no specific article that compares the performance of different configurations of neural networks that serves this purpose. Therefore this chapter aims to develop a technique for PQD detection and classification by using two different ASP techniques for useful feature selection and compare the results of different artificial neural network (ANN) configurations. The robustness of the best-developed model was tested on signals generated with noise. The chapter is further structured as follows. The employed ASP and MLT techniques are briefly presented in Section 11.2. Section 11.3 describes the mathematical models of the targeted PQ events and the feature extraction mechanism, while Section 11.4 addresses comprehensive findings and validation of the PQD approach and detection as well as classification techniques proposed. Finally, Section 11.5 concludes the developed PQD detection and classification approach combining ASP and MLT techniques.

11.2 Methods

This section discusses the employed ASP techniques for the extraction of useful features followed by the MLT technique for effectively detecting and classifying the PQD events.

11.2.1 Advanced signal processing techniques

FT, one of the cornerstones of signal and system analysis techniques, involves the disintegration of the signal into many signals of various frequencies formed by the sum of its constituents' sinusoidal signals [31]. It describes the presence of a particular component of the frequency but does not provide details of time. Another disadvantage of this technique is that it can only be used in analyzing stationary signals but cannot effectively analyze nonstationary signals like PQD [32]. STFT, an upgraded version of FT, can provide both frequency and time information as it uses a sampling window of regular intervals. However, it faces a resolution issue between frequency and time. To deal with the resolution issue, WT and ST methods are widely used.

11.2.1.1 Wavelet transformation

WT has emerged as an effective ASP, where the signals are decomposed into various frequency levels by using a multiresolution signal decomposition technique and the decomposed components hold important characteristics and attributes. Researchers mainly ramify WT into three classes, the first two are represented in a continuous and discrete nature and are respectively referred to as continuous wavelet transform (CWT) and discrete wavelet transform (DWT), while the third class is referred to as WPT [17]. CWT and DWT are widely used in analyzing PQ transients among these three branches. In CWT, a continuous signal $z(t)$ is given by a mother wavelet function constructed from translation and scaling parameters as in Eq. (11.1) [31]:

$$CWT_{\varnothing}\, z(x,y) = W_z(x,y) = \int_{-\infty}^{\infty} z(t)\, \varnothing^*_{x,y}(t)dt \tag{11.1}$$

$$\varnothing_{x,y}(t) = \frac{1}{|x|^{\frac{1}{2}}}\, \varnothing\left(\frac{t-y}{x}\right) \tag{11.2}$$

where x is a scale parameter or change of scale, y is a translation parameter or factor of shift, $\varnothing(t)$ is a mother wavelet function, and $\varnothing^*_{x,\,y}(t)$ is a complex conjugate of $\varnothing(t)$. Conversely, in DWT, the function $z(t)$ of CWT is replaced with a discrete signal $z(k)$ and the variables x and y are replaced by m and n, respectively. The DWT of the function $z(k)$ can be represented by Eq. (11.3):

$$DWT_{\varnothing}\, z(m,n) = \frac{1}{x_0^{\frac{m}{2}}} \sum_k z(k)\, \varnothing\left(\frac{n - ky_0x_0^m}{x_0^m}\right) \tag{11.3}$$

where x_0 is the discrete scale and y_0 is translation factors. In DWT, input signals are passed through a series of high- and low-pass filters. These filters contain outputs that are the detailed coefficient and approximate coefficient, respectively [5]. In this research, the PQD signals are broken down into seven. Hence, they receive one approximate and seven detailed coefficients.

11.2.1.2 Stockwell transformation

Another effective and important ASP tool is ST, which offers enhanced time and frequency analysis of the signals by using the advantages and properties of both STFT and WT [17]. The S-transform of signal $z(t)$ can be deduced upon combining Eqs. (11.4) and (11.5), hence the ST of a continuous signal is defined by Eq. (11.6) [33]:

$$ST(\tau,f) = \int_{-\infty}^{\infty} z(t)\, g(\tau - t,\, f)\, e^{-j2\pi ft}dt \tag{11.4}$$

$$g(t) = \frac{|f|}{\sqrt{2\pi}} e^{\left(\frac{-t^2 \cdot f^2}{2}\right)} \tag{11.5}$$

$$ST(\tau, f) = \int_{-\infty}^{\infty} z(t) \frac{|f|}{\sqrt{2\pi}} e^{\left(\frac{-(\tau - t)^2 \cdot f^2}{2}\right)} e^{-j2\pi f t} dt \tag{11.6}$$

where $g(t)$ is a mother wavelet window function defined by the Gaussian function, f is the frequency, and t and τ are the time parameters. ST on a signal generates a matrix known as the S-matrix whose elements are complex. The discrete form of ST is defined by Eq. (11.7):

$$ST[j, m] = \sum_{m=0}^{M-1} H[n+m] e^{\frac{-2\pi^2 n^2}{m^2}} e^{\frac{i2\pi nj}{M}} \tag{11.7}$$

Eq. (11.8) provides the energy matrix of the same signal:

$$E = |S|^2 \tag{11.8}$$

This research generates four new vectors from the decomposed matrices, namely S_{cmax}, S_{rmax}, $E_{\max}$, and $S_{c\text{-}phase\text{-}max}$, where S_{cmax} and S_{rmax} contain maximum absolute values of columns and rows of the S-matrix, respectively. E_{cmax} contains the maximum absolute values of the columns of the energy matrix. Finally, $S_{c\text{-}phase\text{-}max}$ comprises the S-matrix phase angles linked with the vector S_{cmax}. Detailed information on this topic can be explored in [34].

11.2.2 Artificial neural network

ANNs are made up of a set of algorithms that simulate the evolution of human brains and create electronic brains that are capable of learning and developing like human brains. They consist of nodes that look like the nodes of the human nervous system. These nodes are linked together, and this network knowledge flows through these nodes. In ANN systems, a large number of processors run in parallel, and they are adaptive and can adjust themselves from the initial training [35]. The mentioned capabilities make them excellent candidates for computational technology. Among a variety of structures and learning algorithms, the feedforward neural network (FFNN) will be used in this proposed scheme. In this structure, the network is partitioned into three layers (i.e., input, hidden, and output) and all layers are connected in consecution and node-wise (Fig. 11.1) [36–38].

The connecting route between node i and node j is described by the two-dimensional matrix w_{ij}, called weights. These weights are the backbone of neural networks as they represent the memory of the network. Usually, the

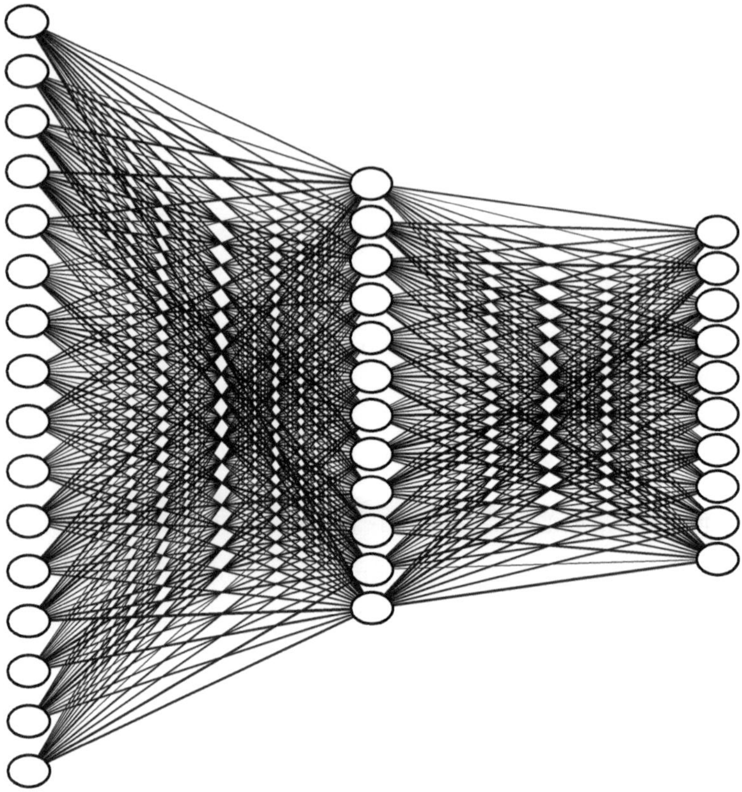

FIGURE 11.1 General representation of feedforward neural network.

output to the node is calculated by multiplying the inputs by the weights plus bias values as denoted by Eq. (11.9):

$$h_j = b_j + \sum_{i=1}^{p} x_i w_{ij} \tag{11.9}$$

where j denotes the jth hidden layer neuron and P denotes the number of inputs. Then, to pass the result from the node to the next node or environment, the output for the node known as activation/squashing will be calculated. ANN models employ several activation/squashing functions, namely linear, sigmoid/logistic, tan sigmoid, ReLU, Leaky ReLU, softmax, and many others [39–42]. Fig. 11.2 illustrates the preceding activation functions graphically.

However, there are two ways of learning: one way is to control the target output with each example, known as supervised learning, and the other way is unsupervised learning where no target output is offered to the network for an example. The training algorithms are a set of techniques used in neural network systems to perform the learning processes. There are many of these

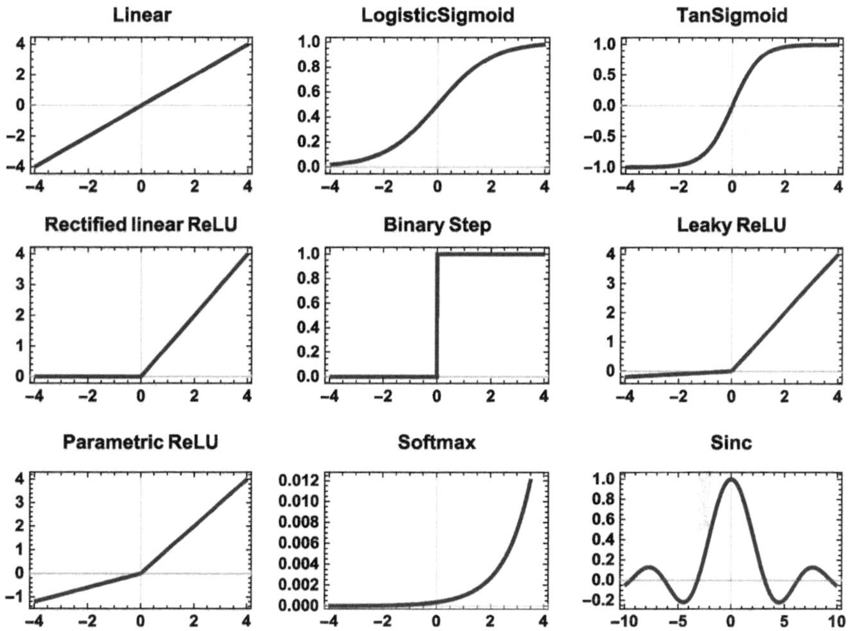

FIGURE 11.2 Activation functions of neural networks.

algorithms and each has pros and cons as measured by speed, storage, and accuracy. In many cases, the neural networks also aim to reduce the mean squared error (MSE) as given by Eq. (11.10):

$$MSE = \frac{\sum (\text{outputs} - \text{targets})^2}{\text{number of instances}} \tag{11.10}$$

As mentioned earlier, each training algorithm has pros and cons. Table 11.1 briefly summarizes the advantages and disadvantages of the employed training algorithms in this chapter.

11.3 Proposed PQD detection and classification scheme

This section discusses the proposed PQD detection and classification scheme that includes PQD modeling, feature extraction using ASP, and training and testing of the MLT.

11.3.1 Mathematical models of PQD

PQD disturbance events are modeled based on parametric equations in the MATLAB platform. Since the parameters of the equations can be easily varied, these models can be used to detect and classify PQD in real time [46].

TABLE 11.1 Artificial neural network training algorithms: pros and cons [43–45].

Training algorithm	Pros	Cons
Gradient descent backpropagation (GD) Resilient backpropagation (RP)	Requires less memory	Requires more iterations
Newton's method	Faster convergence	Computationally expensive
Powell–Beale conjugate gradient backpropagation (CGB) Fletcher–Powell conjugate gradient backpropagation (CGF) Polak–Rieber conjugate gradient backpropagation (CGP)	Faster convergence	Complexity of calculation
Scaled conjugate gradient backpropagation (SCG)	Faster convergence	Better accuracy derivative transfer function
Quasi-Newton backpropagation (BFG) One step secant backpropagation (OSS)	Requires less computation time	Requires larger memory
Levenberg–Marquardt backpropagation (LM)	Faster response	Designed to work with one loss function
Gradient descent with momentum and adaptive backpropagation (GDX)	Large data handling capability	Convergence is case dependent

The proposed approach aims to model 10 types of PQD events as provided in Table 11.2. These signals are given in the order, C_1: sag, C_2: swell, C_3: fluctuation, C_4: interruption, C_5: harmonics, C_6: harmonics + sag, C_7: harmonics + swell, C_8: harmonics + interruption, C_9: notch, C_{10}: transient, and C_0: pure signal. The waveform representation of these PQ events is shown in Fig. 11.3. The fundamental frequency of any signal is 60 Hz and the root mean square (RMS) value of the signal is 230 V.

11.3.2 Feature extraction explanation

In the proposed approach, the voltage signals are continuously recorded for a short window of 0.25 s. Upon using DWT, a seven-level decomposition of the recorded signal with the help of the mother wavelet called Daubechies (db4) was performed. As a result, an approximate and seven detailed coefficients from the DWT were generated. For instance, Fig. 11.4 shows a notch PQD case and associated detailed and approximate coefficients. The method

TABLE 11.2 Mathematical representation of power quality disturbance (PQD) [17], [29,46,47].

Class	PQD	Equation	Parameters
C_0	Pure signal	$Y(t) = A \sin(\omega t)$	$A = \sqrt{2} * 230$ $\omega = 2\pi f$ where $f = 60Hz$
C_1	Sag	$Y(t) = A \sin(\omega t) * (1 - \alpha(u(t-\tau_1) - u(t-\tau_2)))$	$0.1 \leq \alpha \leq 0.9$ $0.1 \leq \beta \leq 0.8$ $0.05 \leq \gamma \leq 0.1$ $0.9 \leq \delta \leq 1$ $T \leq \tau_2 - \tau_1 \leq 9T$ $t_1 \leq t_1$ $u(t) = \begin{cases} 1, & t \geq 0 \\ 0, & t < 0 \end{cases}$
C_2	Swell	$Y(t) = A \sin(\omega t) * (1 + \beta(u(t-\tau_1) - u(t-\tau_2)))$	
C_3	Fluctuation	$Y(t) = A \sin(\omega t) * (1 - \gamma(u(t-\tau_1) - u(t-\tau_2)))$	
C_4	Interruption	$Y(t) = A \sin(\omega t) * (1 - \delta(u(t-\tau_1) - u(t-\tau_2)))$	
C_5	Harmonics	$Y(t) = A (1 + \alpha_1 \sin(\omega t) + \alpha_3 \sin(3\omega t) + \alpha_5 \sin(5\omega t) + \alpha_7 \sin(7\omega t)$	$0.1 \leq \alpha \leq 0.9$ $0.1 \leq \beta \leq 0.8$ $0.9 \leq \delta \leq 1$ $T \leq \tau_2 - \tau_1 \leq 9T$ $\tau_1 \leq \tau_1$ $u(t) = \begin{cases} 1, & t \geq 0 \\ 0, & t < 0 \end{cases}$ $0.05 \leq \alpha_3, \alpha_5, \alpha_7 \leq 0.15$ $\sum \alpha_i^2 = 1$
C_6	Harmonics + sag	$Y(t) = A (1 - \alpha(u(t-\tau_1) - u(t-\tau_2)))$ $(\alpha_1 \sin(\omega t) + \alpha_3 \sin(3\omega t) + \alpha_5 \sin(5\omega t) + \alpha_7 \sin(7\omega t))$	
C_7	Harmonics + swell	$Y(t) = A (1 + \beta(u(t-\tau_1) - u(t-\tau_2)))$ $(\alpha_1 \sin(\omega t) + \alpha_3 \sin(3\omega t) + \alpha_5 \sin(5\omega t) + \alpha_7 \sin(7\omega t))$	
C_8	Harmonics + interruption	$Y(t) = A (1 - \delta(u(t-\tau_1) - u(t-\tau_2)) *$ $(\alpha_1 \sin(\omega t) + \alpha_3 \sin(3\omega t) + \alpha_5 \sin(5\omega t) + \alpha_7 \sin(7\omega t))$	
C_9	Notch	$Y(t) = A(\sin(\omega t) - \text{sign}(\sin(\omega t))) *$ $\sum_{n=1}^{c} K * [u(t-(\tau_1 + 0.02n)) - u(t-(\tau_2 + 0.02n))]$	$0.1 \leq K \leq 0.4$ $0.01T \leq \tau_2 - \tau_1 \leq 0.05T$ $0 \leq \tau_1, \tau_2 \leq 0.5$ $\tau_1 \leq \tau_1$ $u(t) = \begin{cases} 1, & t \geq 0 \\ 0, & t < 0 \end{cases}$ c = number of cycles
C_{10}	Transient	$Y(t) = A(\sin(\omega t) + \text{sign}(\sin(\omega t))) *$ $\sum_{n=1}^{c} [u(t-(\tau_1 + 0.02n)) - u(t-(\tau_2 + 0.02n))]$	

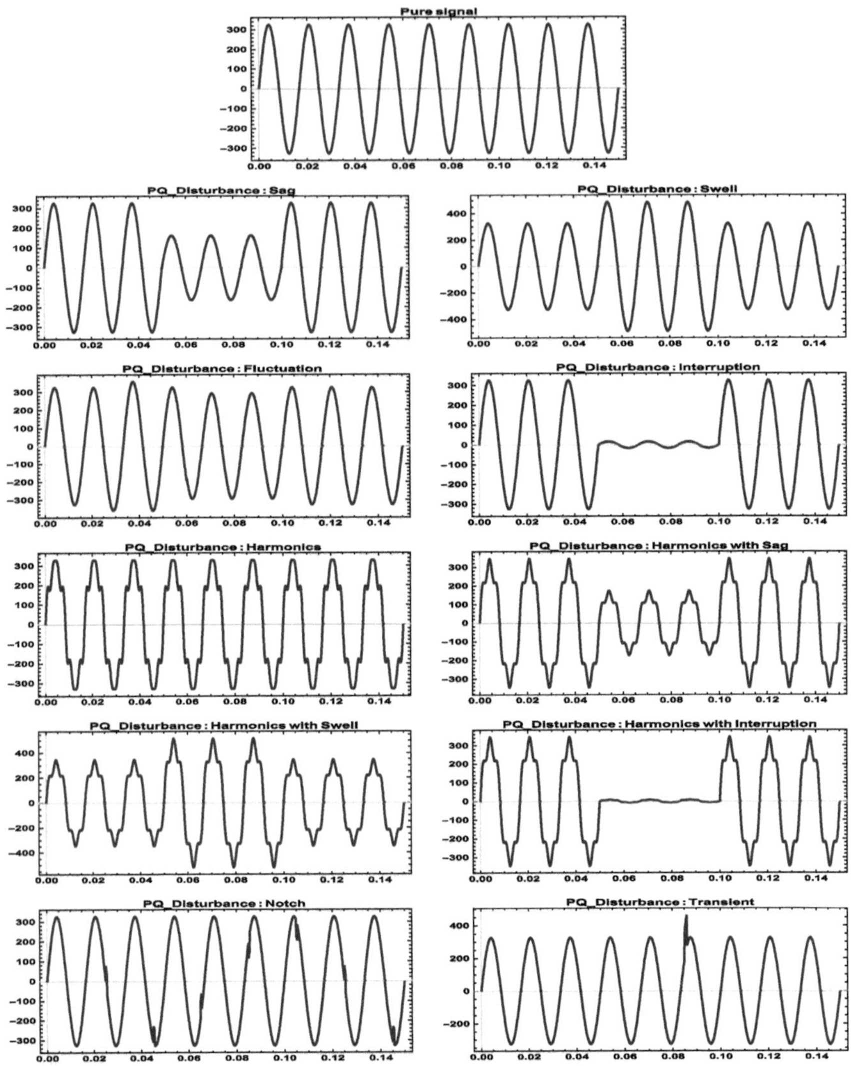

FIGURE 11.3 Modeled pure and power quality disturbance signals.

evaluated six statistical measures in addition to the approximate and detailed coefficients, namely energy, kurtosis, entropy, mean, skewness, and standard deviation. Eventually, this proposed approach extracted and collected 48 (=8 × 6) features for each case. It also kept track of the PQD class associated with the extracted features. Details of DWT-based feature extraction procedures can be found in [48].

Similarly, the proposed PQD detection and classification approach decomposed the recorded voltage signals using ST that produced four

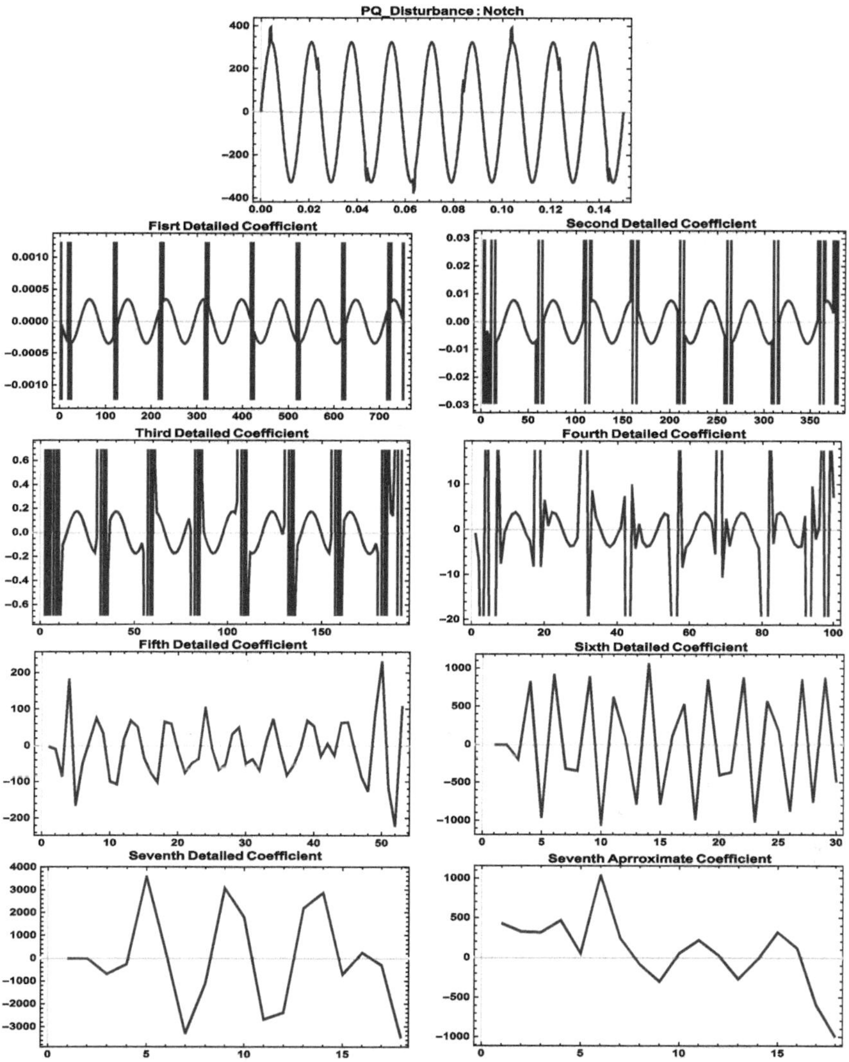

FIGURE 11.4 A power quality disturbance (notch) case and associated discrete wavelet transform coefficients.

independent vectors derived from the S-matrix and E-matrix as discussed in Section 11.2.1. Hence, it extracted seven features, including entropies, kurtosis, maximum and minimum values, standard deviation, and mean values of the generated vectors. Therefore the ST approach ended up with 28 (=4 × 7) features for each case. Details of the ST-based feature extraction procedures can be found in [34]. Finally, this research fetched the DWT and ST extracted features to the FFNNs with a variety of configurations for training, testing, and

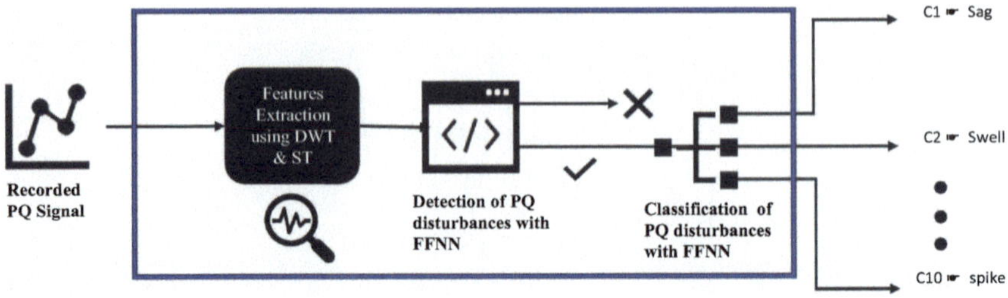

FIGURE 11.5 Proposed power quality disturbance detection and classification scheme. *DWT*, discrete wavelet transform; *FFNN*, feedforward neural network; *PQ*, power quality; *ST*, Stockwell transform.

validation of the proposed PQD detection and classification approaches. Fig. 11.5 presents the proposed PQD detection and classification scheme combining ASP and MLT.

11.4 Results and discussion

Ten different types of PQD were modeled mathematically in a MATLAB/Simulink environment and their mathematical representations were discussed in Section 11.3. This chapter modeled 1500 cases for each PQD type by varying all possible parameters, including their amplitude and duration as mentioned in Table 11.2. Then, it extracted 48 features using DWT and 28 features using ST techniques for each case as discussed in Section 11.3.2. Besides, it also extracted similar features for 5000 cases of pure signal with $\pm 5\%$ variation in the amplitude. The test signals were generated for 15 cycles (0.25 s) with 10 kHz sampling frequency and 60 Hz system frequency, and the RMS of the amplitude of the normal signals was 230 V.

11.4.1 PQD detection results

The proposed PQD detection scheme collected features for 5000 PQD scenarios comprising of 10 classes from the previously mentioned generated cases randomly. The scheme also collected features for 5000 pure signals as generated with $\pm 5\%$ variation of the amplitudes. Then, the proposed approach employed the FFNN to distinguish the PQD cases from their nondisturbed counterparts. The technical parameters of the employed neural networks are summarized in Table 11.3. In this research, the neural network used has been tested with a hidden layer containing 1–15 neurons coupled with the MSE as the objective function, and the recipe that gives the best overall accuracy performance is selected. Besides, it employed various training algorithms for the neural networks, including BFG, CGB, CGF, CGP, GD, GDX, LM, OSS, and RP as discussed in Section 11.2.2. Moreover, the proposed approach chose

TABLE 11.3 FFNN parameters for the PQD detection problem.

Element	Description
Hidden layer neurons	1–12
Activation function	Tan sigmoid (hidden layer), softmax (output layer)
Training algorithm	LM, OSS, SCG, CGP, CGB, CGF, RP, BFG, GD, and GDX
Cost function	Mean squared error
Training set	70%
Testing set	30%

FFNN, Feedforward neural network; *PQD*, power quality disturbance.

tan sigmoid as the squashing/activation function for the hidden layer and softmax for the output layer through a systematic process of trial and error. The MSE errors for the different number of neurons in the hidden layers with Levenberg–Marquardt (LM) training algorithms for PQD detection and classification techniques are tabulated in Table 11.4. As can be observed, the DWT- and ST-based FFNN achieved the best performance for 10 and 9 neurons in the hidden layers, respectively.

Table 11.5 presents the overall confusion matrices for the PQD detection problem employing DWT- and ST-based FFNN approaches. The success and failure detection of PQD and non-PQD cases is presented in the confusion matrix, where bold numbers in the diagonal refer to success and the off-diagonal components refer to failure. As can be seen, the proposed DWT and ST approaches were capable of distinguishing the disturbed situations (cases) from their undisturbed counterparts with accuracies of 98.59% and 99.49%, respectively.

Likewise, this research tested the approach proposed by diversifying the building blocks of neural networks in terms of learning algorithms and changing the number of hidden layers between 1 and 12. Table 11.6 presents the best-performed network configuration in terms of overall accuracy for both DWT- and ST-based approaches. As can be observed, the network with eight neurons in the hidden layer for the ST-based approach showed the best performance. Besides, in most of the cases, the ST-based networks outperformed the DWT-based approaches. For both DWT- and ST-based approaches, the LM training algorithms showed the best performance in terms of overall accuracy (Tables 11.5 and 11.6). Therefore this research tested the obtained best configurations of the neural networks in the detection of the PQD under noisy environments. It applied different levels of additive white Gaussian noises to the generated signals before extraction of the features. Then, it employed those features in those neural networks to investigate their efficacy. Table 11.7 shows

TABLE 11.4 MSE errors for the detection and classification of FFNN with the LM training algorithm.

	MSE (detection)		MSE (classification)	
Number of neurons	DWT-FFNN	ST-FFNN	DWT-FFNN	ST-FFNN
1	5.709E-05	4.065E-05	7.480E-06	1.681E-06
2	1.886E-05	1.316E-05	7.480E-06	1.689E-06
3	7.489E-06	2.408E-06	3.127E-06	2.107E-06
4	7.700E-06	1.817E-06	2.828E-06	1.640E-06
5	6.778E-06	1.547E-06	3.110E-06	2.476E-06
6	7.003E-06	1.389E-06	2.786E-06	1.694E-06
7	5.933E-06	1.621E-06	7.424E-06	1.523E-06
8	4.692E-06	1.155E-06	2.965E-06	1.680E-06
9	5.138E-06	**1.140E-06**	4.785E-06	1.553E-06
10	**4.178E-06**	1.168E-06	3.017E-06	1.548E-06
11	6.045E-06	2.047E-06	3.468E-06	**1.521E-06**
12	4.782E-06	1.148E-06	**2.608E-06**	1.603E-06
13	–	–	3.118E-06	1.693E-06
14	–	–	2.668E-06	1.697E-06
15	–	–	3.194E-06	1.566E-06

DWT-FFNN, Discrete wavelet transform-feedforward neural network; *FFNN*, feedforward neural network; *LM*, Levenberg–Marquardt; *MSE*, mean squared error; *ST-FFNN*, Stockwell transform-feedforward neural network. Bold characters of Table 11.4 specify the minimum MSE values

TABLE 11.5 PQD detection results of the proposed technique with the LM training algorithm.

	DWT (10)		ST (9)	
	PQD cases	Non-PQD cases	PQD cases	Non-PQD cases
PQD cases	4938	62	4977	23
Non-PQD cases	79	4921	28	4972
Overall accuracy (%)	98.59		99.49	

DWT, Discrete wavelet transform *ST*, Stockwell transform; *PQD*, power quality disturbance.

TABLE 11.6 PQD detection results with different training algorithms.

Item	Samples classified successfully													
	RP		BFG		OSS		SCG		CGP		CGF		CGB	
	DWT (4)	ST (9)	DWT (9)	ST (2)	DWT (15)	ST (16)	DWT (4)	ST (9)	DWT (17)	ST (2)	DWT (18)	ST (18)	DWT (18)	ST (8)
PQD cases	4867	4923	4926	4937	4883	4891	4849	4894	4886	4894	4919	4927	4918	4973
Non-PQD cases	4843	4899	4879	4919	4868	4842	4767	4850	4886	4871	4825	4836	4894	4950
Overall accuracy (%)	97.10	98.22	98.05	98.56	97.51	97.33	96.16	97.44	97.72	97.65	97.44	97.63	98.12	99.23

DWT, Discrete wavelet transform; *PQD*, power quality disturbance; *ST*, Stockwell transform.

TABLE 11.7 PQD detection results in noisy environments.

	Samples classified successfully					
	40 dB SNR		30 dB SNR		20 dB SNR	
Item	DWT	ST	DWT	ST	DWT	ST
PQD cases	4892	4965	4856	4954	4816	4942
Non-PQD cases	4939	4938	4934	4925	4904	4911
Overall accuracy (%)	98.31	99.03	97.90	98.79	97.2	98.53

DWT, Discrete wavelet transform; *PQD*, power quality disturbance; *SNR*, signal-to-noise ratio; *ST*, Stockwell transform.

their performance at different levels of signal-to-noise ratio (20, 30, and 40 dB). Results suggest the robustness of the proposed PQD detection approaches under noisy environments. Also, the ST-based approach continued to outperform the DWT-based approach even under noisy conditions. However, the GD and GDX training algorithms showed the worst performance and are not listed in the table.

11.4.2 PQD classification results

Like the PQD detection technique, the classification technique collected DWT- and ST-based features for 1200 cases of each modeled PQD class. Then, it fetched them into different configurations of FFNN to classify the PQD types. Different parameters, including the number of hidden neurons, activation function, and others of the FFNN for the classification problem, are tabulated in Table 11.8. The MSE values of LM training algorithms for the different

TABLE 11.8 FFNN parameters for the PQD classification problem.

Element	Description
Hidden layer neurons	1–15
Activation function	Tan sigmoid (hidden layer), softmax (output layer)
Training algorithm	LM, OSS, SCG, CGP, CGB, CGF, RP, BFG, GD, and GDX
Cost function	Mean squared error
Training set	70%
Testing set	30%

FFN, Feedforward neural network; *PQD*, power quality disturbance.

number of hidden neurons are presented in Table 11.4. As can be observed, the DWT- and ST-based approaches provided the best performance for 12 and 11 neurons in the hidden layers, respectively. Associate confusion matrices for the classification problem for DWT (LM and 12 neurons) and ST (LM and 11 neurons) are presented in Tables 11.9 and 11.10, respectively. Diagonal elements represent properly classified PQD classes and the off-diagonal elements represent the wrong classification. Overall classification accuracies for DWT-FFNN and ST-FFNN were 97.68% and 99.27%, respectively. Table 11.11 presents the correctly classified cases and overall accuracies for other types of training algorithms. As can be seen, the training algorithm SCG with 10 neurons in the hidden layer showed the best performance for the ST-based approach. Also, other ST-based approaches showed better performance over the DWT-based approaches for most of the training algorithms. However, it was observed after a careful comparison of the results of Tables 11.9–11.11 that the LM training algorithms provided the best performance for the classification problem as well. The BFG algorithm provides the least accurate performance compared to its peers. However, the GD and GDX training algorithms showed the worst performance and are not listed in the table.

Finally, the robustness of the best-performing models was investigated under noisy environments by applying different levels of additive white Gaussian noises (20, 30, and 40 dB) to the generated signals before extraction of the features. After collection of the features of the noisy signals, they were fetched to the best configuration of the networks and the obtained results are tabulated in Table 11.12. The results confirm the robustness of the proposed PQD classification approaches under noisy environments. Also, the ST-based approach continued to outperform the DWT-based approach even under noisy conditions.

11.4.3 Validation of the proposed scheme

After the development of the ASP- and MLT-based PQD detection and classification schemes under uncertainty, several independent cases (not seen by the developed schemes during the training and testing phases) were generated and the same features employing DWT and ST were extracted. Then the features were employed for validation purposes of the built models (PQD detection and classification using the neural networks). Table 11.13 presents results obtained from the developed ASP-based MLT approaches for randomly generated representative cases. As can be seen, the developed models (both DWT-FFNN and ST-FFNN) detected and classified the fetched cases successfully. Therefore the proposed techniques have the full potential to be implemented in real electrical power system networks in detecting and classifying PQD.

TABLE 11.9 PQD classification results based on DWT-inspired FFNN (LM).

Class	C1	C2	C3	C4	C5	C6	C7	C8	C9	C10
C1	**1113**	0.000	0048	0035	0.000	0004	0.000	0.000	0.000	0.000
C2	0.000	**1177**	0022	0.000	0.000	0.000	0.000	0.000	0.000	0001
C3	0001	0017	**1157**	0.000	0.000	0002	0.000	0.000	0007	0016
C4	0012	0.000	0.000	**1188**	0.000	0.000	0.000	0.000	0.000	0.000
C5	0.000	0.000	0.000	0.000	**1200**	0.000	0.000	0.000	0.000	0.000
C6	0.000	0.000	0.000	0001	0002	**1163**	0.000	0034	0.000	0.000
C7	0.000	0.000	0.000	0.000	0.000	0.000	**1200**	0.000	0.000	0.000
C8	0.000	0.000	0.000	0.000	0.000	0009	0.000	**1191**	0.000	0.000
C9	0.000	0.000	0017	0.000	0.000	0.000	0.000	0.000	**1174**	009
C10	0.000	0.000	0019	0.000	0.000	0.000	0.000	0.000	0023	**1158**
Overall accuracy 97.68%										

DWT, Discrete wavelet transform; *FFNN*, feedforward neural network; *LM*, Levenberg–Marquardt; *PQD*, power quality disturbance. Bold charecters in Table 11.9 specify the successful classification cases.

TABLE 11.10 PQD classification results based on ST-inspired FFNN (LM).

Class	C1	C2	C3	C4	C5	C6	C7	C8	C9	C10
C1	**1197**	0.000	0.000	0.000	0.000	0003	0.000	0.000	0.000	0.000
C2	0.000	**1196**	0.000	0.000	0.000	0001	0003	0.000	0.000	0.000
C3	0.000	0.000	**1200**	0.000	0.000	0.000	0.000	0.000	0.000	0.000
C4	0.000	0.000	0.000	**1197**	0001	0.000	0002	0.000	0.000	0.000
C5	0002	0.000	0.000	0.000	**1184**	0.000	0.000	0.000	0014	0.000
C6	0.000	0002	0.000	0.000	0.000	**1198**	0.000	0.000	0.000	0.000
C7	0.000	0004	0.000	0002	0.000	0.000	**1194**	0.000	0.000	0.000
C8	0.000	0.000	0.000	0007	0.000	0.000	0.000	**1193**	0.000	0.000
C9	0.000	0.000	0.000	0.000	0023	0.000	0.000	0.000	**1159**	0018
C10	0.000	0.000	0.000	0.000	0.000	0.000	0.000	0.000	0006	**1194**
Overall accuracy 99.27%										

DWT, Discrete wavelet transform; *FFNN*, feedforward neural network; *LM*, Levenberg–Marquardt; *PQD*, power quality disturbance; *ST*, Stockwell transform. Bold charecters in Table 11.10 specify the successful classification cases.

TABLE 11.11 PQD classification results with different training algorithms.

Class	OSS		SCG		CGP		CGB		CGF		RP		BFG	
	DWT (12)	ST (11)	DWT (10)	ST (10)	DWT (10)	ST (12)	DWT (10)	ST (10)	DWT (11)	ST (8)	DWT (11)	ST (9)	DWT (5)	ST (8)
C1	1094	1177	1113	1185	1105	1176	1105	1107	1149	1139	1116	1083	945	960
C2	1134	1188	1148	1188	1149	1184	1149	1132	1145	1139	1147	1088	1178	993
C3	1193	1198	1135	1200	1172	1199	1172	1193	1117	1168	1176	1191	925	1188
C4	1171	1189	1116	1189	1162	1176	1162	1110	1148	1177	1103	1041	1065	1197
C5	1198	1166	1200	1174	1198	1192	1198	1163	1198	1151	1200	1175	1194	1170
C6	1106	1181	1153	1178	1155	1165	1155	1125	1139	1104	1143	1017	998	817
C7	1168	1179	1193	1179	1196	1175	1196	1129	1199	1163	1194	1082	1200	861
C8	1104	1156	1112	1184	1124	1191	1124	1123	1184	1152	1098	1104	977	963
C9	1191	1186	1140	1189	1167	1172	1167	1095	1146	1133	1085	1133	1131	1089
C10	1115	1186	1038	1188	1111	1185	1111	1196	1197	1171	1095	1181	1092	1186
Overall accuracy (%)	95.62	98.38	94.57	98.78	96.16	98.46	96.16	94.78	96.85	95.81	94.64	92.46	89.21	86.87

DWT, Discrete wavelet transform; *PQD*, power quality disturbance; *ST*, Stockwell transform.

TABLE 11.12 PQD classification results in noisy environments.

Item	Samples classified successfully					
	40 dB SNR		30 dB SNR		20 dB SNR	
	DWT	ST	DWT	ST	DWT	ST
C1	1151	1193	1147	1196	1125	1192
C2	1137	1195	1162	1195	1156	1186
C3	1146	1199	1156	1198	1131	1200
C4	1162	1197	1155	1198	1154	1197
C5	1199	1183	1183	1185	1200	1179
C6	1137	1194	1143	1198	1135	1196
C7	1200	1197	1194	1183	1200	1189
C8	1164	1185	1156	1191	1134	1195
C9	1167	1164	1142	1155	1156	1162
C10	1182	1188	1167	1192	1162	1190
Overall accuracy (%)	97.04	99.13	96.71	99.09	96.28	99.05

DWT, Discrete wavelet transform; *PQD*, power quality disturbance; *SNR*, signal-to-noise ratio; *ST*, Stockwell transform.

11.5 Conclusions

This chapter evolved mechanisms for the detection and classification of PDQs combining advanced signal processing and machine learning techniques. In this chapter, 10 different types of PQD were modeled mathematically and their useful features were extracted using discrete wavelet and Stockwell transforms. The features extracted by both DWT and ST were then applied to the FFNNs with different settings. It was observed that the LM training algorithm for both feature extraction-based techniques was the best-performing one in terms of overall accuracies for both detection and classification problems compared to other employed training algorithms. The mentioned training algorithm was capable of distinguishing PQD incidents from their non-PQD counterparts with accuracies of 98.59% and 99.49% for DWT and ST-based approaches, respectively. The LM training algorithm classified the modeled PQD events with accuracies of 97.68% and 99.27% for both DWT and ST-based PQD classification approaches, respectively. The best-performing models showed their robustness under noisy environments. It is worth mentioning that the ST-based approaches outperformed the DWT-based approaches in almost all investigated scenarios. Moreover, the developed PQD detection and classification models detected a few randomly generated

TABLE 11.13 Validation results of the developed PQD detection and classification schemes.

Case number	Actual case name	Predicted results			
		DWT-FFNN		ST-FFNN	
		Detection	Class type	Detection	Class type
# 1	Pure signal	No	–	No	–
# 2	Notch (C9)	Yes	C9	Yes	C9
# 3	Sag (C1)	Yes	C1	Yes	C1
# 4	Swell (C2)	Yes	C2	Yes	C2
# 5	Harmonics (C5)	Yes	C5	Yes	C5
# 6	Interruption (C4)	Yes	C4	Yes	C4
# 7	Harmonics + sag (C6)	Yes	C6	Yes	C6
# 8	Fluctuation (C3)	Yes	C3	Yes	C3
# 9	Harmonics + interruption (C8)	Yes	C8	Yes	C8
# 10	Transient (C10)	Yes	C10	Yes	C10
# 11	Harmonics + swell (C7)	Yes	C7	Yes	C7

DWT-FFNN, Discrete wavelet transform-feedforward neural network; *PQD*, power quality disturbance; *ST-FFNN*, Stockwell transform-feedforward neural network.

representative cases successfully that validated the efficacy of the proposed approaches. Therefore the developed techniques have the full potential to be applied for the identification and classification of PQD incidents in real power system networks.

11.6 List of abbreviations

ANN: artificial neural network
ASP: advanced signal processing
CWT: continuous wavelet transform
DWT: discrete wavelet transform
FFNN: feedforward neural network
FT: Fourier transform
GD: gradient descent
LM: Levenberg–Marquardt
MLT: machine learning tools
MSE: mean squared error
PQ: power quality

PQD: power quality disturbance
ST: Stockwell transform
STFT: Short-time Fourier transform
WT: wavelet transform

Acknowledgments

The authors acknowledge the support received from King Fahd University of Petroleum and Minerals' Center of Research Excellence in Renewable Energy.

References

[1] IEEE Standards Coordinating Committee 22 on Power Quality, IEEE Standards: 1159-1995, IEEE Recommended Practice for Monitoring Electric Power Quality, Nov. 1995.

[2] IEC, "IEC 60050 - International Electrotechnical Vocabulary." .

[3] A. von Meier, Electric Power Systems: A Conceptual Introduction, IEEE Press, 2006.

[4] S.D. Ahmed, F.S.M. Al-Ismail, M. Shafiullah, F.A. Al-Sulaiman, I.M. El-Amin, Grid integration challenges of wind energy: a review, IEEE Access 8 (2020) 10857–10878. Institute of Electrical and Electronics Engineers Inc.

[5] M. Ijaz, M. Shafiullah, M.A. Abido, Classification of power quality disturbances using Wavelet Transform and Optimized ANN, in: 2015 18th International Conference on Intelligent System Application to Power Systems (ISAP), Proceedings of the Conference on, Sep-2015, pp. 1–6.

[6] PTSAdmin, Power Quality and Why It's Important - Pentech, 04-May-2017 [Online]. Available: https://www.pentech.com/power-quality-importantance/. (Accessed June 2020).

[7] H. Wang, P. Wang, T. Liu, Power quality disturbance classification using the S-transform and probabilistic neural network, Energies 10 (1) (Jan. 2017) 107.

[8] T. Taufik, M. Muscarella, D. Sawitri, Power quality analysis of variable frequency drives connected to a reactively compensated mixed load system, in: Proceedings - 2016 International Seminar on Application of Technology for Information and Communication, ISEMANTIC 2016, Institute of Electrical and Electronics Engineers Inc., 07-Mar, pp. 261–266.

[9] P. Kanirajan, V. Suresh Kumar, Power quality disturbance detection and classification using wavelet and RBFNN, Appl. Soft Comput. J. 35 (Jul. 2015) 470–481.

[10] IEEE Recommended Practice and Requirements for Harmonic Control in Electric Power Systems," IEEE Std 519TM-2014, 2014, pp. 1–29.

[11] IEC, IEC 61000-3-2 Electromagnetic Compatibility (EMC)-Limits for Harmonic Current Emissions, fifth ed., 2018.

[12] IEEE and Institute of Electrical and Electronic Engineers, IEEE Std 1159 - IEEE Recommended Practice for Monitoring Electric Power Quality, IEEE, New York, 2009 vol. 2009, no. June.

[13] M.K. Saini, R. Kapoor, B.B. Sharma, in: PQ Events Classification and Detection - A Survey, vol. 2011, IET Conference Publications, 2011, pp. 490–495, 583 CP.

[14] M.K. Saini, R. Kapoor, Classification of power quality events - a review, Int. J. Electr. Power Energy Syst. 43 (1) (Dec. 2012) 11–19.

[15] O.P. Mahela, A.G. Shaik, N. Gupta, A critical review of detection and classification of power quality events, Renew. Sustain. Energy Rev. 41 (01-Jan-2015) 495–505. Elsevier Ltd.

[16] S. Khokhar, A.A.B. Mohd Zin, A.S.B. Mokhtar, M. Pesaran, A comprehensive overview on signal processing and artificial intelligence techniques applications in classification of power quality disturbances, Renew. Sustain. Energy Rev. 51 (2015) 1650–1663.

[17] M. Mishra, Power quality disturbance detection and classification using signal processing and soft computing techniques: a comprehensive review, Int. Trans. Electr. Energy Syst. 29 (8) (Aug. 2019).

[18] G.T. Heydt, P.S. Fjeld, C.C. Liu, D. Pierce, L. Tu, G. Hensley, Applications of the windowed FFT to electric power quality assessment, IEEE Trans. Power Deliv. 14 (4) (1999) 1411–1416.

[19] S. Santoso, W.M. Grady, E.J. Powers, J. Lamoree, S.C. Bhatt, Characterization of distribution power quality events with Fourier and wavelet transforms, IEEE Trans. Power Deliv. 15 (1) (Jan. 2000) 247–254.

[20] Y. Gu, M.H.J. Bollen, Time-frequency and time-scale domain analysis of voltage disturbances, IEEE Trans. Power Deliv. 15 (4) (Oct. 2000) 1279–1284.

[21] J. Barros, E. Pérez, Automatic detection and analysis of voltage events in power systems, IEEE Trans. Instrum. Meas. 55 (5) (Oct. 2006) 1487–1493.

[22] M. Szmajda, K. Górecki, J. Mroczka, Gabor transform, spwvd, gabor-wigner transform and wavelet transform - tools for power quality monitoring, Metrol. Meas. Syst. 17 (3) (2010) 6.

[23] J.G.M.S. Decanini, M.S. Tonelli-Neto, F.C.V. Malange, C.R. Minussi, Detection and classification of voltage disturbances using a Fuzzy-ARTMAP-wavelet network, Elec. Power Syst. Res. 81 (12) (Dec. 2011) 2057–2065.

[24] F. Ucar, O. Alcin, B. Dandil, F. Ata, Power quality event detection using a fast extreme learning machine, Energies 11 (1) (Jan. 2018) 145.

[25] K. Manimala, K. Selvi, R. Ahila, Optimization techniques for improving power quality data mining using wavelet packet based support vector machine, Neurocomputing 77 (1) (Feb. 2012) 36–47.

[26] R. Jalaja, B. Biswal, Power quality event classification using hilbert huang transform, Advanc. Intell. Syst. Comput. 199 (2013) 153–161. AISC. Springer Verlag.

[27] M. Biswal, P.K. Dash, Detection and characterization of multiple power quality disturbances with a fast S-transform and decision tree based classifier, Digit. Signal Process. 23 (4) (2013) 1071–1083.

[28] P.K. Mishra, U. Subudhi, S. Jain, Power quality disturbances classification with deep learning approach, in: Proceedings - 2019 International Conference on Information Technology, ICIT 2019, Institute of Electrical and Electronics Engineers Inc., 01-Dec, pp. 273–278.

[29] Y. Shen, M. Abubakar, H. Liu, F. Hussain, Power quality disturbance monitoring and classification based on improved PCA and convolution neural network for wind-grid distribution systems, Energies 12 (7) (Apr. 2019) 1280.

[30] S. Prakash, S. Purwar, S.R. Mohanty, Adaptive detection of islanding and power quality disturbances in a grid-integrated photovoltaic system, Arabian J. Sci. Eng. (Feb. 2020) 1–14.

[31] E. Fuchs, B. Trajanoska, S. Orhouzee, H. Renner, Comparison of wavelet and Fourier analysis in power quality, in: PQ 2012 8th Int. Conf. - 2012 Electr. Power Qual. Supply Reliab. Conf. Proc, 2012, pp. 93–100.

[32] T.Y. Vega, V.F. Roig, H.B. San Segundo, Evolution of signal processing techniques in power quality, in: 2007 9th Int. Conf. Electr. Power Qual. Util. EPQU, 2007.

[33] N.H.T. Huda, A.R. Abdullah, M.H. Jopri, Power quality signals detection using S-transform, in: Proc. 2013 IEEE 7th Int. Power Eng. Optim. Conf. PEOCO 2013, 2013, pp. 552–557. June.

[34] M. Shafiullah, M. Abido, T. Abdel-Fattah, Distribution grids fault location employing ST based optimized machine learning approach, Energies 11 (9) (Sep. 2018) 2328.

[35] H. Duffau, "Brain Plasticity and Reorganization before, during, and after Glioma Resection," in *Glioblastoma*, Elsevier Inc., 2016, pp. 225–236.

[36] J.A. Anderson, E. Rosenfeld, Talking Nets: An Oral History of Neural Networks, MIT Press, 2000.

[37] D. Hunter, H. Yu, M.S. Pukish, J. Kolbusz, B.M. Wilamowski, Selection of proper neural network sizes and architectures-A comparative study, IEEE Trans. Ind. Informatics 8 (2) (May 2012) 228–240.

[38] S. Maiti, V. Verma, C. Chakraborty, Y. Hori, An adaptive speed sensorless induction motor drive with artificial neural network for stability enhancement, IEEE Trans. Ind. Informat. 8 (4) (Nov. 2012) 757–766.

[39] A. Wuraola, N. Patel, SQNL: a new computationally efficient activation function, in: Proceedings of the International Joint Conference on Neural Networks 2018-July, Institute of Electrical and Electronics Engineers Inc., Oct-2018.

[40] D.L. Elliott, A better activation function for artificial neural networks, ISR Technical Report, University of Maryland 93 (8) (1993) 1–3.

[41] A. Nicolae, PLU: The Piecewise Linear Unit Activation Function, Sep. 2018.

[42] L.H. Eidnes, A. Nøkland, Shifting mean activation towards zero with bipolar activation functions, in: 6th International Conference on Learning Representations, ICLR 2018 - Workshop Track Proceedings, 2018.

[43] H. Karim, S.R. Niakan, R. Safdari, Comparison of neural network training algorithms for classification of heart diseases, IAES Int. J. Artif. Intell. 7 (4) (2018) 185–189.

[44] T.F. Awolusi, O.L. Oke, O.O. Akinkurolere, A.O. Sojobi, O.G. Aluko, Performance comparison of neural network training algorithms in the modeling properties of steel fiber reinforced concrete, Heliyon 5 (1) (Jan. 2019) e01115.

[45] A. Ghaffari, H. Abdollahi, M.R. Khoshayand, I.S. Bozchalooi, A. Dadgar, M. Rafiee-Tehrani, Performance comparison of neural network training algorithms in modeling of bimodal drug delivery, Int. J. Pharm. 327 (1) (2006) 126–138.

[46] K. Deepthi, Power quality assessment using change detection and DFT, Int. J. Recent Technol. Eng. 8 (1) (2019) 100–105.

[47] H.S. Rupal, K.T. Ankit, S.R. Mohanty, N. Kishor, "Detection and classification of power quality disturbances using signal processing techniques," *Asia-Pacific Power*, Energy Eng. Conf. APPEEC 2017-Novem (2018) 1–6.

[48] M. Shafiullah, M.A. Abido, Z. Al-Hamouz, Wavelet-based extreme learning machine for distribution grid fault location, IET Gener., Transm. Distrib. 11 (17) (Nov. 2017) 4256–4263.

Chapter 12

Recent trends and advances in power quality

Subrat Sahoo
ABB Power Grids Research, Vasteras, Sweden

Chapter outline

12.1 Introduction 338
12.2 Business motivation 339
12.3 Measurement methodologies and techniques 340
12.3.1 Flicker measurement 341
12.3.2 Classification techniques using multiresolution methods 341
12.3.3 Alternative architecture design for building power quality monitoring systems 341
12.4 PQ data analytics: examples 343
12.4.1 Case 1: detection of a transient condition 344
12.4.2 Case 2: detection of a voltage sag 345
12.4.3 Case 3: detection of voltage dip in two phases 346
12.4.4 Case 4: detection of a voltage sag—downstream event restored after 1.2 s 347
12.4.5 Case 5: momentary interruption—short circuit cleared in the reclosure 347
12.4.6 Case 6: several reclosure event 348
12.4.7 Case 7: voltage dips in the network leading to overvoltage in the system in an EV environment 350
12.4.8 Case 8: impact on EV charging infrastructure during various PQ events 350
12.5 Recent trends on power quality 350
12.5.1 Impact of modern grid integration framework 351
12.5.2 Impact of supraharmonics 351
12.5.3 Impact of fast voltage variations 352
12.5.4 Developments in monitoring and integration opportunities 353
12.5.5 Data collection format 353
12.5.6 IoT-based framework for power quality solutions 354
12.6 Conclusion 355
References 355

Power Quality in Modern Power Systems. https://doi.org/10.1016/B978-0-12-823346-7.00010-4

12.1 Introduction

Power quality (PQ) has been an acknowledged problem in power systems for decades. However, it has found renewed interest from the utility community in recent times due to the changing generation and consumption dynamics. The generation segment is seeing a disruption from traditional disciplines due to unprecedented growth in renewable sources at different voltage levels. This comes with the technological advances that make the levelized cost of electricity from renewables at par and in some places below that of conventional energy sources. At the same time, global climatic concerns and increased greenhouse gas emissions, originating from fossil-based electricity generation, the transport sector, and other industrial processes, are forcing researchers to look for alternatives such as carbon negative or neutral energy sources. The International Energy Agency (IEA), estimates 300 GW of renewable addition globally every year between 2018 and 2030 to meet Paris Conference of Parties (COP21) commitments [1]. Despite a 7% growth during the last few years, CO_2 emissions from the power sector continue to rise at a record pace, due to rapid urbanization, the growing need for industrialization, globalization leading to increasing traffic flow in all modes of transport, etc.

Penetration of renewable energy at multiple entry points into the voltage network comes with its own share of challenges to cope with existing outdated power system infrastructures (in most parts of the world) and their complexities. PQ therefore becomes a topic of paramount interest, when large-scale generation sources are added to the network with their intermittent availability. A cloud patch can wane a few megawatts of solar park power production in seconds and wind can cease blowing without prior warning. Despite significant advancement in satellite mapping and meteorological research, weather forecast sources have never been accurate at the hours and minutes scale. These volatilities therefore bring challenges to capacity reserve, voltage disturbance suppression, and harmonics balancing to offer guaranteed power to households and industries with a defined quality of service (QoS). The biggest challenge for PQ in renewable energy production comes from power electronics-based devices. These devices have the potential of adding multiple switching frequency transients to the grid, if appropriate power filters are not incorporated.

Consumption patterns have also seen a wide transformation in recent times, due to unsynchronized loads added at multiple entry points and voltage levels. Some examples are electric vehicle (EV) chargers that range from a few tens to hundreds of kW, a host of data centers and their accessories, heating, ventilation, and air conditioning (HVAC) devices, arc furnaces that run by variable frequency drives, switching transformers, etc. that not only add a lot of unwanted harmonics to the grid, but are also responsible for voltage sags, transients due to starting/stopping, as well as intermittent operations, leading to short and long interruptions, brownouts, and flickers. The global projection of EVs integrated into the mainstream by 2030 is 37% in Japan, 30% in North America (United States and Canada), 29% in India, and 22% in the rest of the world. This impacts the charging infrastructure, which was estimated to be 5.2 million at the end of 2018, with an increase of 44% from 2017 [2]. The domestic or private charging point contributes to the lion's share of this number. These charging infrastructures come at different voltages and kW levels, thus impacting the grid at various levels and calling for a very detailed study to prepare the grid to cope with the new energy matrix that is rapidly coming into existence. Small domestic loads, such as light-emitting diode (LED) lamps, laptops, and phone chargers, which have unusual power factor distribution in an effort to reduce active power, are also responsible for PQ problems.

PQ in the utility world is largely referred to as the quality of voltage delivered to the consumer. It is determined by a series of prescribed regulations for magnitude, phase, and frequency. However, by definition, it denotes both voltage and current, where the latter is governed largely by consumer-side activities. The concept and implications are rather widespread depending on the end users. This concept has existed for decades and several guidelines have been followed by both transmission and distribution operators. However, it has gained significance in recent times due to changing generation and consumption equations in the power system, mentioned earlier. Therefore PQ is particularly interesting at the nodal points, where a major dissection between generation and consumption happens. Nevertheless, its contribution at the medium- and high-voltage levels cannot be ignored due to the multi-MW renewable parks feeding directly at such voltages, as well as supercharger networks for EVs that draw power in the MW range from the grid. Additionally, system events due to half-cycle core saturation during energization of transformers, rapid loads, converter station operations, reactor switching into the network, etc. not only cause momentary interruptions and stalling operations but also give a significant suggestion of incipient fault detection [3].

12.2 Business motivation

Power quality problems cost ~$119 billion per year to the industrial facilities in the United States, caused by power supply variations and voltage disturbances, as per an Electric Power Research Institute (EPRI) report [4]. On the other hand, 25 EU states suffer an equivalent of $160 billion (~€150 billion) financial losses per year due to different PQ issues, according to the European Copper Institute [5]. These figures are linked to subsequent downtime and production losses as well as the equivalent of intellectual productivity losses.

Fig. 12.1 summarizes data collected from 24 different distribution utilities in the US by the EPRI. 85% of incidents related to PQ are caused by voltage sags, swells, harmonics, wiring, and grounding problems. Unlike in the United States, the voltage dip is not the single largest PQ problem in Europe and different reasons for PQ issues are rather widespread, as reported in Fig. 12.1.

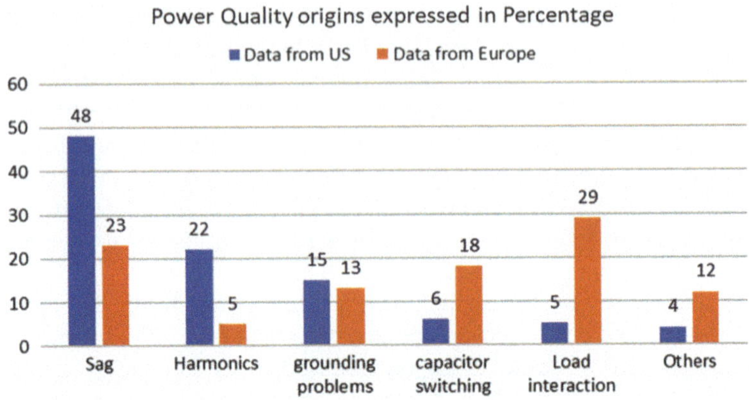

FIGURE 12.1 Causes for power quality problems in the United States and Europe.

The consequences result in load imbalances, low power factor, fast voltage fluctuations, harmonic pollution that results in capital loss, and reduced quality of life and prosperity index. These phenomena also lead to reduced life of equipment, disconnection and failure to reconnect to the grid by renewables, and many unplanned service disruptions. Capacitors, harmonic filters, static VAR compensators, and battery energy storage solutions are some widely used mitigation methods to prevent PQ issues escalating further.

There exists state- and contry-wide regulations which individually set limits for acceptable PQ tolerances. There also exists a grey zone, where the stipulated limits are subject of negotiation between the power producers and consumers. There are several drivers that determine the present-day PQ scenario and these will be more debatable in the coming decades because traditional power systems, which used to be a combination of centralized generation, long-distance transmission, and distributed consumption, are beginning to change. This chapter looks at recent advances in this field and outlines some of the use cases that highlight the importance of PQ parameters that will be subject to constant monitoring mode. Various monitoring trends in PQ are also discussed in this chapter.

12.3 Measurement methodologies and techniques

PQ measurements are implemented in multiple ways, depending on the field of application, academic research, industrial or residential purposes. The degree of applicability also exists in different technology readiness levels, sometimes just showing the conceptual developments of measurement techniques. These techniques derive inspiration from several international standards, namely IEEE 1159 (recommended practice for PQ monitoring) [6], IEEE 519 (harmonic control) [7], IEC 61000-4-30 (testing and measurement techniques) [8], and IEC 61000-4-7 (harmonics and interharmonics assessment) [9]. Different methods for capturing specific aspects of PQ are highlighted next with their respective figures of merit.

12.3.1 Flicker measurement

Flicker is one important aspect for both domestic and industrial consumers as it can impact some of the sensitive equipment, while causing much annoyance. The flicker components can be extracted from the signal as proposed by Cheng et al. [10] by using Pont's method, thus dissociating it from other forms of the disturbance spectrum. However, due to rigorous mathematical complexity of the method, it proves challenging for wider implementation.

12.3.2 Classification techniques using multiresolution methods

Rule-based classification techniques help classify PQ events, such as temporary or permanent disturbances caused by voltage disturbances, while tracing their root cause. This is particularly applicable for real-time systems. Several time–frequency (T–F) analyses can be applied to the rule base. Wavelet and multiresolution *S*-transform have been applied to PQ disturbance signals by several researchers [11,12] for detection and classification of events.

S-transform is a useful technique for T–F domain representation of the signal, initially developed by Stockwell [13] in 1996, which finds application in a variety of fields, including medical science, defense, communication engineering, scientific signal processing, etc. It uniquely combines the frequency-dependent resolution of T–F space along with absolutely referenced local phase information. In such a transformation, the phase characteristics are directly related to the phase of the Fourier transform spectrum, while retaining the localized time. The additional phase information and accuracy of the represented signal in T–F space offers better accuracy compared to continuous wavelet, short-time Fourier transform, and Cohen's class transformations at the cost of a more computationally intensive infrastructure.

12.3.3 Alternative architecture design for building power quality monitoring systems

Depending on the type of architecture involved for monitoring the PQ, multiple design alternatives can be followed.

A data acquisition system proposed by Grigorescu et al. [14] measures voltage, current, frequency, active and reactive power, etc. in a flexible way combining both real-time monitoring and statistical analysis from long-term data acquisition. The functionality of smart meters (SMs) used in automatic metering infrastructure (AMI) architecture offers an ideal platform to integrate or leverage its existing parameters into a larger framework where PQ meters can be integrated. Such a modified approach can be adapted to a newer environment where data is logged only following a sequence of anomalies characterized by the PQ indices, and conversely, in the absence of events, the logging is not carried out for metering purposes. Such a system is called an integrated power quality monitoring system and is widely adopted by many utilities, usually on an experimental basis [15,16]. In these cases, the functionality of SMs is augmented with additional functionalities to be deployed in a smart distribution grid, with added intelligence for depicting voltage quality information capture. Such an infrastructure has the potential of vertical integration to the larger algorithm development, control, and communication framework, such as supervisory control and data acquisition (SCADA) and electric vehicle management system (EVMS), thus leveraging the latter's data throughput system.

By introducing remote computing solutions and transmitting the collected data systematically via a narrow band internet of things (NBIoT) platform, the entire monitoring system chain can be redesigned into a cloud or fog-based web server and pass on useful information directly to the customer in a more integrated fashion. Such a system is commonly referred to as power quality monitoring over internet [17] that uses .NET or ASP Net frameworks and various time synchronization options, including synchronizing the sampled data to a GPS or GLONAS system.

Many experimental frameworks are adopted for quick prototyping and implementation of algorithms in either fixed or reprogrammable chips. Two commonly used frameworks are field programmable gate array (FPGA) and application-specific integrated circuit (ASIC). FPGA-based online PQ monitoring has been proposed by [18] for signal segmentation, using Hilbert–Huang transform techniques, combined with a class-specific weighted random vector functional link network. Furthermore, a fast characterization of voltage variation was demonstrated by [19] using the Hilbert transform and a rule-based system was adopted using an FPGA-based chip. An FPGA-based kit was launched for continuous PQ monitoring using a SPARTAN-3E starter kit as demonstrated in [20].

However, the commercial success of such demonstrations [18–20] remains questionable since the implementations are merely FPGA validation of the transformation techniques with rule-based system implementation on laboratory environment without adequate testing for electromagnetic compatibility (EMC) considerations, the presence of other sources of interference, and wide adoption of the technology. Nevertheless, such implementations facilitate parallel usage of many off-the-shelf technologies for co-demonstration, such as Wi-Fi modules for data transfer and liquid crystal displays for use as a human/machine interface, whereas the processing ability can reside in the FPGA. The operator can thus have the flexibility of accessing any mode in the system grid for checking the voltage quality without going through a centralized analysis of all nodes together.

Francisco et al. [21] proposed a web services architecture for PQ monitoring of SM-monitored data already available in the network, thus expanding the database into a distribution framework. The hope here is to offer a flexible platform, accessible from multiple subsystems. The web services presented are on-demand entities capable of processing information individually, accessed from power meters, SCADA, or databases. Apart from fault detection and diagnosis features, the results of the analysis could be used for fault characterization improvement and further segregation.

An integrated PQ monitoring system from the Bosnian utility [15] talks about collecting data only when there occurs an event and not logging the data during normal circumstances. Such a concept can be less demanding on computational and storage requirements. SMs were used for conducting several laboratory tests and PQ meter data was compared with data from those meters in compliance with IEC standards for PQ measurements [8]. This highlights the benefit of leveraging existing infrastructure and building a figure of merit for an intelligent PQ meter, which in turn depends on these inputs.

Often, combined information on metering accuracy and event classification can be joined together to give a detailed overview of the grid, including network planning, maintenance strategy, and healthy upkeep analysis. Bisceglie et al. used cooperative sensor networks for PQ monitoring, where the sensors were located at various nodes to monitor the power exchange [22]. This collaborative approach of sensors helps to scale up the sensors while organizing them for higher ambition of a network-level PQ management system.

12.4 PQ data analytics: examples

Awareness of PQ started three decades ago with the definition of voltage quality that refers to a constant magnitude, single-frequency sinusoid that is delivered by the utility. However, deviations exist or arise due to the performance of the network as a result of natural or manmade phenomena that impact voltage quality. A complementary definition is current quality that depends on what the consumer receives from the utility. Thus PQ is a mixed phenomenon with contributions from both voltage and current. These definitions are also largely diluted when the power flow becomes bidirectional due to secondary generations feeding back to the grid and active participation of loads that have a notion of feeding back harmonics, imbalance, and voltage variations, thus disturbing the symmetry. The concept of regenerative mechanisms also disturbs this equation but in a more stringent way.

A broad classification of transient and steady-state events is presented in Fig. 12.2, mostly separated by timescale, with a representative image on the top that exemplifies those specific events. This section highlights key PQ events observed at different utility sites. It also contains certain case studies in a simulated environment that have final implications on PQ events. Those case studies are included for the sake of completeness and to give a broad view of all PQ spectrum events. The examples are not exhaustive but are indicative of

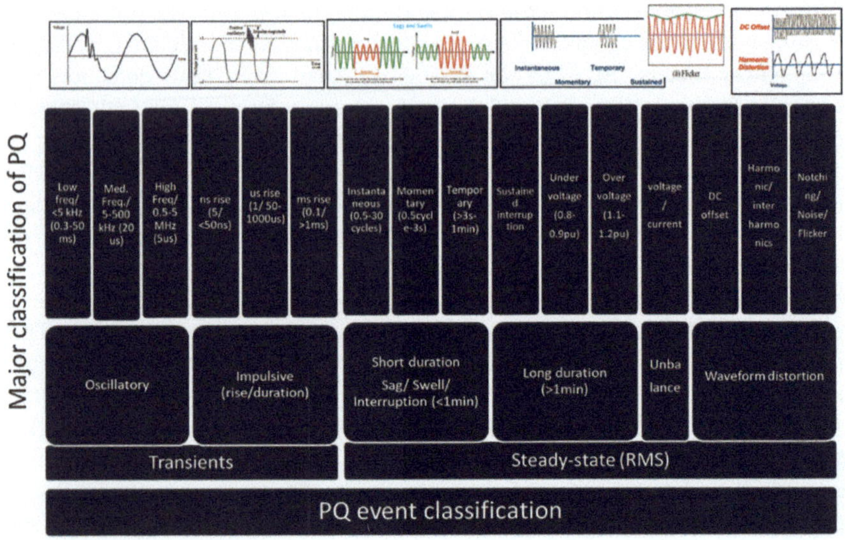

FIGURE 12.2 Classification of power quality (PQ) events in a timescale. *RMS*, Root mean square.

subtle differences existing in the logged events and their reported classifications in the PQ monitors from different vendors. It may be stressed here that the listed events are not fully categorized in the right class and there are elements of human interpretation involved to subjectively verify them with SCADA or manual logs.

12.4.1 Case 1: detection of a transient condition

Fig. 12.3 shows a transient condition that is quickly restored within a cycle, but results in an overcurrent scenario that lasts at least 25 cycles (see Fig. 12.4 and the zoomed part of the initiated transient in Fig. 12.5). Transients usually last less than a cycle. The event is most likely generated due to switching of a

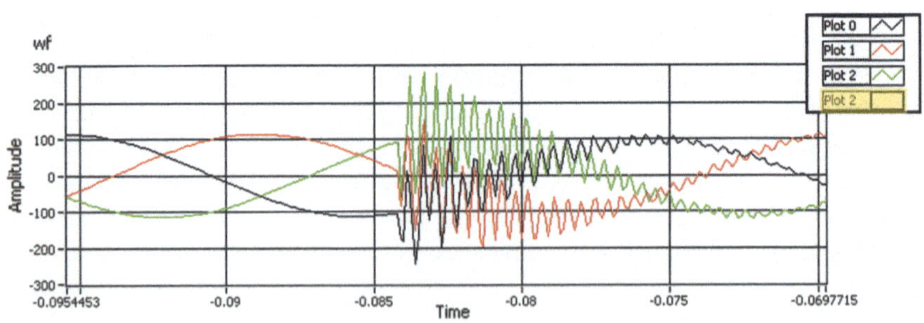

FIGURE 12.3 Transient power quality event detection, reported high-voltage magnitude in kV.

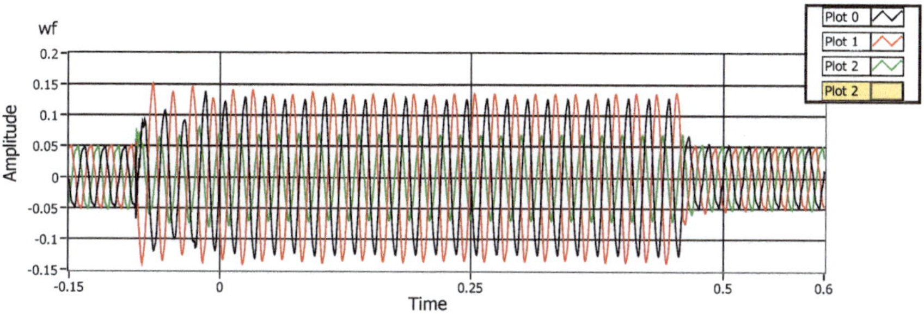

FIGURE 12.4 Transient power quality event detection, reported high-voltage current magnitude in kA.

capacitor bank or an inductor circuit, which tends to oscillate for one to a few cycles before settling down. The voltage might reach a magnitude of 2 pu momentarily during this period and the current also follows the voltage, depending on upstream or downstream contributions.

12.4.2 Case 2: detection of a voltage sag

Fig. 12.6 shows a voltage sag that is almost a textbook quality waveform, where participation of the three phases is identical and the voltage drop of more than 10% of nominal value in one phase is registered as a sag, until the voltage is restored on all three phases. The event is triggered around 0 s for the 0.75 s root mean square (RMS) record observed in Fig. 12.6. The corresponding RMS value of the current is shown in Fig. 12.7. It is noted that the current registers an increased value during the sag, which confirms a downstream event, triggered by load. The nominal current is in the range of 100 A, which increases fourfold, implying a clear case of short circuit, but is restored within a few cycles.

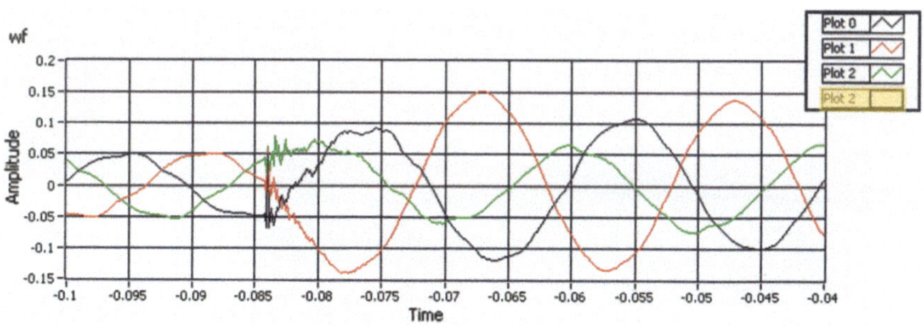

FIGURE 12.5 Transient power quality event detection, reported high-voltage current magnitude in kA, zoomed for two cycles at the beginning of the event.

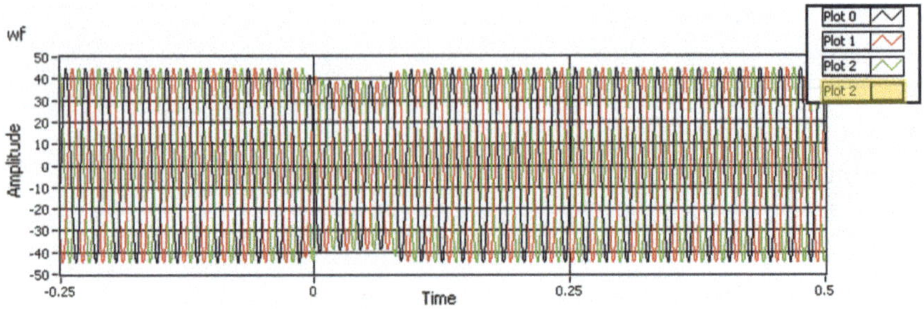

FIGURE 12.6 Voltage sag power quality event detection, reported high-voltage magnitude in kV.

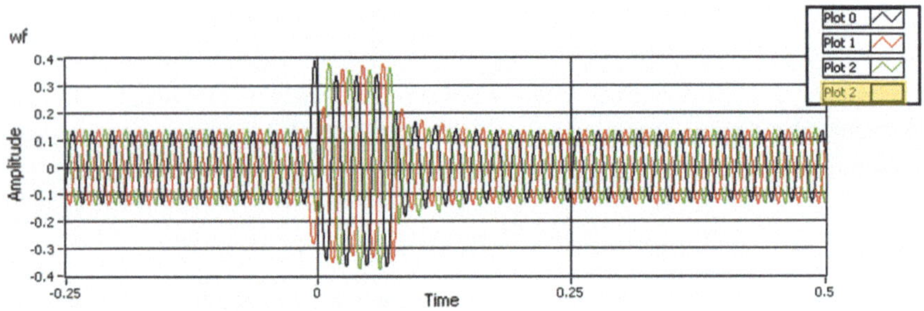

FIGURE 12.7 Voltage sag power quality event detection, reported high-voltage current magnitude in kA.

12.4.3 Case 3: detection of voltage dip in two phases

Fig. 12.8 shows a voltage dip in two phases that leads to a trip after roughly 2 s. Depending on the measurement reading the line-to-line voltage or line-to-neutral voltage, the number of phases participating in an event can be a little tricky. This particular case might actually involve a single line-to-ground fault or a short circuit between two lines.

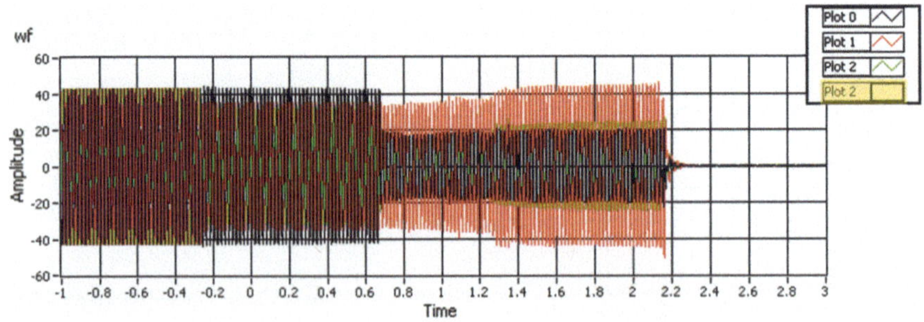

FIGURE 12.8 Voltage sag power quality event detection, reported high-voltage magnitude in kV.

Therefore this additional information needs to be ascertained to verify the exact involvement of the phases. The current magnitude is reported in Fig. 12.9, which confirms this to be a downstream event too, where different phases join the voltage sag event at different times, leading to a short circuit that is not restored at least during the captured cycles of the event (4 s: 3 s postevent and 1 s preevent time). It may be noted here that the fault restoration time cannot be retrieved from this data log as the events are captured for the waveform and RMS values only for a limited time-bound period that depends on the storage ability of the respective monitoring device in question and is preprogrammed and thus cannot be customized for a particular event.

12.4.4 Case 4: detection of a voltage sag—downstream event restored after 1.2 s

Fig. 12.10 shows a voltage sag incident that lasts close to 1.2 s. This is a very common type of event observed in the network where the likely cause is a short circuit due to a dry tree branch falling on a transmission line causing a temporary short circuit. The branch might fall on the ground by itself or burn due to the short circuit, thus clearing the fault in the next reclosure (if there is a trip event, which is not the case in this particular event); also, the protection relay may survive the overcurrent for the said period (as per the protection logic) and normalcy will thus be restored after 1.2 s. Fig. 12.11 shows the current RMS amplitude justifying a downstream event.

12.4.5 Case 5: momentary interruption—short circuit cleared in the reclosure

Unlike the previous case, Fig. 12.12 is a case of momentary interruption that lasts approximately 10 cycles. The most likely cause is a short circuit, tripping the voltage on the high-voltage side. However, the short circuit current of

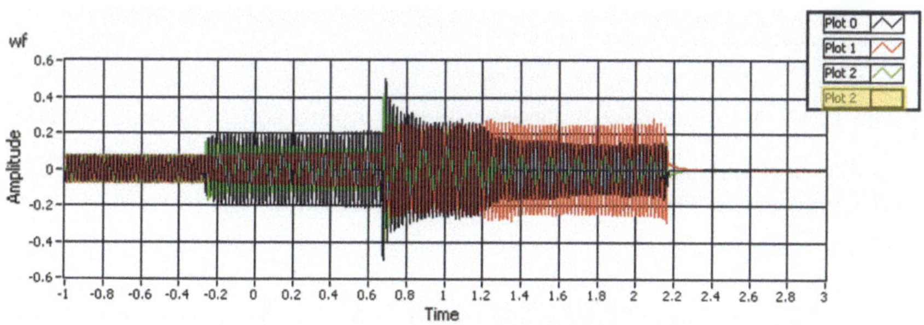

FIGURE 12.9 Voltage sag power quality event detection, reported high-voltage current magnitude in kA.

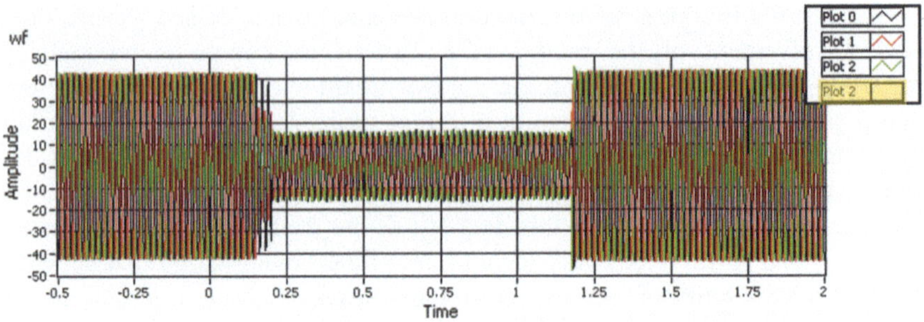

FIGURE 12.10 Voltage sag power quality event detection, reported high-voltage magnitude in kV.

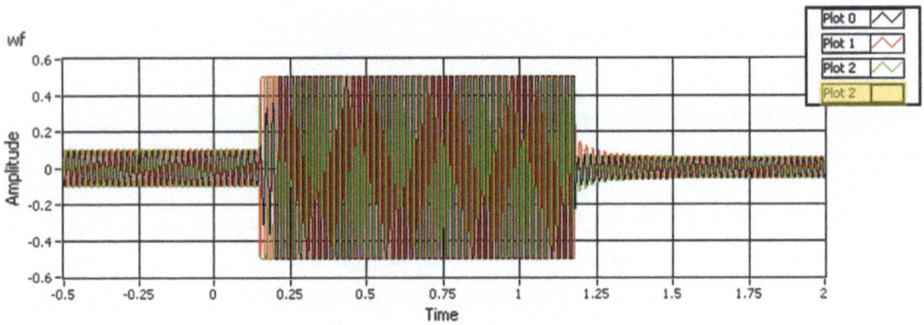

FIGURE 12.11 Voltage sag power quality event detection, reported high-voltage current magnitude in kA.

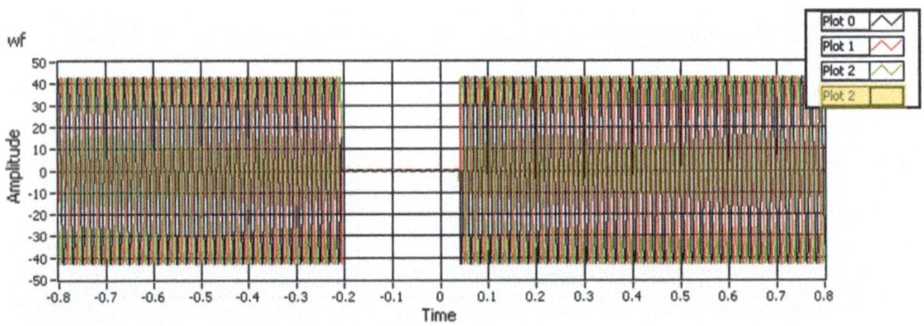

FIGURE 12.12 Voltage sag power quality event detection, reported high-voltage magnitude in kV.

twice the nominal is observed in all three phases in Fig. 12.13 during this downstream event. The fault is cleared by the reclosure, thus restoring voltage after a few cycles.

12.4.6 Case 6: several reclosure event

Fig. 12.14 exhibits a several reclosure incident during the reported period of one voltage sag event. During the event capture, a several reclosure attempt

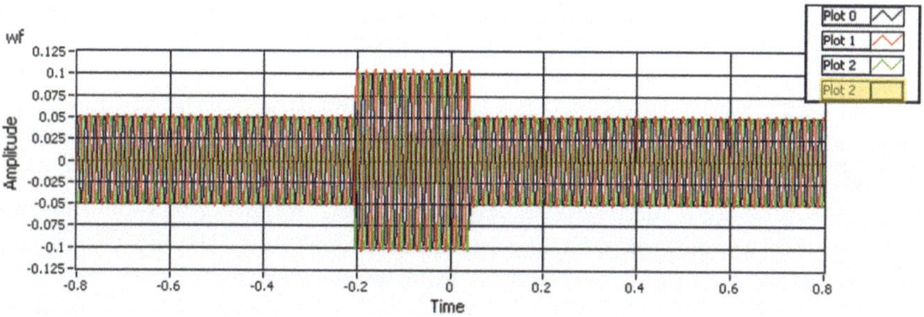

FIGURE 12.13 Voltage sag power quality event detection, reported high-voltage current magnitude in kA.

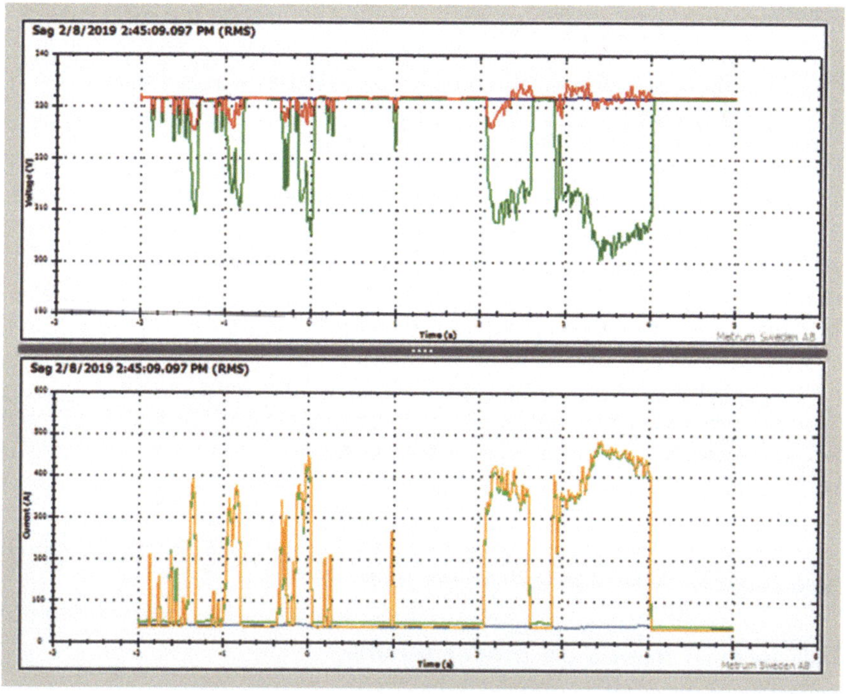

FIGURE 12.14 Voltage sag power quality event detection, (*top*) reported high-voltage magnitude in kV, (*bottom*) reported high-voltage current magnitude in kA.

was made, and each time the relay was rejected by the system. This indicates a real fault in the downstream since the current and voltage are aligned in the opposite direction during each event. Individual phase participation is also very different to the cases discussed so far, thus isolating the green phase as the faulty one.

12.4.7 Case 7: voltage dips in the network leading to overvoltage in the system in an EV environment

An interesting study was argued in [23], where the disconnection of a bulk of coordinated EV chargers, due to a voltage sag event, resulted in an overvoltage. Though this is a simulated environment, where a controlled experiment was analyzed, such studies enforce strong requirements on the EV charging management as well as suggesting a need for delaying the reconnection following a PQ event, such as the one reported here.

The coordinated charging infrastructure is actually pretty much prevalent already in many countries, at least on a test basis, where the vehicle-to-grid (V2G) ability of the electric cars is being authenticated by allowing bidirectional power flow. Several pilot projects are being evaluated in multiple locations [24,25].

12.4.8 Case 8: impact on EV charging infrastructure during various PQ events

Several studies are being carried out globally to verify the impact of EV chargers on the grid, thus making them responsible for PQ events. Another section of one study also focused on the impact of voltage sag on EV onboard chargers. One such study carried out by [26] used different simulation models of six- and 12-pulse rectifiers as well as a pulse width modulated (PWM) rectifier charger and applied different kinds of voltage sags to study which rectifier type is impacted the most. This paper shed light on the working state of the charger also being an important parameter for impact analysis.

Grid disturbance was simulated on different forms of EV interaction into the grid, namely grid-to-vehicle (G2V), V2G, and vehicle-to-home (V2H) modes in [27]. The study found that G2V and V2H modes had a larger impact on the battery charging/discharging current. This is an important step for performing a life cycle analysis of the batteries, when subjecting them to all these different modes during their operation.

12.5 Recent trends on power quality

As mentioned in the introduction, unprecedented changes in renewable energy penetration and the new generation of converters implemented in multiple consumption profiles have caught power system grid code compliance off guard. Policies, regulations, and watchdogs have not been able to keep pace with the rapid developments, thus leading to unforeseen consequences in many parts of the world from a PQ perspective.

CIGRE and CIRED operate a joint working group, JWG C4.24 [28], that addresses PQ and EMC issues related to future electric networks. This JWG also collaborated with IEEE WG on PQ issues concerning grid modernization.

The scope of these different nonprofit bodies is to understand the new trends that are developing in the smart grid framework and find appropriate connections with the PQ scenario. This section highlights recent trends that concern the power world.

12.5.1 Impact of modern grid integration framework

Constant evolution of the grid structure, especially in the low- and medium-voltage network, calls for critical evaluation of the elements that are responsible for impacting PQ. The list is rather exhaustive. A few important considerations in this regard are reported here:

1. Harmonic distortion and voltage imbalance from both the generation and consumption sides of the low-voltage network have to be studied in depth, especially with the involvement of a power electronic converter interface for a closer understanding of the frequency response of the interfacing device. This requirement entails the evolution of new measurement techniques to be adopted in such a situation.
2. Microgrid connections with a provision for islanding create additional PQ challenges during connection and disconnection from the main grid. This is particularly of interest due to the integration of small generation and storage facilities via PWM or other active converters, thus becoming a source of supraharmonics [29–33].
3. Voltage and VAR control is an important step in the large-scale integration of renewable, which has an active participation on demand-side management. This impacts PQ due to participation from consumer-side inverters as well as grid-side converters, especially when there are involvements from EV chargers in the network further contributing to short duration voltage dips, individual rapid voltage changes (RVCs), as well as flicker sensitivities.
4. Power electronics devices invariably create additional switching frequency range interactions between the devices, which is termed supraharmonics. The power electronics equipment on the grid side also contains capacities that have a potential to cause noticeable shift in resonant frequency in the low-voltage network. Such a resonance shift behavior is also noticed due to the recent modification in the distribution network from overhead line to underground cable to achieve better availability of the network due to calamities and natural phenomena.

12.5.2 Impact of supraharmonics

Supraharmonics is a relatively new field of research within PQ. In the absence of power semiconductor devices, distortion of the voltage waveforms was restricted to 2 kHz, commonly referred to as low-frequency harmonics.

Of late, this scenario is changing with the introduction of multiple power electronics devices both in low- and medium-voltage networks. This has the potential of impacting network operation and to some extent energy metering at both voltage levels. This is a phenomenon of harmonics, which is prevalent in the range of 2–150 kHz, and coincides with the operation of power electronics device switching.

The rise of power electronics devices in both these voltage levels can be attributed to unprecedented renewable penetrations at both the domestic and farm levels. The impact of these converters on the grid has been studied to a considerable extent by many researchers [32,33]. The use of multilevel converter topologies has brought technological advancements in modulation strategies bringing enhanced quality of output voltages, thus finding application in various fields of induction motors and drives, active filters, renewable technology interfaces, flexible AC transmission system (FACTS) and high-voltage direct current (HVDC) semiconductor devices, etc. While dealing with high-power applications, the switching frequencies of these power electronics-driven converters operate well above 1 kHz due to the high switching frequencies involved, thus giving rise to an interactive conflict between various devices. An analytical framework was established in [34] to understand the steady-state disturbance between devices in the frequency domain without taking into account their impedance behavior.

12.5.3 Impact of fast voltage variations

Traditional voltage variations in the utility network were classified as short- and long-duration variations that lasted less than 1 min and over 1 min, respectively. Short-duration variations could be classified as sag, swell, and short interruption with further subdivision of instantaneous, momentary, and temporary interruption within the short interruption class. Long-duration variation lasting over a minute is classified as under- or overvoltage and sustained interruption. A classification is illustrated in Fig. 12.2. Besides these traditional voltage variations, the inclusion of renewables has introduced additional variations, which have timeframes between 1 s and 10 min [29].

Similarly, there have been many changes in lighting conditions over time. Compact fluorescent lamps (CFL) and LED lamps sense flicker in an entirely different perspective compared to incandescent lamps, which call for appropriate standard modification for short-term flicker severity (Pst).

The voltage step, which is part of the RVC calculation, also requires a revisit for standard measurements (mostly set on the basis of IEC 61000-3-3 [35] and EN 50160 [36]) as it always lacks a clear, unequivocal definition to be internationally accepted. RVC is a major source of annoyance for flicker severity both for short- (Pst) and long-term (Plt) flicker severity.

Transients in the network are largely classified as oscillatory and impulsive in nature (Fig. 12.2). Oscillatory transients can last from a few microseconds to milliseconds with further subclassification of low-, medium-, and high-frequency oscillatory transients [6], induced by switching operations of capacitor banks, inductive switches, etc., whereas the impulsive transients are clearly a result of micro- to nanosecond impulses that exist well within a short cycle. However, capturing these events is a staggering task, when the measuring equipment (the voltage transformer and current transformer) in operation is not able to register such fast changes due to the inherent limitations in the hardware to operate above a certain kHz due to frequency saturation. Capturing these events would require dedicated hardware, free from low-frequency response limitations. Such transient operations are unavoidable and have become more prevalent in recent times due to global climate change, influencing large-scale thunderstorms, lightning strikes, and the like.

12.5.4 Developments in monitoring and integration opportunities

With the importance of PQ growing every single day, monitoring solutions are also becoming versatile and encompassing part of power and energy metering to offer a wide range of data usability, formatting of data collection that allows cross-examination on vendor neutral platforms, versatility of communication protocols, links to SCADA, network control centers for parallel decision making, and long-term impact study. The following subsections shed light on some emerging fields of interest.

12.5.5 Data collection format

Most of the commercial PQ meters have a similar style of functioning. The data that is intended for the owner's consumption provides many details from different meters presented in proprietary standalone databases without clear interpretation guidelines. The challenges with non-standardized data interpretation are discussed at length in [37]. This forces utilities to use meters from one vendor, so consolidation over a single database becomes easier.

- PQDIF format: In an effort to harmonize this information, IEEE Standard 1159 [6] introduced a file format that is exclusive to PQ, called the power quality data interchange format (PQDIF) that applies to both simulation and measurement data, independent of platforms where they are captured. Additionally, this format uses a bulk of compression algorithm, ensuring dramatic improvement in the file length and storage requirements. Most of the modern PQ meters offer an export option for PQ events over the PQDIF format, to be able to view them elsewhere. Conversely also, some of the meters accept a PQDIF formatted file imported from another source. This feature does not, however, facilitate the bulk comparison of the PQ events or database migration for statistical analysis and interpretation.

- COMTRADE format: Another format of data exchange that is prevalent for transient data storage is called common format for transient data exchange (COMTRADE) [38], which is a more compact form of storing waveforms and widely used by vendors like ABB, Ametek, GE, Qualitrol, Schweitze, Schneider, etc. This format underwent modifications in 2013 by WG H4 of the power system relaying committee, together with IEC 60255-24 Ed 2.0. COMTRADE uses four different file groups, namely ".cfg," ".dat," ".inf," and ".hdr," for exchanging sampled values between a recorder and a reader, with the prime intention of promoting interoperability. The modifications included inclusion of time code, local code, and time quality fields, while offering up to 1 ns resolution of the time stamp. The revision also promised inclusion of the XML format for the CFG section in future upgrades.

12.5.6 IoT-based framework for power quality solutions

Digitalization of a power system is an important step that empowers both the operators and prosumers for a dynamic power management system. Since the market for PQ meters is still evolving and volume-wise is in a very nascent stage, technical developments have to be borrowed from other domains. Bulk deployment in low-voltage networks is dominated by SMs at the consumer end, whereas the niche market segment at the medium- and high-voltage levels is dominated by transient fault recorders and intelligent electronic devices (IEDs). In the former case, due to the bulk nature of the data used by SMs, NBIoT is preferred as the communication medium for handling AMI network data. However, in the latter case, a lot of conservatism is adopted for data transmission due to the criticality of the infrastructure and therefore potential existence of cyber security threats. A dedicated intranet is used for data transmission in these cases between the SCADA system and operator networks.

AMI has seen two secondary applications for upcoming SM deployments in major parts of the world besides traditional energy metering objectives. These are integration of advanced energy management systems via home or wide area network and monitoring of vital system states, such as frequency, phase, and amplitude, under the broader framework of PQ [39]. Such a larger objective thus covers comprehensive tasks of data management, logging, storage, segregation, reporting, and communication, which forms the objective of an IoT-based framework with essential constituents. Though the present versions of SMs are inadequate to fulfill the entire objective of PQ functionality, in many ways they paveway for future development, thus bridging the gap between conventional PQ meters and advanced SMs. Several reports in the literature have inched toward this objective [16,39], albeit at a laboratory scale.

12.6 Conclusion

This chapter aimed at introducing the reader to recent changes in power system dynamics that are unprecedented both from power production and consumption points of view. Therefore maintaining a healthy balance between voltage magnitude, frequency, and phase and its consequent impact on current is becoming unmanageable. This needs deeper investigation of the interacting devices at various PCCs, and thus adoption of correction measures to mitigate them.

The use of power electronics devices in converters in an aid to improve efficiency and also interconnecting multiple systems effectively has the downside of introducing multiple switching frequencies that start to interact with each other. This gives rise to the supraharmonics phenomenon, which is relatively new, and its long-term impact has not been completely understood.

With a lot happening in power systems at the same time, which includes increased generation sources and consumption points, each has the potential to negatively impact the PQ of the overall grid. Though several independent working groups within CIGRE, CIRED, IEEE, and IEC are working together to contribute to effective standards for measuring PQ at different voltage and frequency levels, there is still unfinished business and loopholes in the system that can create economic difficulties, leaving power systems vulnerable and devices failing or being mismanaged, contrary to the grid code compliance.

PCCs for distributed generation in low-voltage and medium-voltage networks are increasing greatly, duc to several federal incentives for promoting clean energy. The changing dynamics in the power system require more physical resources to monitor the PQ at as many entry points as possible, not only to devise mechanisms to contain it at source, but also to find violators who should be subjected to penalties and made liable to satisfy the concerns of the power community.

However, the onus still remains on the utility to prevent the grid from unwanted pollution. As a first measure, most of the distribution transformers in the distribution network are now fitted with a PQ meter to monitor various disturbances and historically observe the trend. The same philosophy should also apply to consumers who can challenge the utility provider in case the latter fails to comply with various stipulated standards. This would open up new provisions in the SM roll-out standard with significant components of PQ indices included in it. All in all, strong vigilance is needed at both ends to maintain a safer, more resilient, and greener smart grid.

References

[1] https://www.iea.org/news/renewable-capacity-growth-worldwide-stalled-in-2018-after-two-decades-of-strong-expansion.

[2] IEA Global EV Outlook, n.d. https://www.iea.org/reports/global-ev-outlook-2019.

[3] S. Sahoo, N. Abeywickrama, T. Bengtsson, R. Saers, Understanding the sympathetic inrush phenomenon in the power network using transformer explorer, in: 2019 IEEE 4th International Conference on Condition Assessment Techniques in Electrical Systems (CATCON), Chennai, India, 2019, pp. 1–5, https://doi.org/10.1109/CATCON47128.2019.CN0031.

[4] S. McNulty, Primen (EPRI), The Cost of Power Disturbances to Industrial and Digital Economy Companies, EPRI IntelliGrid Initiative, June 2001, p. 1006274.

[5] R. Targosz, J. Manson, Pan-European power quality survey, in: 2007 9th International Conference on Electrical Power Quality and Utilisation, Barcelona, 2007, pp. 1–6, https://doi.org/10.1109/EPQU.2007.4424203.

[6] IEEE recommended practice for monitoring electric power quality, in: IEEE Std 1159-2019 (Revision of IEEE Std 1159-2009), August 13, 2019, pp. 1–98, https://doi.org/10.1109/IEEESTD.2019.8796486.

[7] IEEE recommended practice and requirements for harmonic control in electric power systems, in: IEEE Std 519-2014 (Revision of IEEE Std 519-1992), June 11, 2014, pp. 1–29, https://doi.org/10.1109/IEEESTD.2014.6826459.

[8] IEC 61000-4-30 Electromagnetic Compatibility (EMC)- Part 4-30, Ed. 3.0, 2015-02: Testing and Measurement Techniques- Power Quality Measurement Methods.

[9] IEC 61000-4-7 Electromagnetic Compatibility (EMC) Part 4-7: Testing and Measurement Techniques- General Guide on Harmonics and Interharmonics Measurements and Instrumentation, for Power Supply Systems and Equipment Connected Thereto.

[10] Y.C. Chen, C.-N. Chen, C.-I. Chen, A high-resolution technique for flicker measurement in power quality monitoring, in: Presented at the 8th Conference on Industrial Electronics and Applications (ICIEA), Taiwan, 2013.

[11] I.W.C. Lee, P.K. Dash, S-transform-based intelligent system for classification of power quality disturbance signals, IEEE Trans. Ind. Electron. 50 (4) (August 2003) 800–805, https://doi.org/10.1109/TIE.2003.814991.

[12] M.S. Manikandan, S.R. Samantaray, I. Kamwa, Detection and classification of power quality disturbances using sparse signal decomposition on hybrid dictionaries, IEEE Trans. Instrum. & Meas. 64 (1) (January 2015) 27–38, https://doi.org/10.1109/TIM.2014.2330493.

[13] R.G. Stockwell, A basis for efficient representation of the S-transform, Digit. Signal Process. 17 (1) (January 2007) 371–393, https://doi.org/10.1016/j.dsp.2006.04.006.

[14] O.M.G.S.D. Grigorescu, C. Cepisca, A.S. Vinyea, Power quality monitoring systems for smart grid networks, The 8th International Symposium on Advanced Topics in Electrical Engineering, May 23–25 2013.

[15] A.B.M. Music, N. Hasanspahic, S. Avdakovic, E. Becirovic, Integrated Power Quality Monitoring System and the Benifits of Integrating Smart Meters, Public Electric Utility Elektroprivreda of Bosnia and Herzegovina, 2013, pp. 86–91.

[16] M.M. Albu, M. Sanduleac, C. Stanescu, Syncretic use of smart meters for power quality monitoring in emerging networks, IEEE Trans. Smart Grid 8 (1) (January 2017) 485–492.

[17] B.A. Ahmad, H.H. ElSheikh, A. Fadoun, Review of power quality monitoring systems, in: 2015 International Conference on Industrial Engineering and Operations Management (IEOM), Dubai, 2015, pp. 1–8, https://doi.org/10.1109/IEOM.2015.7093825.

[18] M. Sahani, P.K. Dash, FPGA-based online power quality disturbances monitoring using reduced-sample HHT and class-specific weighted RVFLN, IEEE Trans. Ind. Inf. 15 (8) (August 2019) 4614–4623.

[19] M. Valtierra-Rodriguez, D. Granados-Lieberman, J.E. Torres-Fernandez, J.R. Rodríguez-Rodríguez, J.F. Gómez-Aguilar, A new methodology for tracking and instantaneous characterization of voltage variations, in: IEEE Transactions on Instrumentation and Measurement, vol. 65, July 2016, pp. 1596–1604, https://doi.org/10.1109/TIM.2016.2540866, 7.
[20] S. Folea, G. Mois, L. Miclea, Power quality measurement system using FPGAs, in: 2012 13th International Conference on Optimization of Electrical and Electronic Equipment (OPTIM), Brasov, 2012, pp. 1280–1285, https://doi.org/10.1109/OPTIM.2012.6231855.
[21] F.J. Gomez, J. Melendez, S. Herraiz, E. Gonzalez, M. Lopez-Perea, C. Murphy Oconnor, Web services framework for power quality monitoring, in: Smart SysTech 2013; European Conference on Smart Objects, Systems and Technologies, Erlangen/Nuremberg, Germany, 2013, pp. 1–7.
[22] C.G.M. di Bisceglie, A. Vaccaro, D. Villacci, Cooperative sensor networks for voltage quality monitoring in smart grids, in: Presented at the 2009 IEEE Bucharest Power Tech Conference Bucharest Romania, 2009.
[23] S. Kundu, I.A. Hiskens, Overvoltages due to synchronous tripping of plug-in electric-vehicle chargers following voltage dips, in: IEEE Transactions on Power Delivery, vol. 29, June 2014, pp. 1147–1156, https://doi.org/10.1109/TPWRD.2014.2311112, 3.
[24] https://sweden.nissannews.com/sv-SE/releases/release-130188-energi-att-spara-nissan-och-endesa-tecknar-avtal-om-utveckling-av-europas-f-rsta-massmarknadsanpassade-v2g-system#.
[25] https://www.chademo.com/v2g-report-published-identifying-50-projects-around-the-world/.
[26] S. Hui, X. Yonghai, Study on the impact of voltage sags on different types of electric vehicle chargers, in: 2016 China International Conference on Electricity Distribution (CICED), Xi'an, 2016, pp. 1–4, https://doi.org/10.1109/CICED.2016.7576319.
[27] X. Guo, J. Li, X. Wang, Impact of grid and load disturbances on electric vehicle battery in G2V/V2G and V2H mode, in: 2015 IEEE Energy Conversion Congress and Exposition (ECCE), Montreal, QC, 2015, pp. 5406–5410, https://doi.org/10.1109/ECCE.2015.7310420.
[28] JWG C4.24/CIRED Power Quality and EMC Issues Associated with Future Electricity Networks, n.d. http://c4.cigre.org.
[29] M. Bollen, et al., Consequences of smart grids for power quality: overview of the results from CIGRE joint working group C4.24/CIRED, in: 2017 IEEE PES Innovative Smart Grid Technologies Conference Europe (ISGT-Europe), Torino, 2017, pp. 1–6, https://doi.org/10.1109/ISGTEurope.2017.8260116.
[30] M. Bollen, M. Olofsson, A. Larsson, S. Rönnberg, L. Martin, Standards for supraharmonics (2 to 150 kHz), IEEE EMC Mag. 3 (2014) 114–119. Quarter 1.
[31] S.K. Rönnberg, M.H.J. Bollen, H. Amaris, G.W. Chang, I.Y.H. Gu, Ł.H. Kocewiak, J. Meyer, M. Olofsson, P.F. Ribeiro, J. Desmet, On waveform distortion in the frequency range of 2 kHz–150 kHz—review and research challenges, Elec. Power Syst. Res. 150 (2017) 1–10.
[32] A. Novitskiy, S. Schlegel, D. Westermann, Measurements and analysis of supraharmonic influences in a MV/LV network containing renewable energy sources, in: 2019 Electric Power Quality and Supply Reliability Conference (PQ) & 2019 Symposium on Electrical Engineering and Mechatronics (SEEM), Kärdla, Estonia, 2019, pp. 1–6, https://doi.org/10.1109/PQ.2019.8818259.

[33] S.K. Rönnberg, A.G. Castro, M.H.J. Bollen, A. Moreno-Munoz, E. Romero-Cadaval, Supraharmonics from power electronics converters, in: 2015 9th International Conference on Compatibility and Power Electronics (CPE), Costa da Caparica, 2015, pp. 539–544, https://doi.org/10.1109/CPE.2015.7231133.

[34] D. Agudelo-Martínez, C. Garzón, A. Pavas, Interaction of power quality disturbances within 2–150 kHz (supraharmonics): analytical framework, in: 2018 18th International Conference on Harmonics and Quality of Power (ICHQP), Ljubljana, 2018, pp. 1–7, https://doi.org/10.1109/ICHQP.2018.8378859.

[35] IEC 61000-3-3:2013, Electromagnetic Compatibility (EMC) - Part 3-3: Limits - Limitation of Voltage Changes, Voltage Fluctuations and Flicker in Public Low-Voltage Supply Systems, for Equipment with Rated Current $\leq$16 A Per Phase and Not Subject to Conditional Connection.

[36] EN 50160: 2010/A1, European Standard, Voltage Characteristics of Electricity Supplied by Public Electricity Networks.

[37] S. Sahoo, reportImproved Measurement of Power Quality, Energiforsk Report in Swedish, Published 2019, https://energiforsk.se/program/underhall-av-elnat/projekt/forbattrad-matning-av-power-quality/.

[38] IEEE Std C37.111-1999, IEEE Standard Common Format for Transient Data Exchange (COMTRADE) for Power Systems.

[39] E.J. Palacios-Garcia, et al., Using smart meters data for energy management operations and power quality monitoring in a microgrid, in: 2017 IEEE 26th International Symposium on Industrial Electronics (ISIE), Edinburgh, 2017, pp. 1725–1731, https://doi.org/10.1109/ISIE.2017.8001508.

Index

Note: 'Page numbers followed by "f" indicate figures and "t" indicate tables.'

A

Active distribution networks, 202f
  distribution static compensator (D-STATCOM), 204
  electric vehicles and low generation, 212–214, 213f–214f
  high distributed generation and low demand, 208–211, 208t, 209f, 211f–212f, 212t
  Kirchhoff's second law, 202
  line drop compensation, 200
  optimal power flow (OPF), 201
  passive network, 207–208, 207f, 207t–208t
  plug-in electric vehicles (PEVs), 204
  reactive power, 202
  renewable energy sources (RES), 200
  slow charging mode, 200
Advanced signal processing (ASP), 313
  Stockwell transformation (ST), 314–315
  wavelet transformation (WT), 314
Artificial intelligence (AI)-based controllers, 106–107
  artificial neural network (ANN) based controller. *See* Artificial neural networks (ANNs)
  classifications of, 112
  fuzzy logic controller (FLC), 118–119
  system current fast Fourier transform, 123, 128f–129f
  system performance waveforms, 123, 126f–127f
Artificial neural networks (ANNs), 115f, 313
  activation/squashing, 315–316, 317f
  backpropagation, 114
  feedforward neural network (FFNN), 315, 316f
  harmonic separation, 116–118, 117f
  mean squared error (MSE), 316–317
  power quality (PQ), 115, 116f
  pseudocode, 123, 124f
  renewable energy sources (RES), 115, 116f
  structure of, 113, 114f
  sum of squared error (SSE), 114
  supervised learning, 316–317
  training algorithms, 317, 318t
  training states, 123, 125f
  types of, 113, 113f
  validation stage, 123, 125f
Automatic metering infrastructure (AMI), 341–342
Average normalized standing voltage (ANSV), 252

B

Backpropagation, 114
Battery storage system (BSS), 287
  energy management system (EMS), 302
  hybrid storage systems (HSSs), 299, 299f
  modular multilevel converter (MMC), 288, 290f
  rating of, 302, 303t
Bidirectional polarity, 5
Body diode, 249
Brake chopper protection circuits, 137–138
Brazilian grid code, 223–226, 226t, 230–237
Business motivation, 339–340, 340f

C

Capacitors, 246–247, 251f–255f, 257f, 339–340
  charging/discharging modes, 249, 249t
  design, 265
  equivalent series resistance (ESR), 261
  output voltage, 249, 249t
  ripple losses, 265
  switching states, 249, 249t
  ultracapacitor (UC), 287
Cascaded H-bridge (CHB), 246–247, 252, 266t–267t
Chokes, 186
Circuit impedance, 2

Common format for transient data exchange (COMTRADE), 354
Compact fluorescent lamps (CFLs), 166
Computational intelligence-based methods, 145
Conduction losses, 261–264
Continuous wavelet transform (CWT), 314
Converters
  cascaded H-bridge (CHB), 246–247
  configuration of, 248–249, 248f
  electric vehicle supply equipment (EVSE), 151–152
  floating capacitors, 246–247
  neutral point clamped, 246–247
  three-phase six-pulse, 109, 109f–110f
Current-controlled inverters, 136
Current harmonic distortion, 40, 61–62
  commercial system applications, 63
  industrial system applications, 62
  residential system applications, 63
Current mode control (CMC), 299
Current-related problems, 2
Current transformers (CTs), 5
Current unbalance, 33

D

DC offset, 43
Defuzzification, 119
Demand response, 160
Dip aggregation, 225
Dip stratification, 225, 225f
Discrete wavelet transform (DWT), 314
Distributed flexible alternating current transmission system (D-FACTS), 160
Distributed generation (DG), 123, 133–134, 208–211
  grid stability, 134
  large-scale penetration, 193–194
  operating conditions, 134, 134t
  point of common coupling (PCC), 355
  tripping time, 134
  voltage functions *vs.* time, 135
Distribution static compensator (D-STATCOM), 160, 204

E

Electric Power Research Institute (EPRI), 339
Electric vehicles (EV) charging station, 350
  charger types, 152, 155t
  DC charging, 151–152
  demand response, 160
  distributed flexible alternating current transmission system (D-FACTS), 160
  energy storage system (ESS), 159–160
  home/residential charging, 152–153
  plug-in electric vehicles (PEVs), 151–152
  point of common coupling (PCC), 152, 153f–155f
  power quality problems
    harmonics analysis, 157–158, 158f–159f, 159t
    name plate rating, 157, 157f
    supraharmonics, 156
  public charging, 152–153
  state of charge (SoC), 152
  vehicle to grid (V2G), 153
Electric vehicle supply equipment (EVSE), 151–152. *See also* Electric vehicles (EV) charging station
Electromagnetic compatibility (EMC), 342
Electromagnetic interference, 110
Energy management system (EMS), 302
Energy storage + static compensator (E-STATCOM), 305f, 307f
  energy management system (EMS), 302
  grid-connected wind park, 289–291, 290f
  hybrid storage systems (HSSs), 297–301
  modular multilevel converter (MMC), 303t
    battery storage system (BSS), 288, 290f
    circulating current control, 292–294, 294f
    gating signals, 294–295
    inner-loop control, 294, 295f
    leg-voltage controller, 292–294
    low-voltage ride-through capability, 297, 297f
    lumped battery storage system, 288, 289f
    negative sequence current compensation, 296, 296f
    point of common coupling (PCC), 292
    positive sequence current reference, 292
    sequence components, 292, 293f
    submodule capacitors, 294–295, 295f
    ultracapacitor (UC), 288, 290
  squirrel cage induction generator (SCIG), 289–291, 290f
  two-/three-level converter, 287–288, 287f
Energy storage system (ESS), 137–138, 285–287, 286t
  battery, 139
  electric vehicles (EV) charging station, 159–160
  operation, 139, 140f
  supercapacitor, 139

Equivalent series resistance (ESR), 261
European Network of Transmission System Operators for Electricity (ENTSO-E), 135
Even harmonics, 53–57, 54t, 55f–57f

F

Fast Fourier transform, 312–313
Fault ride-through (FRT), 134
  grid codes, 134–135, 135f
  photovoltaic (PV) power plants, 137f
    assessment of, 146–147, 146t
    classification of, 137–138, 138f
    energy storage system (ESS), 138–139, 140f
    flexible alternating current transmission system (FACTS), 138, 140–143
    inverter-resident methods, 143–146
    modified inverter controller (MIC), 139
    power electronics-based method, 139–140, 140f
    voltage drop, 136–137
Feedforward neural network (FFNN), 315, 316f, 322–323, 323t
Fit-and-forget approach, 198–199, 208
FLC. *See* Fuzzy logic controller (FLC)
Flexible alternating current transmission system (FACTS), 137–138
  static compensator (STATCOM), 142, 142f
  thyristor-controlled static compensator (TCSC), 141, 141f
Flickers, 34, 36, 166–169
Form factor (FF), 272
FRT. *See* Fault ride-through (FRT)
Fuzzy logic controller (FLC), 139
  current adaptation and modulation, 121–123, 121f–122f, 122t
  DC link capacitor voltage matching, 119–121, 120f, 121t
  defuzzification, 119
  design parameters for, 119, 119t
  fuzzification, 118
  inference/interpretation, 119
  structure of, 118, 118f

G

Gabor transform, 312–313
Gaussian function, 314–315
Grid codes, 133–135, 135f, 285
  energy storage system (ESS), 285–286, 286t
  static compensator (STATCOM), 285–286, 286t
Grid-connected photovoltaic (PV) systems, 136
Grid side controller (GSC), 145
Grid-to-vehicle (G2V) modes, 350

H

Harmonic factor (HF), 274
Harmonic frequencies, 42
Harmonic order, 42
Harmonics, 61–62, 109, 109f–110f
  artificial neural network (ANN) based controller, 116–118, 117f
  consequences/impacts of, 109–110
  current harmonic spectrum, 93, 95f, 96t, 98f, 99t, 101f, 102t
  distortions, 42, 42f, 62–63
  even harmonics, 53–57, 54t, 55f–57f
  instantaneous current, three phases, 40, 41f, 93, 94f, 97f
  instantaneous voltage, phase to neutral, 93, 94f, 97f
  loads, 78–79, 80f
    single-phase rectifiers, 63–64
    three-phase rectifiers, 65–72
  multitenanted premises, 79–80, 81f
  odd harmonics, 50, 51t, 52f–53f
  phase-to-neutral instantaneous voltage, 40, 41f
  point of common coupling (PCC), 73–77
  residential customers, 167
  resonance, 91–93
  total demand distortion (TDD), 85–87
  total harmonic distortion (THD), 80–85, 96f, 99f
  total rated harmonic distortion (TRD), 87–91
  tripplen harmonics, 57–58, 58f, 59t
Harmonic spectrum, 42
High-gain switched-capacitor multilevel inverters (SCMLIs), 246–247
Home appliances
  modeling methodology, 172
  single-phase AC motors starting, 174–181, 181t, 182f–183f
  single-phase nonlinear loads, 172, 173f–174f, 174t, 175f, 178f
Hybrid storage systems (HSSs), 294
  battery storage system (BSS), 299, 299f
  power segregation, 297, 298f
  state-of-charge (SoC), 300, 301f–302f
  types of, 287
  ultracapacitor (UC), 298, 299f

I

IEC-61000, 312
IEC 61000-2-2, 279
IEC 61000-3-3, 35
IEEE-519, 312
IEEE 1564, 230–237
IEEE Guide for Voltage Sag Indices, 220
IEEE standard 519-2014, 73
IEEE standard 1159–2019, 2
IEEE Standard 1346, 220
IEEE Std. 1547, 134, 134t
Impulsive transients
  negative cycle, 4, 4f
  positive cycle, 4, 5f
Instantaneous power theory (IPT), 111
Integrated power quality monitoring system, 341–342
Interharmonics
  causes, 43
  impacts, 43
  integral multiples, 42
Internet of things (IoT), 354
Inverter-resident fault ride-through (FRT) methods
  computational intelligence-based methods, 145
  grid side controller (GSC), 145
  maximum power point tracking (MPPT), 143, 144f
  modified inverter controller (MIC), 139
  phase-locked loop (PLL), 143–144
  resonance controller, 144
  root mean square (RMS), 145

K

Kirchhoff's second law, 202

L

Laplace transformation, 301
Levenberg–Marquardt (LM), 322–323, 324t
Load-side harmonics
  current harmonic spectrum, customer 1, 76, 78f
  distorted voltage, customer 2, 76, 77f
  harmonic current, customer 1, 76, 76f
  instantaneous current wave shape, customer 1, 76, 77f
  instantaneous phase-to-phase voltage wave shape, 77, 78f
  voltage harmonic spectrum, 77, 79f
Long duration voltage variation
  classifications, 24, 24t
  overvoltages, 27–29
  sustained interruption, 29–30
  undervoltages, 24–26
Longest discharging period (LDP), 265
Long-term flicker, 36, 38, 40t
Low-frequency harmonics, 351–352
Low-voltage ride through (LVRT), 143

M

Machine learning tools (MLTs), 313
MATLAB/Simulink, 164, 172–181
Maximum power point tracking (MPPT), 143, 144f, 284
Mean squared error (MSE), 316–317, 324t
Metal-oxide-semiconductor field-effect transistors (MOSFETs), 246–248, 268
Mitigation techniques, 181–189
Modern grid integration framework, 351
Modified inverter controller (MIC), 139
Modular multilevel converter (MMC), 286, 303t, 306f
  average submodule capacitor voltages, 302–304, 306f
  battery storage system (BSS), 288, 290f
  circulating current control, 292–294, 294f
  DC-link voltage, 302–304, 306f
  gating signals, 294–295
  inner-loop control, 294, 295f
  leg-voltage controller, 292–294
  low-voltage ride-through capability, 297, 297f
  lumped battery storage system, 288, 289f
  negative sequence current compensation, 296, 296f
  point of common coupling (PCC), 292
  positive sequence current reference, 292
  sequence components, 292, 293f
  submodule capacitors, 294–295, 295f
  ultracapacitor (UC), 288, 290
Modulation index (M), 254–256
Momentary interruptions, 347–348, 349f
  causes, 20
  electrical distribution, 21, 22f
  impacts, 21
  root mean square (RMS) voltage variations, 19–20, 21f, 22t, 23f
  voltage wave shape, 22, 23f

Multilevel inverters (MLIs), 106
applications of, 245–246, 246f
renewable energy sources (RES), 119
standing voltage (SV), 252
Multiresolution methods, 341

N

Narrow band internet of things (NBIoT), 342
Neutral blocking filter, 186, 186f
New grid regulations, 133–134
Noise
causes, 45
impacts, 46
neutral current wave shape, 45, 46f
phase current wave shape, 45, 46f
Normalized standing voltage (NSV), 252
Notch depth, 43, 44f–45f
Notching
causes, 44
impacts, 45
notch depth, 43, 44f–45f
three-phase voltage wave shape, 43, 44f

O

Odd harmonics, 50, 51t, 52f–53f
Oscillatory transients, 5, 6f, 352–353
three-phase voltage wave shapes, 5, 7f
Overvoltages, 167, 350
causes, 27–29
impacts
root mean square (RMS), 27, 29f

P

Parallel resonance, 92
Particle swarm optimization (PSO), 139
PCC. *See* Point of common coupling (PCC)
Pearson correlation index, 237
Periodic voltage disturbance, 43
Phase disposition pulse width modulation (PD-PWM), 254–256, 270t
Phase-locked loop (PLL), 143–144
Photovoltaic (PV) power plants, 134–135, 137f
assessment of, 146–147, 146t
classification of, 137–138, 138f
energy storage system (ESS), 138–139, 140f
flexible alternating current transmission system (FACTS), 138, 140–143
inverter-resident methods, 143–146
modified inverter controller (MIC), 139
power electronics-based method, 139–140, 140f
voltage drop, 136–137
Point of common coupling (PCC), 107–108
electric vehicles (EV) charging station, 152, 153f–155f
energy storage + static compensator (E-STATCOM), 292, 307f
load-side harmonics, 76–77
single customer/consumer, 75, 75f
source-side harmonics, 76
two customers/consumers, 75, 75f
voltage dips, 221
Potential transformers (PTs), 5
Power factor correction, 184
Power loss analysis
conduction losses, 261–264
ripple losses, 265
switching losses, 260–261
Power quality data interchange format (PQDIF), 353
Power quality disturbances (PQDs), 311–312
classification, 326–327, 326t, 328t–331t
fast Fourier transform, 312–313
feature extraction explanation, 318–322, 321f–322f
feedforward neural network (FFNN), 322–323, 323t
Levenberg–Marquardt (LM), 323, 324t
mathematical models, 317–318, 319t
noisy environments, 323–326, 326t
training algorithms, 323–326, 325t
validation, 327, 332t
Power quality monitoring systems, 341–343
Programmable protection devices, 188–189, 189f
Proportional integral (PI) controllers, 106–107
Proportional integral differential (PID) controllers, 106–107

R

Rapid voltage changes (RVCs), 351
Renewable energy resources (RERs), 338
large-scale integration, 285
power generation, 284
Renewable energy sources (RES), 123f, 159–160, 193–194
active distribution networks, 200
artificial neural network (ANN) based controller, 115, 116f

Renewable energy sources (RES) (*Continued*)
integration of, 106–107
mathematical modeling
coefficients, 111
instantaneous power theory (IPT), 111
load current, 111
voltage disturbances and harmonics, 110
power quality analysis
harmonics, 109
point of common coupling (PCC), 107–108
voltage fluctuations, 108
voltage transients, 108
voltage unbalance/sag/swell, 108
Residential customers
flickers, 167
frequency deviation, 168–169
harmonics, 167
interruptions, 168
overvoltage, 167
transients, 168
undervoltage, 167
voltage sags, 168
voltage swells, 168
voltage unbalance, 168
Residential systems, 169, 169t
harmonic filter, 184–185, 185f
harmonics, 164
nonlinear load, 165, 165f
MATLAB/Simulink, 164
power factor correction, 184
power quality-based equipment, 186–187
power quality measurement, 170, 170t, 171f, 171t
programmable protection devices, 188–189, 189f
residential customers, 166–169
surge protective devices (SPDs), 187–188, 188f
uninterrupted power supply (UPS), 164
unpredictable disturbances, 165
Residual current devices (RCDs), 170
Resonance
condition for, 91–93
controller, 144
frequency, 91
parallel resonance, 92
series resonance, 91–92
Ripple factor (RF), 272
Ripple losses, 265

S

Series resonance, 91–92
Short-duration root mean square (RMS) variation
classifications of, 6, 8t
momentary interruptions, 19–22
voltage sags, 7–12
voltage swells, 12–19
Short-term flicker, 36, 39t
Short-time Fourier transform (STFT), 312–313
Silicon-controlled rectifiers (SCRs), 65–66
applications, 66
current harmonic spectrum, 68, 68f
harmonic order, 67
Simply current distortion, 40
Single-phase rectifiers
applications, 63
current harmonic spectrum, 64, 65f, 66t
instantaneous current wave shape, 63, 64f
instantaneous voltage wave shape, 63, 64f
Sinusoidal reference waveform, 254–256
Smart meters (SMs), 341–342
S-matrix, 314–315
Solar energy-based power generation system, 284, 284f
Source impedance, 2
Source-side harmonics, 76
Squirrel cage induction generator (SCIG), 289–291, 290f
Standing voltage (SV), 252, 257t
cascaded H-bridge (CHB), 252, 266t
equivalent resistance, 252, 262t
switching pulses, 252, 259t
State-of-charge (SoC), 152, 300, 301f–302f, 308f
Static compensator (STATCOM), 137–138, 142, 142f, 285–286, 286t
Stockwell transformation (ST), 314–315
S-transform, 341
Submodule capacitors, 294–295, 295f
Subsynchronous resonance (SSR), 142
Sum of squared error (SSE), 114
Supervised learning, 316–317
Supervisory control and data acquisition (SCADA), 341–342
Supraharmonics, 156, 351–352
Surge protective devices (SPDs), 187–188, 188f
Sustained interruption
causes, 30

electrical distribution, 30, 30f
impacts, 30
root mean square (RMS), 29, 29f, 31f
voltage wave shape, 30, 31f–32f
Switched-capacitor multilevel inverters (SCMLIs), 253
conduction losses, 261–264
current waveform, 272, 275f
harmonic spectrum, 271, 272f
high-gain switched-capacitor multilevel inverters (SCMLIs), 246–247
output voltage and load current waveforms, 270–271, 270f
ripple losses, 265
switching losses, 260–261
unity-gain switched-capacitor multilevel inverters (UG-SCMLIs), 246–247
voltage waveforms, 271, 271f, 273f
Switching losses, 260–261
Switching pulses, 252, 258, 259t, 260f

T

Three-phase rectifiers
current harmonic spectrum, 68, 70f, 71t
instantaneous current, 68, 69f
instantaneous voltage, 68, 69f
silicon-controlled rectifiers (SCRs), 65–66
applications, 66
current harmonic spectrum, 68, 68f
harmonic order, 67
twelve-pulse rectifier, 68, 71f–74f, 74t
Thyristor-controlled static compensator (TCSC), 137–138
configurations, 141
operations, 141–142
Time–frequency (T–F) analyses, 341
Total demand distortion (TDD)
individual harmonic order, 87, 88t–89t
maximum demand load current, 85, 88t
Total harmonic distortion (THD), 64, 164, 253, 276–279, 277t–278t
current harmonic distortion, 82, 84t, 87f
electrical distribution, 85, 86f
nonlinear loads, 170, 171t
voltage harmonic distortion, 81, 83t, 86f
Total rated harmonic distortion (TRD), 90f
definition of, 87–89
reference point of applicability (RPA), 89
Total standing voltage (TSV), 267, 267t
Traditional voltage control strategies, 196–199, 197f–198f
Transients, 108, 168
classifications of, 3, 3t
detection, 344–345, 344f–345f
impacts, 4
impulsive transients, 4, 4f–5f
oscillatory transients, 5, 6f
single-line diagram (SLD), 5, 6f
sources of, 3, 3t
Tripplen harmonics, 57–58, 58f, 59t
Twelve-pulse rectifier, 68, 71f–74f, 74t
Two-level voltage source converter (VSC), 287–288

U

Ultracapacitor (UC), 287–288, 298, 299f, 304t
Undervoltages, 167
causes, 25
impacts, 25
power distribution, 26, 27f
root mean square (RMS), 24, 24f–25f, 28f
voltage wave shape, 25, 26f
Unity-gain switched-capacitor multilevel inverters (UG-SCMLIs), 246–247

V

Vehicle-to-home (V2H) modes, 350
Voltage control
active distribution networks, 202f
distribution static compensator (D-STATCOM), 204
electric vehicles and low generation, 212–214, 213f–214f
high distributed generation and low demand, 208–211, 208t, 209f, 211f–212f, 212t
Kirchhoff's second law, 202
line drop compensation, 200
optimal power flow (OPF), 201
passive network, 207–208, 207f, 207t–208t
plug-in electric vehicles (PEVs), 204
reactive power, 202
renewable energy sources (RES), 200
slow charging mode, 200
voltage profile, 199, 199f
traditional strategies, 196–199, 197f–198f
Voltage dips
detection of, 346–347, 346f–348f
dip duration, 239, 239f
dip magnitude, 239, 239f, 241f
fault cause, 240, 240f

Voltage dips (*Continued*)
faulted equipment, 237, 237f–240f
grouping faults, 228f
data entry, 228
mitigation recommendations, 229
simultaneity of events, 228
vulnerability area, 229
IEEE 1564/Brazilian grid code, 231f
impact factor (FI), 235, 236f
magnitude *vs.* duration, 233, 234f
monthly stratification, 235, 235t
severity index, 235, 236f
stratification, 232, 232t
indexes stratification, 239, 239f–240f
load shutdowns, 220
low-impedance faults, 219–220
overvoltage, 350
Pearson correlation index, 237
point of common coupling (PCC), 221
power quality monitoring, 229–230
quantification, 224f
Brazilian grid code, 226, 226t
data input, 223
dip aggregation, 225
dip indexes determination, 223
dip stratification, 225, 225f
impact factor (FI), 227, 227t
phase aggregation, 223
severity index, 226–227
weighting factor, 227, 227t
quantification and mapping, 220
sensitive consumer connection, 229, 230f
short-term voltage variations, 220
solid faults, 219–220
vulnerability area, 221, 241, 242f
Voltage fluctuations, 108
causes, 37
definition of, 34
electrical distribution, 38, 38f
flicker, 36
illumination intensity, 34
impacts, 37
long-term flicker, 38, 40f, 40t
short-term flicker, 38, 39f, 39t
Voltage harmonic distortion, 61–62, 81, 83t
Voltage imbalance, 32–34. *See also* Voltage unbalance
Voltage quality requirements
ANSI C84.1, 195, 195t
EN 50160, 195, 195t
Voltage regulation, 196
Voltage-related problems, 2
Voltage sags, 168
causes, 10
depth of, 7–9
detection, 345, 346f, 349f
impacts, 10
peak and root mean square (RMS) voltage variations, 9, 9t, 11f–12f, 12t, 13f, 13t, 14f, 14t, 15f, 15t, 16f, 16t
power distribution, 10, 10f
wave shape, 9, 9f, 12f–16f
Voltage swells, 168
causes, 14
circuit diagram, 16–17, 18f
impacts, 15
peak and root mean square (RMS) voltage variations, 13, 17t, 19f–20f, 21t
voltage oscillogram, 17, 19f–20f
wave shape, 13, 17f
Voltage unbalance, 32, 168
causes, 33
current negative sequence, 34, 37f, 37t
current positive sequence, 34, 36f, 37t
impacts, 33
peak and root mean square (RMS), 32, 32t–33t
power distribution, 33, 34f
voltage negative sequence, 34, 35f, 36t
voltage positive sequence, 34, 35f, 36t

W

Waveform distortion, 38
DC offset, 43
harmonics, 39–42
interharmonics, 42–43
noise, 45–46
notching, 43–45
Wavelet transformation (WT), 314
Wind energy-based power generation system, 284, 285f

Z

Zigzag reactors, 186–187

Kolibri 360 GmbH
Ferdinand-Jühlke-Straße 7
99095 Erfurt, Deutschland
produktsicherheit@kolibri360.de